Proceedings in Life Sciences

VIIIth Congress of the International Primatological Society,
Florence, 7–12 July, 1980

Selected Papers, Part A: Primate Evolutionary Biology
© by Springer-Verlag Berlin Heidelberg 1981

Selected Papers, Part B: Primate Behavior and Sociobiology
© by Springer-Verlag Berlin Heidelberg 1981

Main Lectures: Advanced Views in Primate Biology
© by Springer-Verlag Berlin Heidelberg 1982

Advanced Views in Primate Biology

Main Lectures of theVIIIth Congress
of the International Primatological Society,
Florence, 7-12 July, 1980

Edited by
A. B. Chiarelli and R. S. Corruccini

With 35 Figures

Springer-Verlag
Berlin Heidelberg GmbH 1982

Professor A. B. CHIARELLI
Istituto di Antropologia dell'Universita
Via del Proconsolo, 12
50122 Florence/Italy

Dr. R. S. CORRUCCINI
Istituto di Antropologia dell'Universita
Via del Proconsolo, 12
50122 Florence/Italy
and
Department of Anthropology
Southern Illinois University
Carbondale, IL 62901/USA

For explanation of the cover motive see legend to Fig. 3, p. 152.

ISBN 978-3-642-68302-2   ISBN 978-3-642-68300-8 (eBook)
DOI 10.1007/978-3-642-68300-8

Library of Congress Cataloging in Publication Data. Main entry under title: Advanced views in primate biology. (Proceedings in life sciences). "Proceedings of the main session of the VIII International Congress of Primatology" – P. Includes bibliographical references and index. 1. Primates – Congresses. I. Chiarelli, A. B. II. Corruccini, Robert S. III. International Primatological Society. IV. International Congress of Primatology (8th: 1980: Florence, Italy) V. Series. QL737.P9A38 599.8 81-21545 AACR2.

Offsetprinting and bookbinding: Brühlsche Universitätsdruckerei, Giessen

2131/3130-543210

Preface

The VIIIth International Congress of the International Primatological Society was held from 7 through 12 July 1980 in Florence, Italy, under the auspices of the host institution, the Istituto di Antropologia of the University of Florence. More than 300 papers and abstracts were presented either at the main Congress or in 14 pre-Congress symposia the week earlier (so scheduled to avoid conflicting with either the main invited lectures or the contributed paper sessions).

The texts of these main lectures, plus reports from the organizers on results of the pre-Congress symposia, form this volume. The main lecturers were invited to provide comprehensive coverage of the latest developments in major subfields of Primatology, by the leading researchers in those areas. In this way it was planned that the Congress would provide up-to-date reviews of recent progress, and prospects for future progress, that a biennial International Congress should include as one of its primary functions.

The organizers of individual pre-Congress symposia interspersed their summaries among the main lectures, providing the basic details of existing theories, new data, and whatever consensus could be reached concerning their particular topic.

In keeping with what is probably the most urgent theme of these congresses, we included in this primary volume some extra material in the area of primate conservation to augment the main lecturer's treatment.

We acknowledge all the personnel of the Institute of Anthropology, Florence, and all other volunteers for their assistance.

February, 1982   A.B. CHIARELLI
                                                R.S. CORRUCCINI

Contents

Part A — Main Lectures

Inaugural Address, Congress of Primatology, Florence
G.H.R. von Koenigswald 3

Recent Advances in Molecular Evolution of the Primates
M.L. Baba, L.L. Darga, and M. Goodman (With 9 Figures). 6

Immunogenetic Evolution of Primates
J. Ruffié, J. Moor-Jankowski, and W.W. Socha 28

The Evolution of Human Skin
W. Montagna 35

The Importance of Theory for Reconstructing the Evolution of
Language and Intelligence in Hominids
S.T. Parker and K.R. Gibson (With 2 Figures) 42

Primatology and Sociobiology
J. Wind (With 1 Figure) 65

Dominance and Subordination: Concepts or Physiological States?
E.B. Keverne, E. Meller, and A. Eberhart (With 7 Figures). 81

Sexual Behavior in Aging Male Rhesus Monkeys
C.H. Phoenix and K.C. Chambers 95

Simian-type Blood Groups of Hamadryas Baboons. Population Study
of Captivity-born Animals at the Sukhumi Primate Center —
Preliminary Report
B.A. Lapin, W.W. Socha, and J. Moor-Jankowski. 105

Rhesus Macaques: Pertinence for Studies on the Toxicity of
Chlorinated Hydrocarbon Environmental Pollutants
W.P. McNulty (With 6 Figures). 111

The Role of a Kenyan Primate Center in Conservation
J.G. Else 124

Further Declines in Rhesus Populations of India
C.H. Southwick, M.F. Siddiqi, J.A. Cohen, J.R. Oppenheimer,
J. Khan, and S.W. Ashraf (With 1 Figure) 128

Taiwan Macaques: Ecology and Conservation Needs
F.E. Poirier 138

Prospects for a Self-sustaining Captive Chimpanzee. Breeding Program
B.D. Blood. 143

Part B – Symposium Reports

Miocene Hominoids and New Interpretations of Ape and Human
Ancestry
R.L. Ciochon and R.S. Corruccini (With 7 Figures) 149

Infanticide in Langur Monkeys (Genus *Presbytis*): Recent Research
and a Review of Hypotheses
G. Hausfater and C. Vogel (With 2 Figures) 160

Recent Advances in the Study of Tool-Use by Nonhuman Primates
W.C. McGrew 177

Primate Communication in the 1980s: Summary of the Satellite
Symposium on Primate Communication
C.T. Snowdon, C.H. Brown, and M.R. Petersen 184

Primate Locomotor Systems: Summary of Results of the
Pre-Congress Symposium in Pisa
H. Ishida, R.H. Tuttle, and S. Borgognini-Tarli 200

Results of the Pre-Congress Symposium on "Methods and Concepts
in Primate Brain Evolution"
D. Falk and E. Armstrong. 206

The Effects of Drugs and Hormones on Social Behavior in Nonhuman
Primates
A. Kling and H.D. Steklis 212

Report on Symposium Entitled: "Comparative Biology of Primate
Semen"
K.G. Gould 217

Chromosome Banding and Primate Phylogeny. Inaugural Address
H.N. Seuànez 224

Comparative Psychology Symposium: Introduction
P.A. Bertacchini 236

The Present and Future Status of Comparative Psychology:
Proceedings of the Corigliano Calabro Symposium
J.T. Braggio 250

Contributors

You will find the addresses at the beginning of the respective contribution

Armstrong, E. 206
Ashraf, S.W. 128
Baba, M.L. 6
Bertacchini, P.A. 236
Blood, B.D. 143
Borgognini-Tarli, S. 200
Braggio, J.T. 250
Brown, C.H. 184
Chambers, K.C. 95
Ciochon, R.L. 149
Cohen, J.A. 128
Corruccini, R.S. 149
Darga, L.L. 6
Eberhart, A. 81
Else, J.G. 124
Falk, D. 206
Gibson, K.R. 42
Goodman, M. 6
Gould, K.G. 217
Hausfater, G. 160
Ishida, H. 200
Keverne, E.B. 81
Khan, J. 128

Kling, A. 212
Koenigswald, G.H.R. von 3
Lapin, B.A. 105
McGrew, W.C. 177
McNulty, W.P. 111
Meller, E. 81
Montagna, W 35
Moor-Jankowski, J. 28, 105
Oppenheimer, J.R. 128
Parker, S.T. 43
Petersen, M.R. 184
Phoenix, C.H. 95
Poirier, F.E. 138
Ruffié, J. 28
Seuànez, H.N. 224
Siddiqi, M.F. 128
Snowdon, C.T. 184
Socha, W.W. 28, 105
Southwick, C.H. 128
Steklis, H.D. 212
Tuttle, R.H. 200
Vogel, C. 160
Wind, J. 65

Part A
Main Lectures

Inaugural Address, Congress of Primatology, Florence

G.H.R. von KOENIGSWALD [1]

Ladies and Gentlemen,
Dear Colleagues,

I have the privilege to open the VIII International Congress of Primatology in Florence, the famous town of Arts and Sciences, with a short scientific address. You will not find me on the program. I have taken the place of Prof. Tobias, who unfortunately fell ill and was not able to come, and it is only since yesterday that I knew that I had to speak today. I had expected the Congress to be opened by a special performance of the Florence City Ballet: "Four chromosomes on a lazy afternoon in September," but instead you have to listen to me. But this is a pleasure and an honor for me.

I was born before the discovery of Heidelberg Man, and by a number of fortunate circumstances I could follow intimately the whole history of subsequent discoveries. My best friend was the youngest son of the anthropologist Rudolf Martin, and already as a schoolboy my interests were influenced by our science. Fifteen years old, I visited with a friend the sand pit of Mauer near Heidelberg, where Heidelberg Man had been discovered not yet 10 years before, hoping to make a similar discovery. I naturally failed, but a kind workman presented me with a lower molar of a fossil Rhinoceros that he just had found. This very personal contact with a fossil Rhino, the schoolboy's feeling of having a Rhino in his pocket, gave me my enthusiasm for fossil mammals, and this fascination still is unchanged.

So I followed the birth of Primatology as a special branch on the tree of science after the war. I attended the first congress in Frankfurt, and subsequent congresses in Zurich and Portland. Every time a new side-branch was added, and now, here in Florence, there are so many that a lot of "satellite symposia" became necessary to deal with the flood of lectures and information. Mainly interested in the Evolution of Man, I want to thank Professor Chiarelli not only for the organization of such a special symposium before the congress, but also congratulate him as founder and editor of the "Journal of Human Evolution", now in its ninth year, with its main office here in Florence.

For this opening address I have chosen a subject from my own field of research, which adds a romantic touch to the bones and skulls, and at the same time is of highly important scientific value: the inside story of the discovery of Peking Man.

1 Senckenberg Museum, 6000 Frankfurt, FRG

The story begins in a Chinese drugstore in Peking. The most typical animal in Chinese art is the dragon, a mythical animal with supernatural power, ruler of the East. That it really existed is evident from the fossil bones and teeth found in many places in China. It is adorned with the antlers of a Miocene deer, *Cervocervus*, not rare in Central China. The remains of such mighty dragons must also be very powerful, and so dragon bones (lung ku) and dragon teeth (lung tse) had been introduced into Chinese medicine. Documents go back to the T'ang and Sun Dynasties, but their use might go back to prehistoric times. This medicine excavated from bone beds or caves and rock fissures, can be found also in practically every Chinese drugstore outside of China: Indonesia, Malaya, Philippines, Thailand and even in America. Just a few years ago, during the last Pan-Pacific Congress in Vancouver, we found typical Chinese drugstores not as independent shops, but in a far corner in the basement of modern department stores, complete with dragon teeth and other fossils.

As the dragon bones are less expensive than the teeth, skulls are broken into pieces and even the teeth were damaged to show the calcite crystals in the pulpa cavity, as a sign that they are of good quality. They are sold in small quantities, ground and swallowed with some alcohol. They are sold according to weight and very expensive as is every medicine which is supposed to keep elderly gentlemen young.

Already in 1870 some teeth from the drug market in Shanghai reached the British Museum, but it was not before 1900, when during the "Boxer War" German troops marched into Peking, that a larger collection of dragon teeth was assembled. The collection was made by Dr. Haberer, a pupil of the famous German paleontologist von Zittel, for the Museum of Munich, and described by Max Schlosser in 1903. This, at the same time, was the first contribution toward the fossil mammalian fauna of China. Most animals were of Miocene age — typical guide fossil *Hipparion*, a three toed horse, of which the teeth are abundant. I myself have seen close to 50,000 *Hipparion* teeth in drugstores. The most important specimen in the Haberer collection was a very worn upper molar of human type, perhaps of a Miocene hominid! This tooth came from a drugstore in Peking, and it is that little tooth that touched off the search of Early Man in the surroundings of Peking.

In the 20s, the Swedish geologist and archaeologist Gunnar Anderson tried to find suitable sites for the discovery of remains of Early Man near Peking. With the help of Chinese colleagues he discovered Chou Kou Tien at the edge of the Western Hills, and the first excavations were made by Zdansky. Of human remains he only found two isolated teeth, unpacking his material in Uppsala. But later Davidson Black, anatomist of the Peking Union Medical College took over, and the Rockefeller Institution gave the money for large-scale excavations. On 16 October 1927, Dr. Birger Bohlin found a fine lower molar; he directly went to Peking to hand the tooth over to Dr. Black. He recognized the importance of that find — the tooth was different from all human teeth described till then — and he made it the type-specimen of his "Sinanthropus pekinensis". Many colleagues would not believe that from a single tooth such conclusions could be drawn, but later in 1928, when Dr. Pei found the first skulls at Chou Kou Tien, it was evident that Black had been right.

Peking Man was a typical primitive hominid, with a flattened skull and a strong supraorbital ridge, formed like a ledge above the eyes. He was not only anatomically a man, but also a tool maker and had known the use of fire. He was a great hunter with ample game available, but in spite of that a cannibal, most probably for magic reasons. All skulls were opened to extract the brain, all long bones were split to extract the marrow. All lower jaws were broken, perhaps to silence the victims?

The discovery of Peking Man at the same time helped us to solve the riddle of the Javanese *Pithecanthropus.* 1889 Dr. Dubois, a Dutch physician, went to the East Indies. Influenced by Darwin, he wanted to find the "missing link between Man and the anthropoids". He must not have had too clear ideas about what to expect, because when in October 1891 he found in an excavation at Trinil on the banks of the Benga-wan Solo in Central Java a skull cap, he thought that he had discovered a fossil chim-panzee *(Anthropopithecus).* Only three years later he called his find "Pithecanthro-pus", using a name given by Haeckel to a hypothetical form linking Man (Anthropus) with the anthropoids (Pithecus).

What Dubois really had found was only the roof of a skull, all the diagnostic tem-poral parts missing. So his interpretation aroused a lot of discussions, and later even Dubois changed his mind and regarded his *Pithecanthropus* as a kind of giant gibbon. But what was it really? A man, an ape, a man-ape, an ape-man, a bastard between man and ape or an artificially deformed skull? There are hundreds of publications about this problem. Then, 40 years after its discovery, in 1931 Davidson Black could show that *Sinanthropus* and *Pithecanthropus* are closely related — and that the much dis-puted *Pithecanthropus* was really an early human being.

But Dubois did not want to accept this result. To him *Pithecanthropus* remained an ape, and *Sinanthropus* was just a kind of degenerated Neanderthal Man. But fortu-nately, only 6 years later, in 1937, we found Pithecanthropus II in Java. This find was complete enough to show that Davidson Black had been right.

Most unfortunately the complete material of Peking Man, remains of a dozen skulls and more than 100 teeth, were lost during the war. But the site still is there and in no way exhausted, and a number of new finds, not many, have already been made by our Chinese colleagues.

My own work was mainly concentrated on the Chinese drug market in the south, Hong Kong and Canton. Here a most interesting Pleistocene fauna from caves and rock fissures from the Kwangtung and Kwangi Provinces can be found. No *Hipparion,* but a very different assemblage including *Elephas, Stegodon,* panda, orang, the still myste-rious *Gigantopithecus* — the first teeth came in 1935 from drugstores, long before this interesting being was found by Woo in excavations — and also human teeth. Some oversized first lower premolars are giving away a southern Peking Man, others are modern. Among these we find the same type of Haberer's tooth from the Peking drug-store, with the same red matrix noted by Schlosser. Father Teilhard de Chardin looked over my collection and recognized them as coming from mesolithic strata. So a modern tooth helped us to discover a fossil Man!

I wish you all a good Congress! I thank you for your patience.

Recent Advances in Molecular Evolution of the Primates

M.L. BABA [1], L.L. DARGA [2], and M. GOODMAN [3]

Introduction

In the last three decades investigations into the processes underlying molecular evolution have expanded our knowledge concerning the phylogenetic history of the primates. Advances in molecular evolution have coincided with exciting paleontological discoveries, especially the recent australopithecine and ramapithecine finds that question anew the events and timing leading to the origins of the hominine species. The molecular approach has yielded important insights into phylogenetic relationships among primates by sharpening our understanding in several areas. The possibility of correct placement for the enigmatic *Tarsier* with respect to prosimian and anthropoid lineages, and understanding the evolutionary history of the gibbon and siamang with respect to the other ape lineages has been strengthened. Further insights into relationships within Old and New World monkey groups have also been gained.

Most importantly, methods developed for analyzing the data of molecular evolution have clarified the processes underlying molecular change. Maximary parsimony analysis of amino acid sequence data has revealed a pattern of acceleration and deceleration in the evolution of primary structure for several proteins. Such nonuniform rates of change suggest that divergence dates calculated from amino acid sequence data by use of a molecular clock will not be precisely accurate.

Parsimony analysis of rates of change for structural-functional regions within proteins involved in energy metabolism has provided strong evidence that positive and stabilizing selection is the most important force orchestrating change at the molecular level.

In this chapter we will review the current state of the science of molecular anthropology, including its implications for primate phylogeny, for interpretation of the fossil record, and for the processes underlying molecular evolution.

1 Department of Anthropology, College of Liberal Arts, Wayne State University, Detroit, Michigan, USA
2 Children's Hospital, Detroit, Michigan, USA
3 Department of Anatomy, School of Medicine, Wayne State University, Detroit, Michigan, USA

Biochemical Methods for Investigating Phylogeny

Maximum Parsimony Method Applied to Amino Acid Sequence Data

Continuing progress in amino acid sequence analysis has yielded a relatively large body of primary structure data for several proteins. Substantial structure data now exist for alpha hemoglobin (83 species), beta hemoglobin (103 species), myoglobin (65 species), fibrinopeptides A and B (47 species), cytochrome C (94 species) and alpha lens crystallin (46 species).

The maximum parsimony method is a powerful tool for unraveling evolutionary relationships among a set of contemporary amino acid sequences. A thorough discussion of the theoretical foundations and algorithms enabling parsimony analysis can be found elsewhere (Moore 1976, Goodman 1976, Goodman et al. 1979). By directly comparing the primary structure of a single protein found in many extant species, the parsimony algorithm generates a dendrogram depicting those ancestral mRNA sequences and that branching network which, in total, requires the fewest number of nucleotide replacements to account for observed sequence differences between lineages. The parsimony method has the advantages of conforming to both Occam's Razor and Hennigian principles. Occam's Razor is satisfied, since the parsimony procedure constructs dendograms solely on the basis of evolutionary pathways for which molecular evidence exists. The parsimony method also adheres to the principles of phylogenetic systematics (Hennig 1966) in that primitive similarities and similarities due to convergence are distinguished from shared, derived features. The accuracy of the method is also unaffected by varying rates of molecular change within and between lineages.

Recent advances in the application of the parsimony method have sharpened its resolving power. An augmentation procedure (Baba et al. 1981) corrects for the underestimation of nucleotide replacements in regions of the dendogram which are represented by only a few sequences.

Recently, a procedure was described by Goodman et al. (1979) to correct for the problem which results when excessive numbers of parallel and back mutations in a single protein cause incorrect groupings of sequences to resemble one another in a tree of lowest nucleotide replacement (NR) length. Beginning with the assumption that the gene phylogeny may differ from the species phylogeny, Goodman et al (1979) noted that it is possible to improve the effectiveness of the search for the correct tree by taking into account additional evolutionary events at the genetic level. The gene phylogeny for a set of operational taxonomic units (OTU's), based upon single protein analysis, is defined as the ancestral order of branching of phylogenetically related genes leading to the expressed genes in the extant OTU's. The species phylogeny is the evolutionary branching of lineages descending to the modern species. When the putative gene phylogeny depicted in the tree of lowest NR length violates well established evidence on the cladistic relationships of a species group, our problem is to decide whether the discrepancy is due to an error in the branching arrangement of species or to gene duplication and gene expression events. This problem can be solved by calculating the minimal number of hypothetical gene duplication (GD) and gene expression (GE) events necessary to account for observed differences between the putative gene phylogeny depicted in the tree of genic changes. The tree with the lowest NR+GD+GE score thus becomes the most parsimonious tree.

A species phylogeny derived independently from molecular data may be obtained by combining amino acid sequence data in a tandem alignment, as if each protein sequence were part of a giant polypeptide chain encoded by a single gene. By increasing our sample of the genome, dendrograms generated by parsimony analysis of such a tandem alignment correct for the problems of convergence and gene duplications encountered in a single-protein analysis, and thus more closely approximate the true species phylogeny.

The Immunological Approach

Before a substantial body of literature on primary sequences had accumulated, biochemical approaches to primate evolution relied heavily upon immunological techniques such as immunodiffusion, microcomplement fixation and radioimmune assay. An advantage of the immunodiffusion method, thoroughly described by Goodman and Moore (1971), is that different proteins diffuse through the agar at different rates and form separate precipitin lines when they react with antibodies. The use of whole serum in the production of antisera therefore yields comparative results on several proteins simultaneously. This system thus permits visualization of evolutionary divergence for several individual proteins for each species compared.

On the basis of set theory logic, quantitative results of immunodiffusion investigations are translated into antigenic distance tables. The data in these tables are converted to a dissimilarity matrix which gives the antigenic distance between any pair of species in the collection. A divergence tree depicting the order of ancestral branching for the total collection of species is then generated from the dissimilarity matrix by Sokal and Michener's (1958) unweighted pair-group method.

Moore demonstrated by mathematical proof (Moore 1971) that dendrograms so produced on the basis of immunological distance data approximate cladograms, provided that species represented in the dissimilarity matrix evolved antigenically at roughly comparable rates. The branching order depicted in the dendrogram approximates the true cladogeny of species more closely when antigenic distance values are derived from a broad spectrum of proteins rather than from one or two proteins only.

DNA Hybridization and Nucleotide Sequence Data

Evolutionary change among species is perhaps most precisely reflected through comparison of their genetic material, deoxyribonucleic acid (DNA). DNA can change during evolution in a variety of ways, and several methodologies have been developed to measure this change (Kohne 1976). DNA hybridization techniques, using both repeated and nonrepeated sequences of DNA, first demonstrated the value of DNA studies for the assessment of phylogenetic relationships among primate taxa (Hoyer et al. 1964, Hoyer and Roberts 1967, Kohne et al. 1972, Hoyer et al. 1972, Bonne et al. 1980). More recently, advances in methods used to determine nucleotide sequences have permitted direct comparison of nucleotide bases among ever larger groups of species (Jukes 1980).

Mitochondrial DNA (mtDNA) is particularly well suited to the study of evolutionary relationships; it is small, relatively simple to analyze by several methods, and has the same gene content and gene order among all animal species (Brown et al. 1980). Furthermore, since mtDNA evolves at a rate 5–10 times faster than single copy nuclear DNA, it can provide a uniquely sharp focus on the phylogeny of closely related species (Ferris et al. 1980). Recently, the method of restriction endonuclease cleavage mapping has yielded especially high resolution data pertaining to relationships within the Hominidae (Brown et al. 1980).

Biochemical Evidence on the Phylogeny of Primates

Maximum Parsimony Analysis of Amino Acid Sequence Data

Tandem Alignment of Amino Acid Sequences. An assessment of phylogenetic relationships within Primates most closely approximating the true cladogeny is generated by maximum parsimony analysis of a tandem alignment of amino acid sequence data.

Figure 1 is the parsimony cladogram for six proteins, including alpha and beta hemoglobin sequences, myoglobin, fibrinopeptides A and B, cytochrome *C*, and alpha lens crystallin. Within this divergence tree Primates form a monophyletic assemblage,

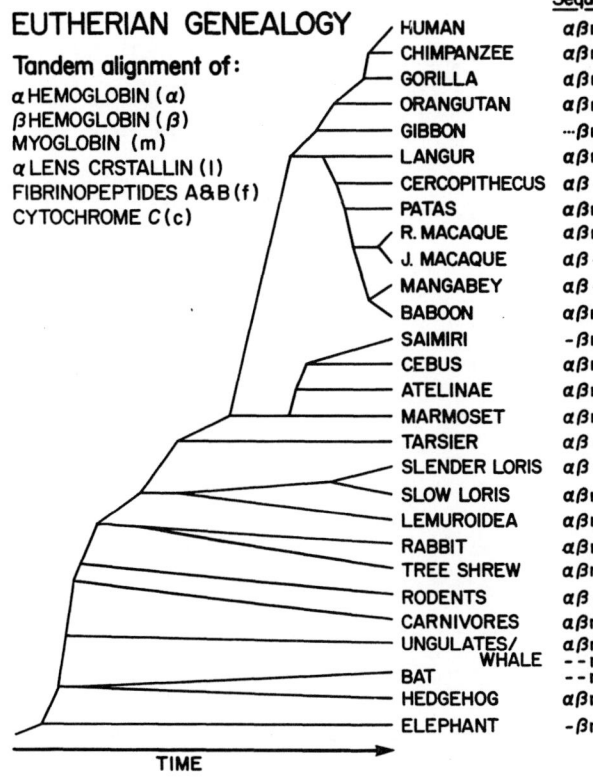

EUTHERIAN GENEALOGY

Tandem alignment of:

α HEMOGLOBIN (α)
β HEMOGLOBIN (β)
MYOGLOBIN (m)
α LENS CRSTALLIN (l)
FIBRINOPEPTIDES A&B (f)
CYTOCHROME C (c)

	Sequences
HUMAN	αβmlfc
CHIMPANZEE	αβm-fc
GORILLA	αβm-f-
ORANGUTAN	αβm-f-
GIBBON	···βm-f-
LANGUR	αβm---
CERCOPITHECUS	αβ--f-
PATAS	αβm---
R. MACAQUE	αβm̶lfc
J. MACAQUE	αβ----
MANGABEY	αβ--fᶜ
BABOON	αβm---
SAIMIRI	-βm---
CEBUS	αβm-fc
ATELINAE	αβm-fc
MARMOSET	αβm---
TARSIER	αβ----
SLENDER LORIS	αβ----
SLOW LORIS	αβmlfc
LEMUROIDEA	αβml--
RABBIT	αβmlfc
TREE SHREW	αβml--
RODENTS	αβ-lfc
CARNIVORES	αβmlfc
UNGULATES/WHALE	αβmlfc
	--ml-c
BAT	--ml-c
HEDGEHOG	αβml--
ELEPHANT	-βmlf-

TIME

Fig. 1. Scientific names and taxonomic units of lineages depicted in this dendrogram are as follows: human *(Homo sapiens)*; chimpanzee *(Pan troglodytes)*; gorilla *(Gorilla gorilla)*; orangutan *(Pongo pygmaeus)*; gibbon *(Hylobates* sp.); langur *(Presbytis)*; cercopithecus *(Cercopithecus)*; patas *(Erythrocebus patas)*; R. macaque *(Macaca mulatta)*; J. macaque *(Macaca fuscata)*; mangabey *(Cercocebus)*; baboon *(Papio* sp.); saimiri *(Saimiri* sp.); cebus *(Cebus* sp.); atelinae *(Ateles* sp.); marmoset *(Saguinus* sp.); tarsier *(Tarsius)*; slender loris *(Loris tardigradus)*; slow loris *(Nycticebus coucang)*; Lemuroidea; rabbit (Lagomorpha); tree shrew *(Tupaia* sp.); rodents (Rodentia); carnivores (Carnivora); ungulates-whale (Ungulata/Cetacea); bat (Chiroptera); hedgehog (Insectivora); elephant (Proboscidea)

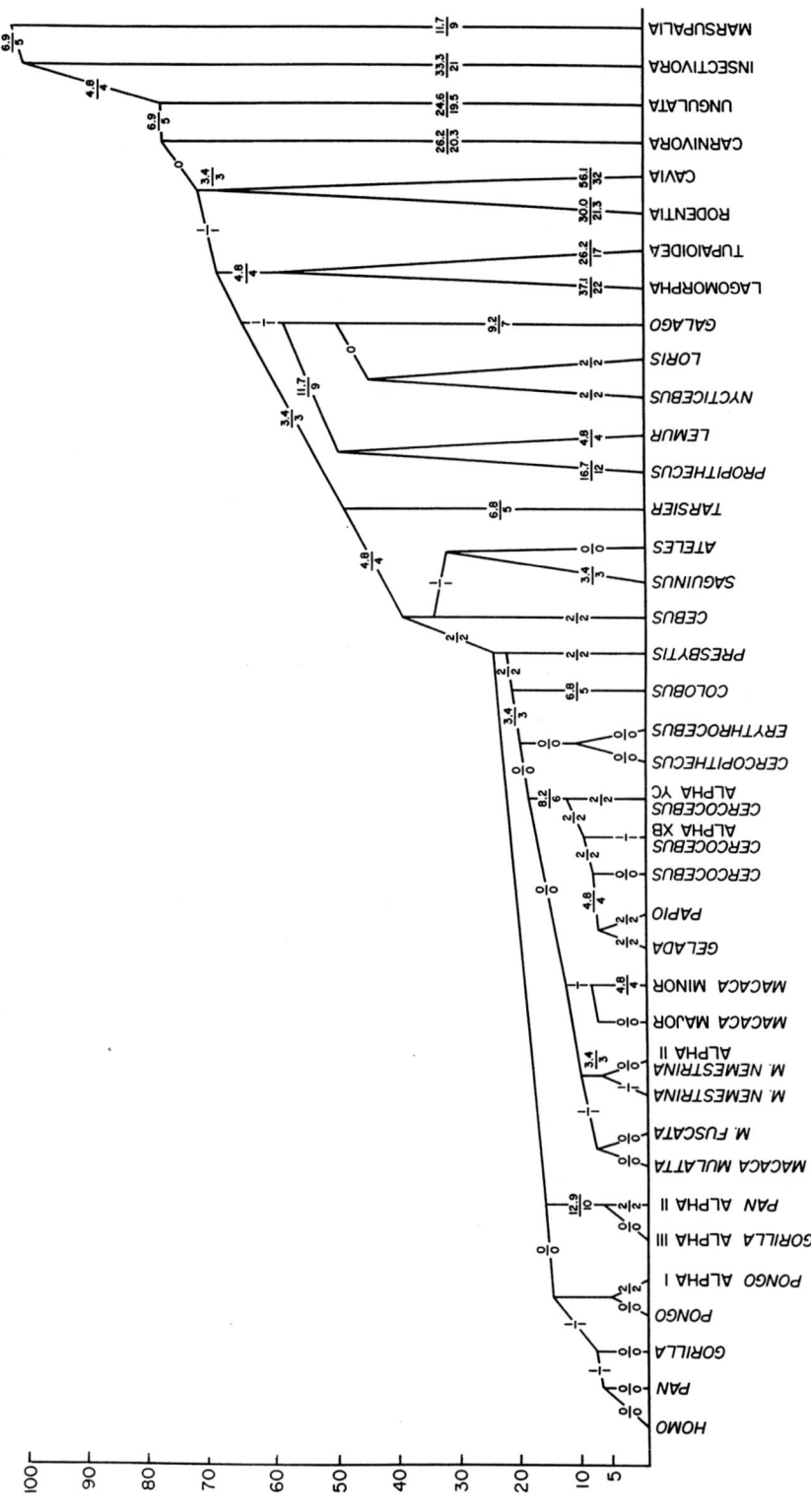

Fig. 2. Most parsimonious tree for alpha hemoglobin. Species depicted in this dendrogram are as follows: *Homo* (*H. sapiens*); *Pan* (*P. troglodytes*); *Gorilla* (*G. gorilla*); *Pongo* (*P. pygmaeus*); *Pongo* Alpha-I, *Gorilla* Alpha-III; *Pan* Alpha-III; *Macaca mulatta*; *M. fuscata*; *M. nemestrina*; *M. nemestrina* Alpha-II; *Macaca* major (*M. irus* major alpha); *Macaca* minor (*M. irus* minor alpha); *Gelada* (*Theropithecus gelada*); *Papio* (*P. anubis*); *Cercocebus* (*C. atys*); *Cercocebus* alpha XB (*C. atys* alpha XB); *Cercocebus* Alpha YC (*C. atys* YC); *Cercopithecus* (*C. Aethiops*); *Erythrocebus* (*E. patas*); *Colobus* (*C. badius*); *Presbytis* (*P. entellus*); *Cebus* (*C. apella*); *Saguinus* (*S. fusciollis*); *Ateles* (*A. Geoffroyi*); *Tarsier* (*Tarsius syrichta*); *Propithecus* (*Propithecus* sp.); *Lemur* (*L. fulvus*); *nycticebus* (*N. coucang*); *Loris* (*L. tardigradus*); *Galago* (*G. senegalensis*); *Lagomorpha* (*Oryctolagus cuniculus*); *Tupaoidea* (*Tupaia glis*); Rodentia (*Mus musculus* C57 B1 mouse, *M. musculus* NB mouse, *Ondatra zibethecus, Lemmus sibiricus, Clethrionomys rutitus, Microtus xanthongnathus, Rattus* sp.); Carnivora (*Canis familiaris* duplicated alpha, *C. latrans, C. familiaris, Urocyon cineroargenteus, Thalarctos maritimus, Meles meles, Procyon lotor, Nasua narica, Felis felis, Panthera leo*); Ungulata (*Ovis aries, capra* sp., *Capra* 2-alpha duplicated, *Bos taurus, Llama* sp., *Camelus dromedarius, Sus scrofa, Equus equus, Equus asinus, E. equus* fast alpha 24 Y); Insectivora (*Erinaceus europaeus*); Marsupalia (*Macropus cangaru, Didelphis* sp. fast alpha)

Unaugmented and augmented link length values are given for each link between internal nodes or between an internal node and a terminal point. For each link the unaugmented value is the number shown below the line, augmented above the line. Augmented link lengths are the number of nucleotide replacements (fixed mutations) between adjacent ancestor and descendant sequences corrected for superimposed fixations by an augmentation algorithm. The ordinate is a time scale in millions of years before present with major ancestral nodes fixed by palaeontological views concerning ancestral separations of the organisms from which the alpha hemoglobin came. Where fossil evidence was unsufficient to establish ancestral branch points, a heuristic procedure was used in which times were estimated from the magnitude of the link lengths in these areas of the tree and by interpolation betwen the points which were placed on the basis of palaeontological evidence most closely related to the *Tupaia* – Lagomorpha grouping. Primates divide to form Strepsirhini (lorises and lemurs) and Haplorhini (*Tarsier* and Anthropoidea). Within Haplorhini, Anthropoidea is a monophyletic unit bifurcating to form the sister groups Platyrrhini and Catarrhini. Platyrrhini splits to form Callitrhicidae and Cebidae; Catarrhini then divides into Cercopithecoidea and Hominoidea. Cercopithecoidea contains a monophyletic Colobinae, and the papionine cluster of *Papio* and *Cercocebus*, which is most closely related to *Macaca*.

The tree depicts the Hominoidea as a monophyletic unit, with hylobatids diverging first, followed by *Pongo*. Members of the Hominidae *(Homo, Pan,* and *Gorilla)* are more closely related to one another than to the Asiatic apes. The parsimony cladogram depicted in Fig. 1 suggests a closer relationship between *Homo* and *Pan* than between *Pan* and *Gorilla* or *Homo* and *Gorilla.*

Parsimony Analysis of Individual Proteins

Maximum parsimony analysis of amino acid sequence data has generated phylogenetic trees for alpha hemoglobin, beta-type hemoglobin, myoglobin, cytochrome *C*, fibrinopeptides A and B, carbonic anhydrase I and II, and alpha lens crystallin.

Alpha Hemoglobin Sequences. Figure 2 depicts relationships within Primates based upon parsimony analysis of 83 alpha hemoglobin sequences. The overall branching order of lineages within this dendrogram closely resembles that displayed in Fig. 1.

Within the Old World monkey grouping, where several species are represented, the leaf-eating Colobinae *(Presbytis* and *Colobus)* separate from the Cercopithecidae. Relationships within the Cercopithecidae provide evidence to support the concept of a monophyletic Papionini; *Cercocebus, Papio,* and *Gelada* cluster to form a single assemblage, most closely related to *Macaca*. All macaque species cluster into a single grouping.

Hominoidea follows the branching arrangement expected from traditonal taxonomies, with the exception of an allelic variant of *Gorilla* alpha which emerges separately from the hominoid stem. The Asiatic ape *Pongo* separates from the African apes and humans, with *Homo* and *Pan* sharing identical alpha hemoglobin sequences.

Beta-type Hemoglobin Sequences. The dendrogram shown in Fig. 3 was produced from parsimony analysis of 103 beta and delta hemoglobin sequences. The relationships between higher primate taxa suggested by this phylogenetic tree are similar to those indicated by the tree in Fig. 2. The branching order within Cercopithecidae agrees closely with that suggested by parsimony analysis of alpha hemoglobin sequences. Colobine lineages leave the main Old World monkey branch first, followed by the cercopithecines. Again a monophyletic unit comprised of *Cercocebus, Papio, Gelada,* and the macaques provides evidence for a papionine tribe.

Beta hemoglobin sequences in *Pan* and *Homo* are identical to one another, and to the putative ancestral sequence of the human-African ape ancestor. Thus gorilla, human, and chimpanzee lineages form a trichotomous relationship which does not clearly demonstrate which of the two are closer to one another.

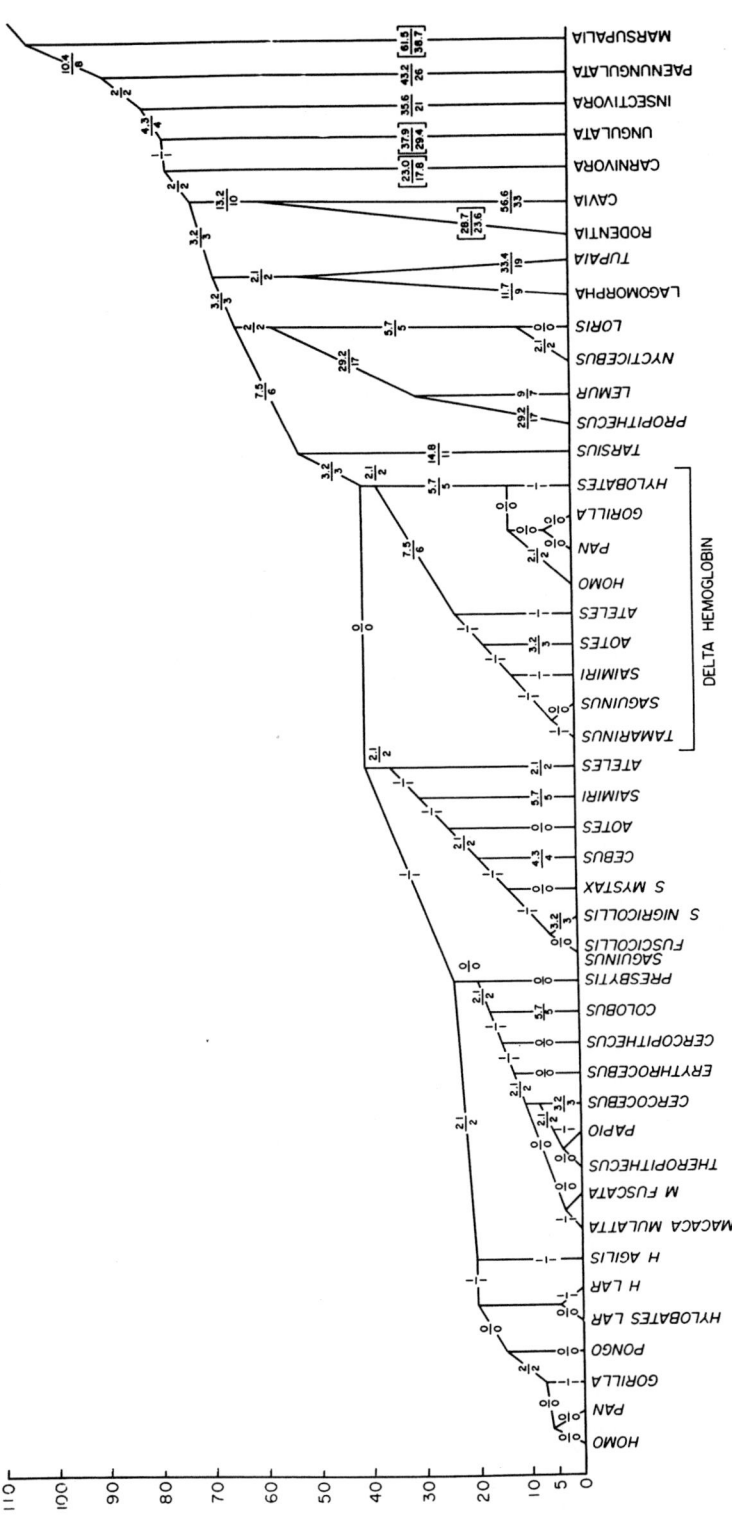

Fig. 3. Most parsimonious tree for beta-type hemoglobin. Scientific names of species and groups of species depicted in this dendrogram which differ from those listed in the legend of Fig. 2 are as follows: *Aotes* (*Aotus* sp.); *Saguinus* (*S, mystax* delta); Rodentia (*Mus musculus* C57 B1 mouse, *M. musculus* AKR mouse, *M. cervicolor* major D-like, *M. cervicolor* major S-like, *M. musculus* minor, *Rattus* sp., *Lemmus sibricus, Ondatra zibethecus, Clethrionomys ruticus, Microtus yanthognathus*); *Cavia* sp., Carnivora (*Canis familiaris, C. latrans, Orocyon cinereoargenteus, Thalarctos maritimus, Meles meles, Procyon lotor, Nasus narica, N. narica* slow beta, *N. nasua, Felis felis, Panthera leo, Felis catus* slow beta); Ungulata (*Ovis aries, Capra* sp., *Ovis aries* C, *Capra* sp. C, Barbary sheep C, *Bos taurus* A, *B. taurus, B. taurus* fetal, *Ovis aries* fetal, *Llama* sp., *Camelus dromedarius, Sus scrofa, Equus equus*); Insectivora (*Erinaceus europaeus*); Paenungulata (*Elephas maximus*), Marsupalia (*Macropus cangaru, M. cangaru* two, *Didelphis* sp.). Unaugmented and augmented link length values are displayed as in Fig. 2. Divergence dates for ancestral nodes were established by the procedure described in the legend of Fig. 2

The genealogy shown in Fig. 3 depicts a single duplication of beta-type genes in the early Anthropoidea, descending to form ceboid delta and hominoid delta branches. The ceboid delta and hominoid delta branches thus appear to be more closely related to one another than to the ceboid beta or hominoid beta branches, respectively. Phylogenetic relationships within the hominoid delta grouping demonstrate the clearly hominoid affinity of *Hylobates*.

Myoglobin Sequences. The phylogeny of 17 primate myoglobin sequences is presented in Fig. 4. This dendrogram was based upon investigation of 65 myoglobin lineages. With this dendrogram the strepsirhine branch depicts the classical Lorisiformes and

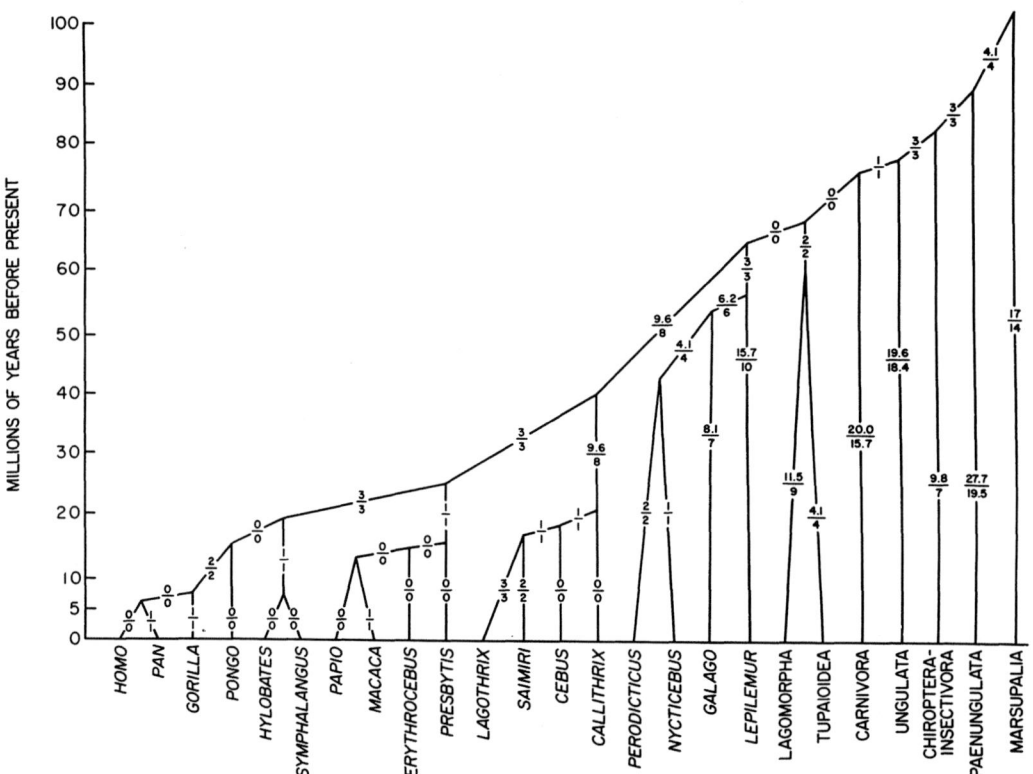

Fig. 4. Most parsimonius tree for myoglobin. Scientific names of species and groups of species depicted in this dendrogram which differ from these listed in the legend of Fig. 2 are as follows: *Symphalangus (S. syndactylus; Lagothrix (L. lagothrix); Saimiri (S. sciureus); Callithrix (C. jacchus); Perodicticus (P. potto); Galago (G. crassicaudatus); Lepilemur (L. mustelinus);* Carnivora *(Phoca* sp., *Eumetopias* sp., *Canis familiaris, Urocyon cineroargentus, Lycaon pictus, Meles meles);* Ungulata *(Sus scrofa, Cervus elaphus, Bos taurus, Ovis aries, Phocoenoides dalli, Delphinus delphis, Orcinus orca, Globicephala melaena, Balaenoptera acutorostrata, B. physalus, Megatpera nouaengliae; Eschrichtius gibbosus, Kogia sinus);* Ungulata *(Equus equus, E. zebra);* Chiroptera-Insectivora *(Rousettus aegyptiacus, Erinaceus* sp.); Paenungulata *(Elephas maximus, Loxodonta africana);* Marsupalia *(Macropus cangaru, Didelphis* sp.). Unaugmented and augmented link length values are displayed as in Fig. 2. Divergence dates for ancestral nodes were established by the procedure described in the legend of Fig. 2

Lemuriformes as coequal sister groups, with the lorisid genera *Galago* and *Perodicticus* forming a cluster more closely related to one another than either is to *Lemur*. *Homo*, *Pan*, and *Gorilla* split as a trichotomy from the node of their common ancestor, with *Pan* and *Gorilla* each displaying one NR divergence from the ancestral condition.

Fibrinopeptides A and B and Cytochrome C Sequences. Parsimony analysis of 47 sequences of fibrinopeptides A and B yielded the order of ancestral branching among ten primate lineages represented in Fig. 5. The outlines of this dendrogram conform closely to those suggested by the globin gene phylogenies.

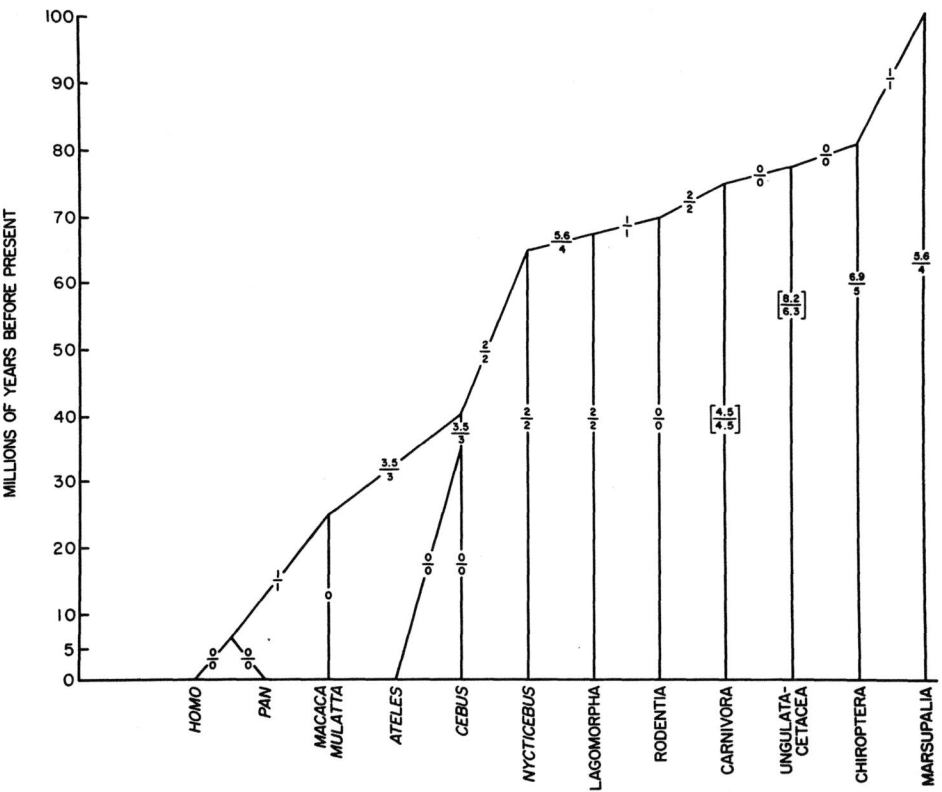

Fig. 5. Most parsimonious tree for fibrinopeptide A and B. Scientific names of species and group of species depicted in this dendrogram which differ from those listed in the legend of Fig. 2 are as follows: *Symphalangus (S. syndactylus)*; Papio *(P. leucophaeus)*; Lagomorpha *(Oryctolagus cuniculus)*; Rodentia *(Rattus* sp.); Carnivora *(Urocyon* sp., *Canis familiaris, Felis felis, Panthera leo)*; Ungulata *(Cervus elaphus, C. canadensis, C. nippon, Muntiacus muntjak, Odocoileus hemionus, Rangifer tarandus, Bison bonansus, Bos taurus, Bubalus bubalus, Syncerus caffer, Capra* sp., *Ovis aries, Gazella* sp., *Giraffa camelopardalis, Antilocapra americana, Llama* sp., *Vicugna vicugna, Camelus dromedarius, Sus scrofa, Equus asinus, E. caballus* X *asinus, E.* sp.zebra, *E. equus, Tapirus terrestris, Dicerus simus)*. Paenungulata *(Elephas maximus)*; Marsupalia *(Macropus cangaru)*. Unaugmented and augmented link length values are displayed as in Fig. 2. Divergence dates for ancestral nodes were established by the procedure described in the legend of Fig. 2

Figure 6 shows the most parsimonious arrangement of 16 eutherian lineages of cyto-chrome *c* based upon a parsimony analysis of 94 sequences. Within this dendrogram, Primates is part of a eutherian assemblage which also includes Lagomorpha and Roden-tia. The primate-lagomorph-rodent unit appears to be most closely related to Carnivora. Only four primate sequences are represented in this phylogenetic tree, and these dis-play relationships identical to those suggested by classical taxonomy. *Nycticebus* leaves the main primate stem first, followed by *Ateles* and finally a *Homo-Macaca* grouping.

Alpha Lens Crystallin. Parsimony analysis of 18 sequences of alpha lens crystallin is the basis for the dendrogram in Fig. 7. As seen in Fig. 6, Primates is part of a larger eutherian cluster including the Lagomorpha, Rodentia, and also Tupaioidea. The four primate sequences are arranged in two groupings of two lineages each, forming strep-sirhine and catarrhine branches.

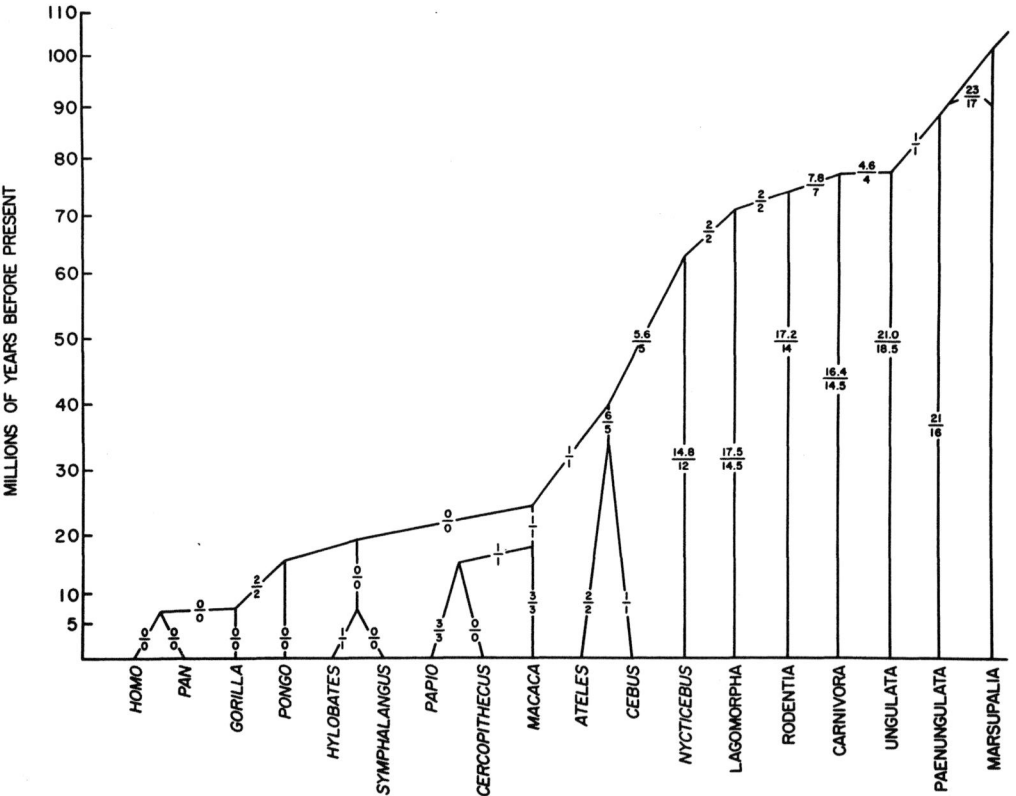

Fig. 6. Most parsimonious tree for cytochrome *C.* Scientific names of species and species groups depicted in this dendrogram which differ from those listed in the legend of Fig. 2 are as follows: Lagomorpha *(Oryctolagus cuniculus);* Rodentia *(Mus musculus);* Carnivora *(Canis familiaris, Phoca* sp.); Ungulata-Cetacea *(Equus asinus, E. equus Ovis aries, Camelus dromedarius, Hippopotamus* sp., *Sus scrofa, Eschrichtus glaucus);* Chiroptera *(Rousettus* sp.), Marsupalia *(Macropus cangaru).* Unaugmented and augmented link length values are displayed as in Fig. 2. Divergence dates for ancestral nodes were established by the procedure described in the legend of Fig. 2

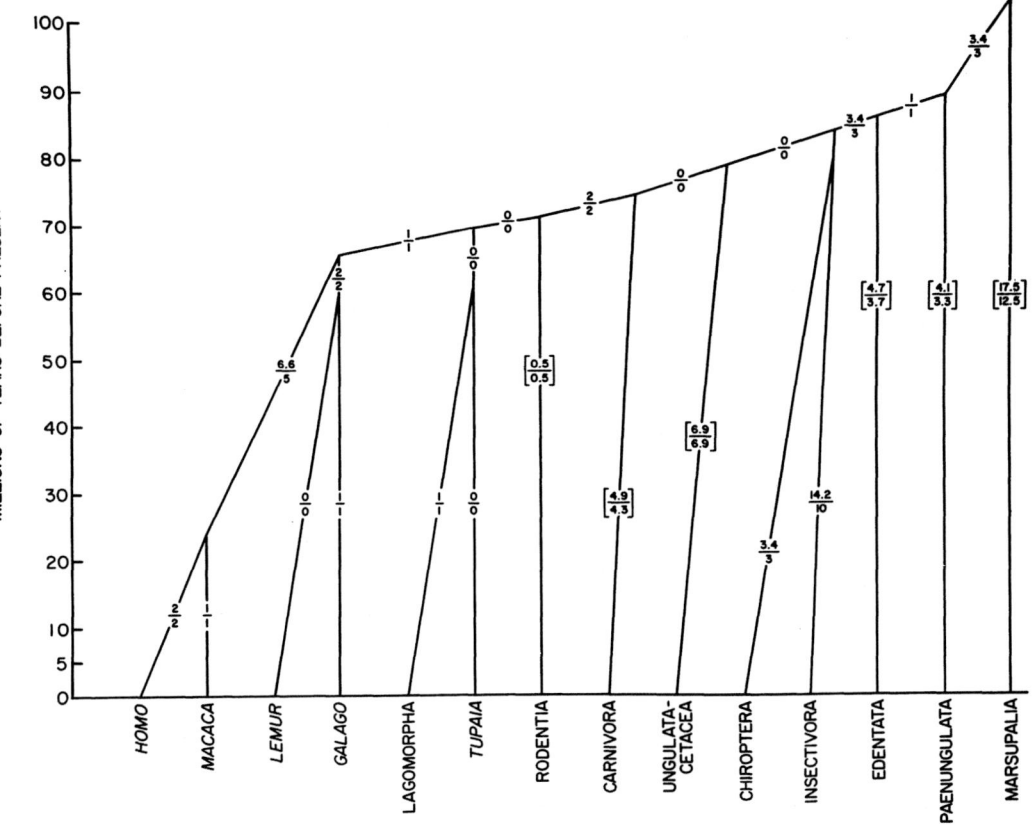

Fig. 7. Most parsimonious tree for alpha-lens crystallin. Scientific names of species and groups of species depicted in this dendrogram which differ from those listed in the legend of Fig. 2 are as follows: Lagomorpha *(Oryctolagus cuniculus, Ochotona princeps)*; Tupaia *(T. glis)*; Rodentia *(Cavia* sp., *Rattus* sp.); Carnivora *(Canis familiaris, Mustela* sp., *Phoca* sp., *Felis felis, Manis pentadactyla)*; Ungulata-Cetacea *(Tapirus terrestris, Dicerus simus, Equus equus, Giraffa camelopardalis, Bos taurus, Camelus dromedarius, Sus scrofa, Balenoptera aeutoristrata, Phocoenoides dalli)*; Chiroptera *(Artibeus* sp.), Insectivora *(Erinaceus europaeus)*; Edentata *(Bradypus tridactylus, Tamandua* sp., *Choloepus* sp.); Paenungulata *(Procavia capensis, Orycteropus afer, Trichechus manatus, Loxodonta africana)*; Marsupalia *(Macropus rufus, Didelphis marsupialis)*. Unaugmented and augmented link length values are displayed as in Fig. 2. Divergence dates for ancestral nodes were established by the procedure described in the legend of Fig. 2

Immunodiffusion Data

Over 10,000 trefoil Ouchterlony plate comparisons have been carried out using rabbit and chicken antiserum to whole serum or purified serum proteins of primate, tree shrew, and elephant shrew species. Computer processing of the Ouchterlony data generates the divergence tree shown in Fig. 8. Major phylogenetic relationships among primate taxa depicted in this dendrogram closely parallel that revealed by parsimony analysis of amino acid sequence data.

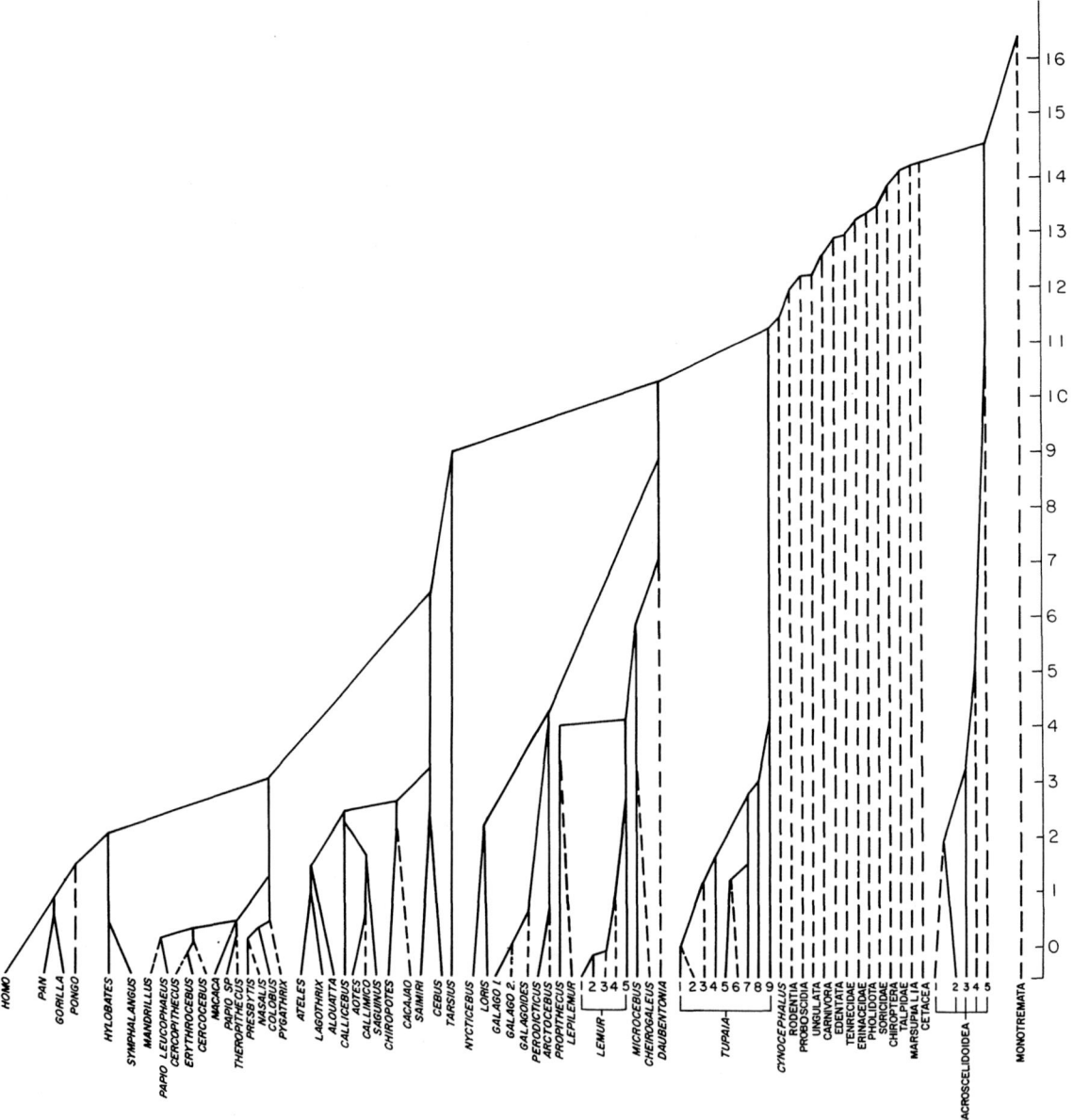

Fig. 8. Divergence tree of primates from immunodiffusion plate comparisons

Primates appears as a monophyletic assemblage, more closely related to the Tupai-oidea than to nonprimate eutherian mammals.

The primate assemblage forms two major lineages; the Strepsirhini, including lori-soid and lemuroid branches, and the Haplorhini. Haplorhini bifurcates into *Tarsier* and Anthropoidea. The latter is composed of two coequal sister groups, the ·Catarrhini and the Platyrrhini. Two catarrhine lineages, the Hominoidea and Cercopithecoidea, diverge to form monophyletic assemblages.

Within Platyrrhini, the callithricid genera *Callimico* and *Saguinus* do not unite most closely, suggesting, as have some parsimony analyses, that Cebidae is polyphyletic. *Cebus* and *Saimiri* join, providing independent evidence for the subfamily Cebinae; the clustering of *Ateles, Alouatta* and *Brachyteles* supports the notion of common ancestry for ateline species and *Alouatta.*

Cercopithecoidea divides to form the classical Colobinae and Cercopithecidae. The concept of a monophyletic Papionini does not gain support from the Ouchterlony data, however. Rather, *Papio, Theropithecus,* and *Macaca* form one phylogenetic cluster, *Cercopithecus* and *Erythrocebus* join *Cercocebus* to form a second cluster, while *Mandrillus* and *Papio leucophaeus* comprise a third assemblage.

Within Hominoidea, classical relationships are confirmed, with hylobatid lineages representing the most ancient splitting followed by the other Asian ape *Pongo. Homo* is closely related to the African apes, but since *Pan* and *Gorilla* join one another, it is not possible to determine which is most closely related to the human line.

DNA Hybridization and Nucleotide Sequence Data

The DNA hybridization technique permits measurement of homologies in both repeated (Hoyer and Roberts 1967) and nonrepeated nucleotide sequences (Kohne 1970; Kohne et al. 1972). Early findings confirmed the phylogenetic relationships among major primate taxa suggested by classical taxonomy and supported by immunological methods. Humans and African apes were found to be more closely to one another than either are to *Hylobates;* hominoid lineages as a whole were more closely related to catarrhine than to platyrrhine monkeys; members of Anthropoidea were found to cluster together when compared to members of Strepsirhini.

Bonner et al. (1980) measured genetic divergence among primate taxa with greater precision by using single copy DNA, which contains more genetic information than the repetitious sequences used in earlier studies. These DNA hybridization results provided further support for the Strepsirhini-Haplorhini bifurcation of Primates, as *Tarsius* was found to be closer genetically to anthropoids than to lemurs and lorises. Bonner's work with single copy DNA (personal communication to M.G. 1981) also suggests that the mouse lemur *(Microcebus* and *Cheirogaleus)* has phylogenetic affinities with Lemuroidea, rather than Lorisoidea.

Most recently, mitochondrial DNA analysis has permitted finer resolution of phylogenetic relationships with Hominoidea, due to the exceptionally rapid rate of mtDNA evolution. In a study of mtDNA for 11 primate species, Brown et al. (1980) confirmed the monophyly of Hominidae; within the Hominidae, *Pan* and *Gorilla* formed a sister group, with *Homo* then joining the African ape assemblage.

Processes of Molecular Change

Analyses of change at the molecular level, in additon to clarifying phylogenetic relationships, can reveal the relative importance of natural selection, neutral mutation and random genetic drift, processes which underly evolution. Maximum parsimony analysis of individual proteins may be used to investigate both the tempo of evolutionary change for different lineages and the degree of uniformity in rates of change for different lineages. Understanding how and when molecular change occurs can reveal to what extent we may rely upon models, such as a molecular clock, for describing and timing evolutionary events.

Since the maximum parsimony method reconstructs putative ancestral mRNA sequences for each internal node within the branching network of a dendrogram, it is possible to calculate rates of molecular change for individual lineages during particular periods of time in the past. These rates of change are calculated as nucleotide replacements per 100 codons per 100 million years.

Rates of Evolution

A detailed investigation of rates of evolutionary change has been conducted for the respiratory enzyme, cytochome c (Baba et al. 1981) and for the globin genes (Goodman et al. 1975).

In the study of 87 cytochrome c sequences, rates were calculated from the time of procaryote-eucaryote divergence to the present. Three major periods of alternating acceleration and deceleration in rates of cytochrome c evolution were found (see Fig. 9). Caution was exercised in interpreting rates of change for the period of early eucaryote divergence, due to paucity of evidence for dates of divergence in this time period. Early rates of change between the time of the divergence of *Tetrahymena* vs other eucaryotes and the time of divergence of plants from animals (1400 MyBP to 900 MyBP) was determined to be 8.2%, twice as rapid as the overall rate of change in the vertebrate lineages during the last one-half billion years (4.8%). The rate rises to 14.9% (triple the vertebrate average rate) during the first period of acceleration from the plant vs animal split to the arthropod-snail-starfish vs vertebrate divergence (900 MyBP to 680 MyBP). A succeeding drop in rate is followed by a second peak in the evolutionary period between the divergence of shark vs other jawed vertebrates to teleosts vs tetrapods and then continues to climb to 11.3% between teleost vs tetrapods to frog vs amniotes. A drop in the rate of change to half the vertebrate average occurs between 340 MyBP and 300 MyBP (frog vs amniotes to bird-reptile vs mammals); in the line from the bird-reptile vs mammal ancestor to the basal eutherian ancestor it falls to 1.4%.

Of special interest is the dramatic speedup in rates of cytochrome c evolution in the line leading from the basal eutherian ancestor to the primates. The rate of change between the divergence point of primate-rabbit-rodent vs other eutherians to the spider monkey vs catarrhine divergence point (90 MyBP to 40 MyBP) is 17.3%, the fastest rate calculated for any evolutionary period. The rate slows once more in the line leading to humans.

Maximum parsimony analysis of globin genes also revealed distinct periods of accelerated and decelerated rates of change (Goodman et al. 1975). Two separate accelerations and subsequent decelerations in the evolution of these proteins have been found; the first acceleration occurs between the vertebrate and bird-mammal ancestors; the second, between the eutherian and anthropoid ancestors.

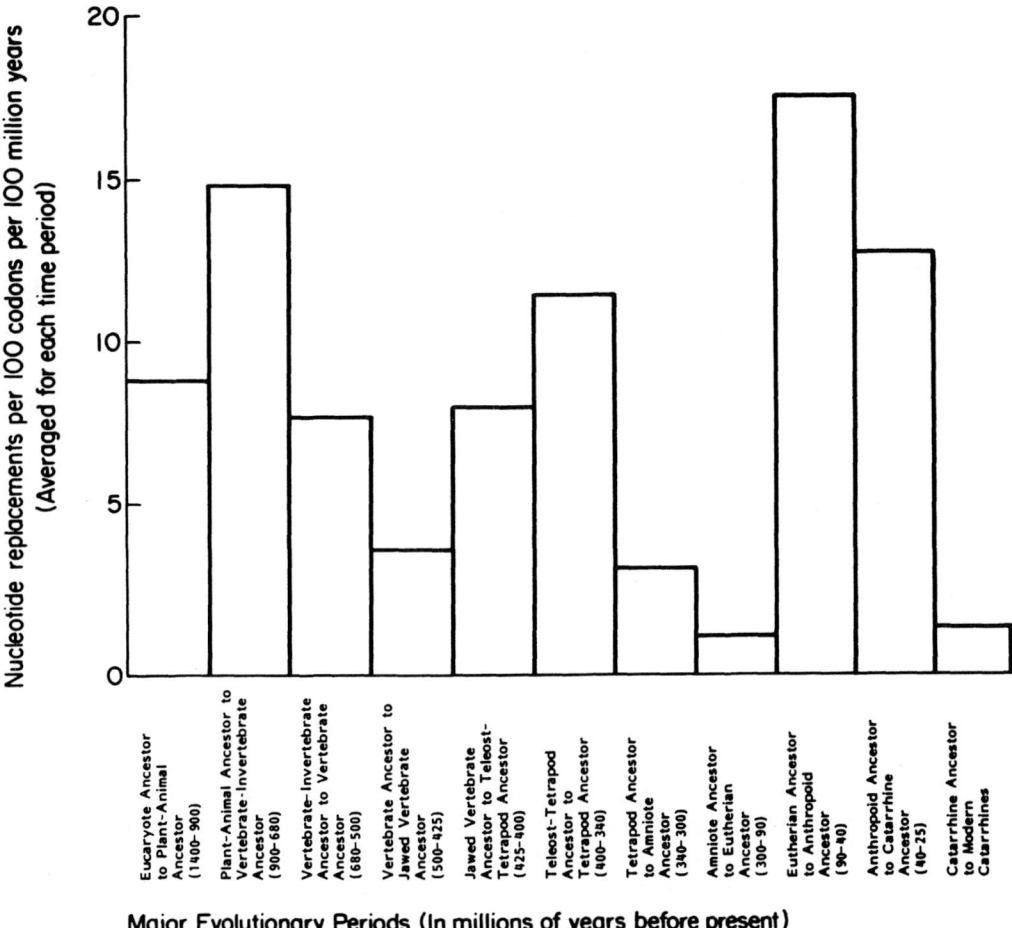

Fig. 9. Fluctuations in the tempo of evolutionary change for cytochrome *c*

Table 1 summarizes and compares rates of change in four molecules during the past one-half billion years. The evolution of alpha and beta hemoglobin and cytochrome *c* depicted in this table reflects not only nonuniform rates of change over time, but also a remarkably synchrony in the tempo of change for these four protein chains.

Rates of nucleotide replacements accelerated for all of the globin genes during the early vertebrate period, and continued to be elevated for all four proteins between the time of the emergence of the gnathostome ancestor and the time for the emergence of the amniote ancestor. The second acceleration for all of these proteins begins with the emergence of the eutherian ancestor and persists until the emergence of the anthropoid ancestor from the strepsirhine branch. Finally, change in all of the proteins decelerated following the divergence of the anthropoid ancestor, with the slowdown continuing until the present.

Table 1. Rates of evolution in alpha and beta hemoglobin, myoglobin and cytochrome c

Evolutionary period	Age (MyBP) [a]	α hemoglobin NR% [b]	β hemoglobin NR%	Myoglobin NR%	Cytochrome c NR%
Vertebrate to gnathostome ancestor	500–425	96.7	97.4	88.0	4.4
Gnathostome ancestor to present	425–0	28.5 (20.3–36.1)	28.1 (20.4–34.3)	27.4 (19.7–30.3)	5.0 (2.9–13.0)
Gnathostome to amniote ancestor	425–300	62.4	54.8	65.9	8.5
Amniote ancestor to present	300–0	15.0 (8.9–25.1)	17.2 (8.7–24.8)	12.3 (4.8–15.8)	3.9 (1.3–15.7)
Amniote to eutherian ancestor	300–90	10.5	13.0	10.6	1.4
Eutherian ancestor to present	90–0	25.7 (14.0–37.0)	31.2 (16.8–49.9)	17.8 (6.0–27.9)	6.0 (1.1–17.2)
Eutherian to Anthropoidea ancestor	90–40	36.9	43.8	22.2	19.0
Anthropoidea ancestor	40–0	5.3	8.6	14.7	7.0

[a] Millions of years before the present
[b] Nucleotide replacements per 100 codons per 100 Myr

Such a synchronized acceleration-deceleration pattern of evolution for molecules involved in energy metabolism is to be expected if certain adaptive radiations placed greater selective pressure on energy-associated proteins to develop new or altered functions. If the selectionist model is correct, decelerations represent the action of stabilizing selection holding newly perfected functions constant. During periods of relatively slow evolution, stochastic processes may play a relatively greater role in the actual change of the molecule, with those amino acid residue positions not involved in known functions being the primary areas of change. Neutral mutations may play a relatively greater role in the shaping of amino acid change during periods of slow evolution, since functional areas would be held constant by stabilizing selection, and most change expected to be the result of random drift.

Bonner et al. (1980) recently provided additional support for the concept of non-uniform rates of change at the molecular level by demonstrating that rates of DNA evolution have been markedly slower among some primate lineages. Hybridization studies of single copy DNA revealed that the DNA of Malagasy primates has evolved at a rate one-half as fast as lorises and tarsiers since the time of the strepsirhine-haplorhine divergence.

Structural-Functional Analysis of Molecular Change

It is possible to test the selectionist model of evolutionary change by comparing rates of evolution for different structural-functional areas within a protein during different

periods of time. If selection is the major force directing change at the primary structure level, then periods of rapid evolution should be characterized by elevated rates of
amino acid substitutions in regions of a molecule associated with critical functions,
relative to rates of change for residue positions of unknown function. Accelerated evolution in functional areas is to be expected during periods of rapid protein evolution,
as positive selection acts to alter or improve the functioning of molecules in species
radiating into new ecological zones.

Structural-functional analysis of the cytochrome c protein (Baba et al. 1981) suggests that during those periods of accelerated overall rates of change, the oxidase-reductase area of the molecule showed an elevated rate of nucleotide replacement relative
to the rate displayed by exterior positions of no known function. During the time
period between the plant-animal-fungi ancestor and vertebrate-invertebrate ancestor
the average amount of change for the oxidase-reductase area (number of nucleotide
replacements in this area divided by the number of amino acid positions in this area)
exceeds the average amount of change for all positions. A rapid rate of change for the
protein in general, and for the oxidase-reductase area in particular, would be expected
during the emergence of the Metazoa when changes in energy utilization by the cells
was taking place. There was also, at this time, a more rapid rate of change in residue
positions associated with the heme group than is expected for those sites during later
time periods.

Evidence for elevated rates of change in the oxidase-reductase area is even stronger
for the time period of the primate radiation, especially before the emergence of the
anthropoid ancestor, when the overall rate of evolution for cytochrome c was rapid.
In contrast, nonprimate animals (including birds, reptiles, and mammals) show reduced
rates of change in the oxidase-reductase area during this time period. In the early primate line the rate of change in this critical functional area is 4 times greater than the
average rate of change in the molecule as a whole, and 13 times greater than the rate
for exterior positions with no known function. Such rapid change in the oxidase-reductase area during early primate evolution may account for the differences in cytochrome c activity found between higher primates and nonprimate mammals (Borden
et al. 1978).

The action of stabilizing selection is seen in the time period between the anthropoid ancestor and the catarrhine ancestor; the molecule was then characterized by an
overall slow rate of molecular change, and there were no nucleotide replacements in any
functional area of the molecule during that period of time. In each of the lines leading
to extant primate species *(Homo, Pan, Macaca,* and *Ateles)* there is no change in any
of the functional areas, and a slow rate of change in the molecule as a whole. Such
a deceleration in rates of change within functional areas supports the hypothesis
(Goodman 1976) that stabilizing selection slows the rate of molecular evolution once
new or altered functions have been perfected.

Structural-functional analysis of molecular evolution among globin genes provides
further evidence for the selectionist model of primary structure change (Goodman
et al. 1975). Functional analyses of nucleotide changes in globin genes during the time
of rapid evolution found in the early vertebrate period reveals an elevated rate of substitutions in the prospective $A_1 B_1$ subunit contact sites compared to external sites
without known function. Immediately after the gene duplication event resulting in

alpha and beta chains, elevated nucleotide substitutions occurred in those amino acid residue positions associated with cooperative functions, A_1 B_1 contacts, the Bohr effect, and DPG binding. These changes can reasonably be attributed to adaptive substitutions.

During the following period of decelerated globin change, residue positions which had previously acquired functions became highly conservative.

The period of acceleration during early primate globin evolution was accompanied by an elevated rate of change at those positions involved in A_1 B_1 contact sites. Evolution of functional sites, and change in the molecules as a whole, then became quite conservative in the lineage leading to the higher primates. This analysis of the evolution of the globin genes also suggests a prominent role for positive and stabilizing selection.

The Clock Model of Evolutionary Change

Nonuniformity in rates of nucleotide replacement for several proteins analyzed thus far suggests that a molecular clock cannot generate accurate divergence dates from amino acid sequence data. This hypothesis has been tested by applying a clock model to primary sequence data from eight proteins, including alpha and beta hemoglobin, myoglobin, fibrinopeptides A and B, cytochrome c, alpha lens crystallin, and carbonic anhydrase I and II.

Clock dates for major divergence events within Primates were calculated for eight individual proteins (Tables 2 and 3). The clock for each protein was calibrated at the point of earliest splitting within the Eutheria. The legends of Tables 2 and 3 explain the rationale for calibration dates assigned to each protein. Clock dates were calculated for both unaugmented and augmented nucleotide replacement values.

Average clock dates for each major divergence point were also determined. Table 4 displays average clock dates calculated by weighing data from each protein equally, i.e., straight averaging of the individual protein dates shown in Tables 2 and 3. Another set of average clock dates was produced by weighing the value of each nucleotide replacement equally, i.e., nucleotide replacement values were summed before clock times were calculated. Average clock dates produced by this later method and average dates shown in Table 4 displayed no significant differences. The average clock dates shown in Table 4 are far too recent given well established fossil evidence.

Parsimony analysis of rates of evolution for several proteins (Goodman et al. 1975, Baba et al. 1981) has demonstrated that molecular change is characterized by alternating periods of acceleration and deceleration. Since the relative magnitude of the decelerations has tended to increase toward the present, divergence dates generated by the clock model become more unreasonable the more recent the evolutionary event. This explains why the clock dates for the *Homo-Pan-Gorilla* and *Homo* vs *Pan* splits are so inaccurate, since recent discoveries in East Africa suggest that *Australopithecus* existed at least 5 million years ago. The clock dates presented here will be even more in error if *Ramapithecus* is ultimately placed in the ancestry of *Homo* rather than that of *Pan*.

Obviously, the longer two species have been diverging from their most recent common ancester, the greater will be the genetic distance between them. Such amounts

Table 2. Clock dates for major divergence events calculated by maximum parsimony analysis of amino acid sequence data for individual proteins. (Calculations based on *unaugmented* nucleotide replacement values)

Divergence events	Clock dates for individual proteins (MyBP)							
	β Hb	Myo	Fib A & B	α lens crys.	α Hb	Cyt c	CA I	CA II
Paenungulata vs remaining Eutheria [a]	90	90	90	90				
Chiroptera-Insectivora vs remaining Eutheria	84.7	78.4		50.9	80.5	69.9		
Ungulata-Cetacea vs remaining Eutheria	74.1	68.2	86.3	49.9	69.9	69.9	69.9	69.9
Strepsirhini vs Haplorhini	45.9	61.0	45.2	43.8	41.6	58.9		
Playrrhini vs Catarrhini	17.8	29.5	19.5		25.2	40.3		14.1
Cercopithecoidea vs Hominoidea	13.9	13.0	13.3	14.6	23.3	7.9	18.0	2.9
Hylobatidae vs Hominidae	4.9	6.7	8.7					
Pongo vs Hominidae	4.8	8.1	7.5		3.3		11.0	
Homo-Pan-Gorilla	0.9	2.7	0		1.8			
Homo vs *Pan*	0	2.0	0		0	0	4.0	

[a] Calibration date based on fossil evidence

The clock dates shown in this table were calculated on the basis of a calibration date of 90 million years ago, representing the earliest splitting within the Eutheria. Maximum parsimony analysis of data sets including *Proboscidea* sequences depict *Proboscidea* as the most ancient lineage within Eutheria, thus the point of divergence for Paenungulata (a superorder including the order *Probiscidea*) vs remaining Eutheria was set at 90 MyBP. Clock dates for other early divergence points within Eutheria (e.g., Chiroptera-Insectivora) vs remaining Eutheria and Ungulata-Cetacea vs remaining Eutheria) were initially calculated from data sets calibrated at 90 MyBP for the Paenungulata vs remaining Eutheria ancestor. Such clock dates were then used to calibrate the clock for those data sets which did not include *Proboscidea*

Table 3. Clock dates for major divergence events calculated by maximum parsimony analysis of amino acid sequence data for individual proteins. (Calculations based on *augmented* nucleotide replacement values

Divergence event	Clock dates for individual proteins (MyBP)							
	β Hb	Myo	Fib. A & B	α lens crys.	α Hb	Cyt. c	CA I	CA II
Paenungulata vs remaining Eutheria [a]	90	90	90	90				
Chiroptera-Insectivora vs remaining Eutheria	85.3	78.4		62.2	83.5	72.7		
Ungulata-Cetacea vs remaining Eutheria	75.4	69.2	86.4	59.7	72.7	72.7	72.7	72.7
Strepsirhini vs Haplorhini	45.6	62.8	44.6	45.2	43.8	58.7		
Platyrrhini vs Catarrhini	15.4	28.2	18.5		27.6	43.6		12.7
Cercopithecoidea vs Hominoidea	12.1	11.9	12.1	12.8	24.9	7.5	17.3	2.6
Hylobatidae vs Hominidae	4.7	6.2	7.9					
Pongo vs Hominidae	4.2	7.4	6.8		2.7		10.2	
Homo-Pan-Gorilla	0.8	2.5	0		1.5			
Homo vs *Pan*	0	1.8	0		0	0	3.9	

[a] Calibration date based on fossil evidence

Calibration dates for data sets represented in this table were determined according to the procedure described in the legend of Table 2

Table 4. Clock date averages for major divergence events

Divergence event	Clock date averages (MyBP)		
	Unaugmented NR values	Augmented NR values	Protein data used [b]
Paenungulata vs remaining Eutheria [a]	90	90	$\beta mF\alpha$-ℓ
Chiroptera-Insectivora vs remaining Eutheria	72.8	76.4	βm α-$\ell\alpha c$
Ungulata-Cetacea vs remaining Eutheria	69.9	72.7	$\beta mF\alpha$-$\ell\alpha$ I II
Strepsirhini vs Haplorhini	49.4	50.1	$\beta mF\alpha$-eqc
Plathyrrhini vs Catarrhini	24.4	24.3	βmF αc
Cercopithecoidea vs Hominoidea	13.4	12.7	$\beta mF\alpha$-$\ell\alpha c$ I II
Hylobatidae vs Hominidae	6.8	6.3	βmF
Pongo vs Hominidae	7.0	6.3	βmF α I
Homo-Pan-Gorilla	1.4	1.2	βmF α
Homo vs *Pan*	1.0	1.0	βmF αc I

[a] Calibration date based on fossil evidence
[b] β = β hemoglobin; m = myoglobin; F = fibrinopeptides A & B; α-ℓ = α lens crystallin; α = α hemoglobin; c = cytochrome c; I = carbonic anhydrase I; II = carbonic anhydrase II

of genetic divergence can be averaged and correlated positively with time elapsed since divergence. But to date, such correlations are meaningful only in so far as they are general and relative; precise divergence dates cannot be generated from amino acid sequence data. However, advances in nucleotide sequencing may demonstrate a greater utility for the molecular clock in the future. Nucleotide sequence data, unlike amino acid sequence data, are able to capture the full extent of nucleotide substitutions, as they reveal the presence of silent mutations. Since silent mutations seem to constitute a large proportion of the total amount of nucleotide change (Jukes 1980), clock dates generated on the basis of nucleotide sequence data may more closely approximate those expected from analysis of the fossil record. This assumes that generation-time will not be found to greatly influence nonuniform rates of molecular change. However, the theoretical implications of the clock concept can still be misleading. Proponents of the clock have generally viewed molecular evolution as a process directed by the random walk of mutations. Structural-functional analysis of amino acid sequence change in the globins (Goodman et al. 1975) and cytochrome c (Baba et al. 1981) have demonstrated that positive and stabilizing selection is the most important process governing the change of molecules over time.

References

Baba ML, Darga LL, Goodman M, Czelusniak J (1981) Evolution of cytochrome *c* investigated by the maximum parsimony method. J Mol Evol 17:197–213

Bonner TI, Heinemann R, Todaro GJ (1980) Evolution of DNA sequences has been retarded in Malagasy primates. Nature (London) 286:420–424

Borden D, Ferguson-Miller S, Tarr G, Rodriquez D (1978) Fed Proc 36 (6):1517

Brown WM, Cann RL, Ferris SD, George M, Wang A, Wilson AC (1980) The assessment of genetic variation and evolutionary relationships using mitochondrial DNA analyses. Abstr 2nd Int Cong Syst Evol Biol, Univ British Columbia, Vancouver, p 34

Ferris SD, Brown WM, Wilson AC (1980) Mitochondrial and nuclear DNA diversity in apes and humans. Abstr Int Cong Syst Evol Biol, Univ British Columbia, Vancouver, p 196

Goodman M (1976) Toward a genealogical description of the primates. In: Goodman M, Tashian RE (eds) Molecular anthropology. Plenum Press, New York, pp 321–353

Goodman M, Moore GW (1971) Immunodiffusion systematics of the primates. Part I. The Catarrhini. Syste Zool 20:19–62

Goodman M, Moore GW, Matsuda G (1975) Darwinian evolution in the genealogy of hemoglobin. Nature (London) 253:603–608

Goodman M, Caelusniak J, Moore GW, Romero-Herrera AE, Matsuda G (1979) Fitting the gene lineage to the species lineage; a parsimony strategy illustrated by cladograms. Syste Zool 28:132–163

Hennig W (1966) Phylogenetic systematics. Univ Chicago Press, Chicago

Hoyer BH, Roberts RB (1967) Studies of nucleic acid interactions using DNA-agar. In: Taylor H (ed) Molecular genetics, part II. Academic Press, London New York, pp 425–479

Hoyer BH, McCarthy BJ, Bolton ET (1964) A molecular approach to the systematics of organisms. Science 144:954–967

Hoyer BH, van de Velde, NW, Goodman M, Roberts RB (1972) J Hum Evol 1:645–649

Jukes JH (1980) Silent nucleotide substitutions and the molecular evolutionary clock. Science 210:973–978

Kohne DE (1970) Evolution of higher organism DNA. Q Rev Biophys 3:327–375

Kohne DE (1976) DNA evolution data and its relevance to mammalian phylogeny. In: Goodman M, Tashian RE (eds) Molecular anthropology. Plenum Press, New York, p 249

Kohne DE, Chiscon JA, Hoyer BH (1972) Evolution of primate DNA sequences. J Hum Evol 1:627–644

Moore GW (1971) A mathematical model for the construction of cladograms. Mimeogr Ser No 731. North Carolina Univ Inst Statistics

Moore GW (1976) In: Goodman M, Tashian RE (eds) Molecular anthropology. Plenum Press, New York

Sokal RR, Michener CD (1958) A statistical method for evaluating systematic relationships. Kans Univ Sci Bull 38:1049–1438

Immunogenetic Evolution of Primates

J. RUFFIÉ [1], J. MOOR-JANKOWSKI [2], and W. SOCHA [2]

Introduction

The comparative immunogenetic studies of blood groups in primates, including man, which are the fruits of 15 years of collaboration between the French groups at College de France, Paris, and the American group at the Laboratory for Experimental Medicine and Surgery in Primates (LEMSIP) of the New York University School of Medicine, have allowed a better understanding of certain important aspects of the evolutionary mechanisms.

Evidently, it would be qualified as naive to consider that the major blood group systems of man (classified by us as "human types") have appeared de novo in our own species. We must assume that complex organs, such as the brain of man or camerular eye did not appear all at once, but have developed in successive steps. Their development required a very long time and stepwise sequence of formation. We find them — sometimes still in a primitive state — in the species which preceded us and their study should allow us to follow up the development and adaptative meaning of various human organs, and thus lead us to a better understanding of their structure. The immunological systems are no exception to this "rule of parentage". They underwent a long phylogenic preparation prior to their arrival at the state of the systems presently defined in man and in the higher primates. It is the analysis of blood groups of the simian type that reveals some of the past phylogenic steps. We shall present here some salient points resulting from these studies.

Paleosequences and Neosequences

When one compares the blood group system in monkeys and in man (for example, the M-N or Rh systems which are the best known) one recognizes two classes of blood factors: *paleosequences* and *neosequences*.

The former category is represented by the following factors of the M-N-blood system: N, N^{vg}, M, H^e, Mi^a; and following blood factors of the Rh-Hr system: Rh_0 (D),

1 Laboratoire d'Anthropologie Physique, Collège de France, Paris, France
2 LEMSIP, NY School of Medicine, 550 First Avenue, NY 10016, USA

R^c, and hr' (c) (Moor-Jankowski and Wiener 1972). These factors (or fractions of blood factors) which are found in a number of closely related species of *Hominoidea* (Socha and Moor-Jankowski 1979) must have existed already in the common precursor species of those presently existing in nature, but have passed through numerous steps in their development without undergoing notable modifications. They represent the paleosequences of DNA which have not changed since early stages of evolution.

The neosequence factors on the other hand are not found but in a single species, or in a few closely related species. These are, for example:

In the M-N system: the series S/s of man (Wiener et al. 1972)
the series A^c, B^c, D^c of anthropoid apes
In the Rh system: the series rh''/hr'' (E/e) of man (Moor-Jankowski et al. 1973)
the series C^c, E^c and F^c of chimpanzees

It seems that we are dealing here with more recent sequences, i.e., neosequences, each of which has evolved separately in a different phylum. Nevertheless, they all reveal the same phylogenic origin. They are solely witnesses of a diversified evolution which, starting from a common branch, provided origins for a number of more or less specialized zoological groups.

Polymorphism

Proceeding from the transspecific to intraspecific level, one can ask the question *"What is the significance of the very high genetic polymorphism demonstrated by the study of blood groups in primates?"* One could also ask *"What can we learn as to the actual mechanisms of speciation from information provided by the study of blood groups of primates?"*

It is noteworthy that for the most ancient blood group systems, as for instance A-B-0, we find the same allels in zoological groups which are considered far apart from each other (Wiener et al. 1974). This situation seems to indicate that a new species did not result from a single mutant or from an ancestral pair of mutants as it is proposed by the neo-darwinists; it rather implies that the evolution of an entire population was at the basis of the processes of speciation. Actually, if a species were to descend from a single ancestor (the Adam theory of hominization) or from a pair of individuals, such species would not inherit but a small part of polymorphism of the ancestral population. Observations, however, of the populations which are presently developing into new species (as for instance the African complex of *Papio,* or the *Drosophila* of Central America, studied by Dobzhansky 1967 and Ayala 1978) reveal such a high level of polymorphisms that the origins of these groups from single individuals must be excluded.

In practice, the process most probably occurs as follows: A population becomes isolated at the periphery of a zone covered by a species. The ecological condition of the isolate are at the limit of its adaptability, and the maximal selection pressures favor development of new genetic combinations which are better adapted to the conditions existing in the peripheral niche inhabited by the isolate. If the genetic patrimony of the isolated group is sufficiently diversified (polymorphic) to tolerate a transformation imposed by the ecological conditions, the separated population will multiply and be able to expand from the original zone of habitation and colonize new ecological niches. At the same time, because of its marginal situation, the newly emerging population will show a tendency of being sexually isolated because the gene flow which it receives from other populations is too weak to counterbalance the divergent genetic trends due to selection. In that phase, all modifications which bring about isolation of the population on its way to speciation (in other words, all that prevents the external gene flow from hindering the newly achieved genetical structure, and — consequently — all that fosters adaptation to the new ecologic niche) must be considered selectively favorable. In reality, this conquest of the new niches diminishes the competition and provides access to the new natural resources. It is at that moment that we believe chromosomal recombination may intervene which will result in acceleration and strengthening of the sexual isolation of a population on its way to speciation.

We must remember that in a normally equilibrated and homogenous population, the spontaneously appearing chromosomal abnormalities — even though not rare — will be eliminated because their selective value is, by and large, negative. They would create difficulties in the course of normal meiosis and would diminish or destroy the *fertility* of their carriers. Contrariwise, in a population which is conquering a new ecological niche, a chromosomal recombination which strengthens sexual isolation and thus protects the acquired differences must be considered as a favorable phenomenon.

Certain authors (Lejeune 1968, Turleau et al. 1972, Chiarelli 1967) suggested that speciation is due to chromosomal recombination. Certainly, such a possibility cannot be excluded theoretically. However, it could in no way explain the existence of the very high polymorphism of immunological characteristics which has been depicted in numerous species considered to be of relatively recent origin. We think that it is not chromosomal recombination which directly causes the speciation, at least not in the majority of cases; it only strengthens the already initiated speciation and renders it irreversible.

Sequential Duplications

The last point, and an essential one, demonstrated by immunogenetics of primates, is the *role of the sequential duplications in the diversification of species through evolution.* These duplications were mentioned first in 1935 by Bridges, and most recently have been pointed out by Ohno. It is the redundancies that we find, for instance, in the formation of the molecule of haptoglobin, hemoglobin, and many other active molecules, which often consist of the nearly end-to-end repetitions of the same sequences, more or less modified. The blood group systems are not an exception to this rule and many of them (M-N, Rh, Kell) are presently assumed to be of multiple-loci origin. Comparative study of these blood group systems in primates throws light on the nature and significance of these multi-loci characteristics (Moor-Jankowski and Socha 1980). In order to better understand them, we must return to the notion of paleosequences and neosequences.

Table 1. Relationship between human and chimpanzee blood groups. (Reproduced from J Hum Evol 8:453, 1979)

Human			Chimpanzee
Blood group system	Specificities present on the red cells		Blood group system
A-B-0	AB, B, H	A_1, A_2		A-B-0
Rh-Hr	rh', rh'', hr'', etc.	hr', Rh_0 = R^c, c^e	c_1^c, C^c, E^c, F^c	R-C-E-F
M-N-S-s	S, s, Hu, U, etc.	M, N^v, Mi^a = V^c	A^c, B^c, D^c, X^c, Y^c	V-A-B-D
I-i	I	i		I-i
Other blood group systems			N^c, H^c, K^c, M^c, T^c, O^c, etc.	Unrelated simian-type specificities

The Rh and M-N blood group systems, for example, are composed, as stated above, of paleosequences [Rh_0 (D) and hr' (c) for the system Rh; N^{vg}, N, n, H^e, Mi^a for the M-N system] which are almost identical in all the species in which they are detected. One cannot help thinking that they were first to appear in the common ancestor. Since, among anthropoid apes, the chimpanzee has been studied most extensively for its blood groups, comparison of chimpanzee and human blood groups is used to show the immunogenetic relationships between both species. Table 1 lists blood group specificities that are common for both man and chimpanzee (most probably paleosequences), and those which are detectable in one species only (thus, neosequences).

We may be dealing here with multiplication of the one and the same chromosomal segment, but the newly formed chromosomal zones followed in each species a different evolutionary pathway. If the natural selection had not only retained that scheme, but also generalized it, this seems to indicate that the process in question represented a certain selective advantage.

Multiplication of the same locus produces "new" genetic material which is able to undergo mutation and be submitted to the pressures of natural selection, without, however, interfering with the initial gene, the activities of which may be necessary for maintaining the biological equilibrium of the cell.

The evolution may thus proceed on multiple levels without endangering the fundamental balance of the genome. This may explain the progressive enrichment of the hereditary material which one observes when proceeding from the simplest forms of life to more complex ones. For all the above reasons, the study of blood groups of primates provides us with a model capable of elucidating, in a quite original way, certain important aspects of the mechanism of evolution. To conclude, the following four tables will be shown, which summarize the present status of the knowledge of the blood groups of nonhuman primates (Tables 2, 3, 4, and 5).

Table 2. Blood groups of chimpanzees

Class of blood groups	Blood group system	Blood specificities	Blood types
Human-type (Determined with reagents used for testing human blood)	A-B-0	A, H also in secretions	A_1, $A_{1,2}$, A_2 or 0
	M-N	M, N	M, or MN
	Rh-Hr	Rh_0, hr'	All Rh_0, and hr'-positive
Simian-type Determined with reagents prepared for chimpanzee red cells)	V-A-B	V^c, A^c, B^c, D^c	v.0, v.A, v.B, v.D, v.AB, v.AD, v.BD, V.O, V.A, V.B and V.D
	R-C-E-F	R^c, C^c, E^c, F^c, c^c, c_1^c	rc_1, rc_2, rCF, $rCFc_1$, $rCFc_2$, Rc_1, Rc_2, RC, RCc_1, RCc_2, RCE, $RCEc_1$, $RCEc_2$, RCF, $RCFc_1$, $RCFc_2$, RCEF, $RCEFc_1$, $RCEFc_2$
	Unrelated specificities	G^c, H^c, K^c, M^c, N^c, O^c, T^c	Positive or negative

Table 3. Blood groups of anthropoid apes other than chimpanzee

Species	Human blood groups defined by reagents prepared against human red cells		Simian-type blood groups defined by reagents prepared against primate red cells	
	System	Blood types	System	Blood types
Gorillas (lowland and mountain)	A-B-0	Exclusively B (in secretions only)		
	M-N-S	N, or MN	V-A-B	V.0 exclusively
	Rh-Hr	All Rh_0 - and hr'-positive	R-C-E-F	RC, RCF or $R_{(var)}$
Gibbons	A-B-0	A_1, A_2, B, A_1B, or A_2B All secretors		
	M-N-S	M, N, or MN	V-A-B	Not reactive with available reagents
	Rh-Hr	hr'-positive only	R-C-E-F	Not sufficient data
Orangutans	A-B-0	A_1, A_2, $A_{1,2}$, B, A_1B, or A_2B Almost all secretors		
	M-N-S	M or m	V-A-B	Nonspecific reactions with available reagents
	Rh-hr	Nonspecific reactions with available reagents	R-C-E-F	Nonspecific reactions with available reagents

Table 4. Blood groups of macaques

Species	Human-type A-B-0 specificites defined by saliva inhibition and serum tests	Simian-type blood groups defined by hemagglutination tests using iso- or crossimmune sera produced in rhesus
Rhesus monkey *(Macaca mulatta)*	Mostly B (A and AB observed although extremely rare. Gene 0 presumably present)	D^{rh} *graded blood group system:* D_1, D_2, D_3 and d *Unrelated specificities:* $A^{rh}, B^{rh}, C^{rh}, F^{rh}, G^{rh}, J^{rh}, L^{rh}, M^{rh}, N^{rh}, O^{rh}, P^{rh}$
Pig-tailed macaque *(Macaca nemestrina)*	0, A, B and AB	D^{rh} *graded blood group system:* $D_1, D_2, D_3, D_4,$ and d *Unrelated specificities:* $A^{rh}, B^{rh}, C^{rh}, F^{rh}, G^{rh}, J^{rh}, L^{rh}, M^{rh}, N^{rh}, O^{rh}, P^{rh}$
Crab-eating macaque *(Macaca fascicularis)*	0, a, B and AB	D^{rh} *graded blood group system:* D_3, D and d *Unrelated specificities:* $A^{rh}, B^{rh}, C^{rh}, F^{rh}, G^{rh}, J^{rh}, M^{rh}, N^{rh}, O^{rh}, P^{rh}$
Stump-tailed macaque *(Macaca arctoides)*	Exclusively B	D^{rh} *graded blood group system:* D_2(var), D_3 and d *Unrelated specificities:* Not polymorphic with available reagents
Barbary macaque *(Macaca sylvanus)*	Exclusively A	D^{rh} *graded blood group system:* D_2 and D_3 *Unrelated specificities:* Not polymorphic with available reagents
Bonnet macaque *(Macaca radiata)*	A, B and AB	Not polymorphic with available reagents

Table 5. Blood groups of baboons and other Old World monkeys except macaques

Species	Human-type A-B-0 groups defined by saliva inhibition and serum tests	Simian-type blood groups defined by hemagglutination tests using iso- and crossinmmune sera of primate origin
Baboons *(Papio)* various species	0 (very rarely), A, B and AB	B^p *graded blood group system:* B_1, B_2, B_3 (D_2^{rh}), B_4 and b *Unrelated specificities:* $A^p, C^p, G^p, E^p, N^p, O^p, S^p, T^p, U^p, V^p, L^p, M^p$, OC, ca hu (monomorphic in *P. hamadryas* and *P. ursinus* but polymorphic in other species of baboons)
Geladas *(Theropithecus gelada)*	Exclusively 0 (often with irregular isoagglutinins)	3 yet unnamed factors
Vervet monkeys *(Cercopithecus pygerythrus)*	A (mostly), B and AB	Not polymorphic with available reagents
Celebes *(Cynopithecus niger)*	0, A, B and AB (very rarely)	Not polymorphic with available reagents
Patas *(Erythrocebus patas)*	Exclusively A	Not polymorphic with available reagents

References

Ayala F (1978) Les mécanismes de l'évolution. Evol Sci 13

Bridges CB (1935) Salivary chromosome maps. Sci Hered 26:60−64

Chiarelli B (1967) Caryological and hybridological data for the taxonomy and phylogeny of the Old World Primates. Turin

Dobzhansky Th (1967) *Drosophila pavlovskiana,* a race or a species. Am Midl Nat 78 (1):244

Lejeune J (1968) Adam et Eve ou le monogénisme. Nouv Rev Théol 90/191

Moor-Jankowski J, Socha WW (1980) Immunogenetic markers in primate animals and their use in breeding and standardization. Proc 16th Gen Meet Int Assoc Biol Standard, San Antonio, Texas, 15−20 September 1979. Dev Biol Standard 45:35−43

Moor-Jankowski J, Wiener AS, Socha WW, Gordon EB, Kaczera Z (1973) Blood group homologues in orangutans and gorillas of the human Rh-Hr and chimpanzee C-E-F systems. Folia Primatol 19:360−367

Ohno S (1970) Evolution by gene duplication. Springer, Berlin Heidelberg New York

Socha WW, Moor-Jankowski J (1979) Blood groups of anthropoid apes and their relationship to human blood groups. J Hum Evol 8:453−465

Turleau C, Grouchy de J, Klein M (1972) Philogénie chromosomique de l'homme et des primates hominiens. Essai de reconstitution de l'ancêtre commun. Ann Gene 15 (4):225

Wiener AS, Gordon EB, Moor-Jankowski J, Socha WW (1972) Homologues of the human M-N blood types in gorillas and other nonhuman primates. Haematologia 6:419−432

The Evolution of Human Skin

W. MONTAGNA [1]

Man's skin is unique among land mammals in that it appears to be largely hairless. Whereas our skin bears millions of hairs, some so small as to be nearly invisible, hairs generally grow vigorously and prominently on men's faces, and particularly on the scalp, axillae, mons, and anogenital areas of both sexes. Except, as we shall see, for sensory perception, most of the hairs on the human body serve no discernible function (those on the head can protect the scalp from the elements), and since all body hairs are to some degree sustained by androgens, their main purpose is likely to be ornamental and epigamic. Lacking adequate protection from hair, human skin has developed adaptive structural changes that give it greater strength, resilience, and high sensibility.

With notable exceptions, the topographic, anatomical, and physiological properties characteristic of human skin are not found in the skin of other primates. The differentiation of man's scalp and forehead, the facial disc, axillae and, for that matter, the skin over the entire body, are all uniquely human. Our skin is thicker than that of most other primates, tougher, more taut, and more elastic. Tunneling through this thick skin is a vast and intricate network of arteries, veins, and capillaries, with a blood supply far in excess of its own biological needs. The skin of all nonhuman primates we have studied, including that of the great apes, is by contrast relatively ischemic.

The characteristic features of human skin change constantly during a lifetime, more so than in other primates; beginning with its differentiation in utero, skin undergoes uninterrupted changes into and through old age. In addition, human skin has distinctive sexual dimorphic differences which are much more conspicuous than they are in other primates.

Since all nonhuman primates are covered with fur, the adaptations that have occurred in their skin differ only slightly if at all from those in other furred mammals. Like the latter, the primates from birth onward are buffered from the environment, and their skin, thus protected, need not be adapted to the environment. And it follows that all the major changes that have taken place in man have occurred concurrently with the loss of hair cover, hence are truly geared to the environment.

If the uniqueness of man's skin is to be appreciated, the cutaneous system has to be considered not only in toto but in specific detail because it does not possess a single major structure that is not also found in the skin of some other primate. Viewed

1 Department of Cutaneous Biology, Oregon Regional Primate Behavior Research Center, 505 N.W., 185th Avenue, OR 98006, USA

anatomically and uncritically, human skin, like that of other primates, has hair follicles and sebaceous glands, nails, sweat glands, and apocrine glands. This is an apparent contradiction to what I have said above, but it is the specific properties of each of these structures that constitute the uniqueness of human skin.

Man's relative hairlessness, or the miniaturization of the hair over much of his body, marks his major divergence from the other primates, and all of the differences found in his skin are in some way related to this fact. And, these departures must have coincided with man's attainment of an erect bipedal posture and locomotion, and be interwoven and related to the total biological needs of an erect body; a "hairless" skin must have accompanied bipedality pari passu. The connection is fairly logical. Man is a large animal whose body mass must be balanced vertically to prevent its toppling over. His skeletal muscles are continuously expending energy to maintain attitudinal, postural, and righting reflexes. This energy in turn creates heat that must be dissipated lest we perish. The loss of hair cover eminently accommodated this need, but it left the body exposed to environmental hazards. To cope with these imperilments, the human organism had need of a superlative tactile sensibility that would keep it constantly informed of external conditions. This acute cutaneous sensory system in turn could develop only with the acquisition of a large brain that could accommodate all the signals coming to it and that could determine instantly which course of action to pursue. Thus, nakedness, vascularity, and increased modalities of sensibility cannot be regarded as separate phenomena but as necessary and related adjuncts of the gestalt of the evolving human skin.

Bipedality in hominoids is relatively ancient, going back more than 3 million years according to recent paleontological findings (Leakey and Hay 1979). But even though human skin has been adapting to bipedality for a long time, only the broadest accommodations have been achieved, leaving some of the finer adjustments still in a makeshift manner.

Man maintains his orientation in space by virtue of the continuous action of interrelated postural reflexes, which supported by two pillars are far more complex than those in quadrupeds which are supported by four. Even when a human being is standing still, sitting, squatting, or lying prone, most of the skeletal muscles of the trunk, limbs, and head must maintain a degree of tone by working synergistically lest the body collapse. Therefore, except during unconsciousness, much of our muscle mass is continuously using energy and generating heat that must be dissipated. And it is to this very task that human skin has been primarily tailored. Dissipation of heat is the function that most conspicuously distinguishes human skin from that of all other mammals.

The loss of hair cover has placed on human skin the burden of performing many functions not demanded of hairy skin. To begin with, the epidermis, which is thicker in man than in all other primates, has a substantially thicker horny layer than that in furry animals. Moreover, over the entire body it is criss-crossed by many lines whose characteristic patterns reflect the direction of pull and stretch to which the skin is subjected. These congenital wrinkles seem to have two purposes: they expand the body surface and allow the skin to be stretched without reaching a breaking point too quickly. Since none of the other primates have such surface imprints, it is assumed that they do not need them. On its inner surface or underside in contact with the dermis, the epidermis is sculptured in such a way as to reflect its topographic location and its corresponding adaptation to that location. Structurally, the uneven topography of the epidermal underside accounts for the great variations in thickness and thus provides a richer source of keratinocytes for producing the thick horny layer than if it were flat. We considered this complex structural characteristic uniquely human until we looked at the epidermis of the anthropoid apes. The epidermis of the gorilla and chimpanzee both show a degree of sculpturing, in contrast with the underside of the epidermis of other primates, which is flat like that of other furred animals. In the other primates, discrete and characteristic structures on the underside of the epidermis are found only on the glabrous skin: the lips, margins of the eyelids, palms, soles, and digits, proximal nail folds, ischial callosities, and anogenital surfaces. The departure from this rule by chimpanzees and gorillas, who are heavily furred, seems to anticipate the structural patterns that have occurred in man.

Because ultraviolet light, though essential, can damage the naked skin of man exposed to too much of it, some of it must be absorbed. This is the function of melanin, a yellow to black pigment produced by melanocytes, found mainly between and underneath the basal cells of the epidermis. Melanocytes produce melanin and pass it on to the keratinocytes. The development of heavily pigmented skin among all human inhabitants of the tropics cannot be dismissed as a lucky coincidence, even though melanin may be produced as a by-product by melanocytes. (In addition to their obvious role of filtering noxious ultraviolet rays, melanocytes perform other functions about which little has been known up to now.) Despite the heavy fur that characterizes all nonhuman primates, their skin has melanocytes that, depending on the species, are numerous or sparse, large or small, very active or inactive. For example, melanocytes are active in the skin of bush babies and rhesus monkeys during late fetal life, but gradually become amelanotic at birth and during early infancy and remain so during life, except for the face, which becomes darker. Conversely, in Celebes apes and chimpanzees, and on the faces of rhesus monkeys, melanocytes are almost amelanotic at birth but become progressively more active during early infancy; in adult Celebes apes and chimpanzees the entire skin is black. The curious incidence of an occasional white-skinned animal among black-skinned chimpanzees suggests that ultraviolet light has little relevance to the amount of melanin pigment in the skin of heavily furred primates.

A striking difference between human and nonhuman primate skin is the abundance of elastic tissue in human skin and its unique architecture. Except around hair follicles and sweat glands, elastic fibers in the dermis of other primates are sparse. (About the significance of elastic tissue we know little except that it attaches smooth muscles to hair follicles and dermis, acts as a support for glands, and may toughen naked skin.) (Montagna and Parakkal 1974.) Chimpanzee and gorillas are the exception to this rule: though heavily furred, the content of elastic fibers, their organization and architecture in their skin are similar to those of man. Here, then, is still another indication that these animals are anatomically closer to man than the other nonhuman primates we have studied.

Although hair follicles on the human body are vestigial, the sebaceous glands accompanying them are larger and more numerous than in any other primate. It can even be said correctly that man is a sebaceous animal. Because sebum is secreted in great amounts around all of the body orifices, it would appear that it acts at least as an

emollient, but its main function, if any, is not known for certain. To it is due much of
our characteristically human, as well as our individual, odor. In nonhuman primates,
sebaceous glands are, with the notable exceptions of lemurs, relatively small except on
the face and perineum and wherever they have become specialized as scent organs, e.g.,
in the sternal pit of spider monkeys, the inguinal glands of marmosets and tamarins,
and the brachial glands and external genitalia of lemurs and lorises.

Man sweats profusely from his 2 to 5 million glands. Eccrine sweat glands physiologically fall into two distinct categories: those on the palms and soles, which respond
principally to psychogenic stimulation, and those on the rest of the body surface,
which respond mainly to thermal stimulation. Structurally similar, these two kinds of
glands have different developmental and phylogenetic histories. Those on the palms
and soles, probably the phylogenetically most ancient, are the first cutaneous glands
to develop in the embryo and are present on the pes and manus of all other primates
and other plantigrade and digitigrade mammals except lagomorphs. They are numerous
on the prehensile surface of the tails of howler, spider and woolly monkeys, and on
the knuckle pads of chimpanzees and gorillas. The original function of sweat glands
may have been to respond to psychogenic stimuli. Furthermore, these glands secret
a small, but constant, amount of water that keeps the thick horny surfaces soft and
sensitive. Except for tupaias and tarsiers, the prosimians and most of the Ceboidea lack
eccrine sweat glands over the rest of the body. There are, however, variable numbers
of them in the hairy skin of all Old World monkeys. Apocrine glands, tubuloalveolar
structures that have nothing to do with sweating in primates, predominate on the hairy
skin of prosimians and New World monkeys. They are more numerous than eccrine
glands in Old World monkeys, gibbons, and orangs but less numerous in chimpanzees
and gorillas. This morphological progression, however, is not accompanied by an
obvious functional one since, although they appear to be structurally competent, the
sweat glands secrete only minimally. Monkeys have dry skin, chimpanzees and gorillas
only a little less so. If they sweated as profusely as man, their pelage would be constantly soaked in a tropical climate and they would be forced to live almost constantly
in wet blankets of fur. Even in man not all eccrine sweat glands are functional and
nearly every individual has his or her own peculiar sweat pattern. Furthermore, the
distinction between psychic and thermal stimulation is not all that clear-cut. Some of
the glands in the axilla and in undetermined areas on the body can also respond to
psychic stimuli. It is significant that although the glands in the palms and soles
respond mainly to adrenomimetic drugs and those on the rest of the body to cholino-
mimetic ones, all can be stimulated to secrete with different doses of either drug.
We all know that men sweat more than women, and that some individuals sweat very
little whereas others profusely. Thus, the several million glands on the human body act
principally as heat regulators, but this function is perhaps too recent to be totally
efficient.

On the human body, apocrine glands are found in the external auditory meatus, the
eyelids, the perineum, and sometimes the mons pubis; they are numerous only in the
cavum axillae. Old World nonhuman primates indicate that apocrine glands were gradually replaced by eccrine glands. Ontogenetically, rudiments of glands appear tran-
siently nearly everywhere in the skin of 5- to 6-month old human fetuses from the
upper part of hair follicles above the sebaceous gland anlagen but later disappear over

most of the body; in adults a few can be found almost anywhere, especially on the face and scalp, where they are considered to be ectopic.

In the human axilla, the axillary organ is designed precisely to produce and disperse odorous secretions. The entire axillary bed is an aggregate of large apocrine and eccrine glands on a one-to-one ratio. One large sebeceous gland and one large apocrine gland are attached to each hair follicle. Both apocrine and sebaceous glands secrete into the pilary canal small amounts of viscid substances, which are diluted by the watery eccrine sweat and distributed over the surface of the epidermis and the entire length of the hairs; everywhere on these surfaces, microorganisms break down the secretion to make it fetid. We become aware of this additional body odor during stress, strenuous exercise, or in hot environments because sweat is the vehicle that distributes the odorous substances. A dry axilla has little odor. Thus, every detail of the axilla is precisely tailored to produce and ventilate substances that give human beings their own distinctive aura. Among the nonhuman primates, only the chimpanzee and the gorilla have an axillary organ that is anatomically and functionally like that of man; analogous aggregates of glands which secrete odorous substances are found in lorises and lemurs.

Among the many varieties of primate skin, only human skin displays striking sexual dimorphism. Women's skin is more supple, apparently more fatty, and often more turgid than men's. Smaller body hairs and a less deeply creased surface impart a smoothness and a more velvety appearance than is found in male skin. The bony eminences of the female body are gently contoured, and the breasts, buttocks, and external genitalia are overtly swollen. These features are not shared by other primates; the only clue to their origin seems to be the sex skin of Old World primates, which during the reproductive years undergoes structural and chemical changes controlled by the hormonal changes that accompany the ovarian cycle (Parker 1974). Whereas no such overt skin tumescence and detumescence that accompany the ovarian cycle occur in human skin, there is some water retention in the skin during the follicular phase and the breasts of young women undergo periodic changes in volume (Milligan et al. 1975) and in sensitivity (Robinson and Short 1977) during the cycle. It is not yet known whether comparable changes occur elsewhere in the skin.

Work in progress shows that the swelling and detumescence of the sex skin of female macaques are largely due to increases and decreases in the amounts of water binding mucopolysaccharides which can be brought about only by cyclic changes in the activity of fibroblasts (Carlisle and Montagna 1979). Because fibroblasts do become alternatively active and dormant during swelling and deswelling, they must contain estrogen receptors that control these fluctuations. Although such receptors have been convincingly demonstrated in other tissues by other investigators, they have not yet been seen in skin fibroblasts. I have just suggested that sex skin is the precedent of enlarged breasts and buttocks and labia. Let me go one step further and suggest (1) that all of a woman's skin is homologous to sex skin and (2) that fibroblasts in the breasts and labia contain the largest number of estrogen receptors. Concerning the first point, most women experience water retention, weight gain, and even skin edema during the follicular phase of the ovarian cycle (Lauritzen and van Keep 1978). These transitory changes are reversed during the luteal phase. We do not know about the second suggestion; time will tell if it is correct. This hypothesis may also be relevant to the syndrome of testicular feminization. People with this aberration are genotypic males whose estrogen and androgen levels are normal or slightly elevated. However, they have an inherited insensitivity to endogenous and exogenous androgens, apparently because of decreased cytosol androgen receptor activity, as a result of which they are phenotypically women. Further studies on nonhuman primate sex skin, testicular feminization and on normal human female skin may provide a clue to the characteristic properties of women's skin. If this sounds preposterous, there is as yet no better explanation.

This presentation would not be complete without some reference to baldness. Highlighting the major facts may help to counteract some of the foolishness that continues to be written on this subject. Since human culture has proclaimed from time immemorial that scalp hair is the crown of human beauty, baldness has always been regarded as an affrontery and a biological mischief. Yet, notwithstanding its cataclysmic aspects, the biological aspect of baldness is at once exciting and perplexing since in both sexes scalp hairs, at least in the early years of life, grow longer than those in any other part of the body. To understand baldness, we must first realize that despite gross appearances to the contrary, the human scalp and forehead are structurally indistinguishable. It is the hairline that defines and accentuates the two, but as daily experience shows, is a fey and vague demarcation, never fixed, forever shifting.

If allowed to, scalp hair follicles could grow uninterruptedly for years and produce hairs of such length that they could become a liability. Hair and beard five or more feet in length would be disastrous: they could be tripped over or become mixed in our food and libations, and entangled in environmental obstacles. But we have hands that turn this handicap into an ornament. In most nonhuman primates, head hairs are generally only slightly longer than those elsewhere on the body. In a few species — male Hamadryas baboons, lion-tailed macaques, stump-tailed macaques, some marmosets, and a few others — the crown hairs are variously long. On the human scalp, hair grows vigorously and in many persons densely during the early years, then in most men gradually thins after the age of 30. But baldness is not a reduction in the number of hair follicles: it is a gradual, systematic involution of these follicles until they are similar to those found during the fetal stage, and produce insignificant hair.

This process begins in all human beings during late gestation. Until the 7th month, the hair on the forehead of the human fetus is as long as that elsewhere on the head, and there is no demarcation between scalp and forehead. During the 8th month the follicles on the forehead either remain small or undergo a gradual involution, while those over the rest of the scalp continue to grow so that by birth the forehead appears naked. After birth, the hairs of the upper forehead continue to diminish until by middle or late childhood a clear hairline has been established. But the insidious process continues, in most cases imperceptibly, throughout life. It becomes obvious in those who express male-pattern baldness. Now the follicles on the frontal and parietal scalp involute just as those on the forehead had done during gestation and beyond. In widespread cases of baldness, the upper occipital scalp also becomes involved; except in extreme old age, the lower occipital and the temporal areas rarely become involved. Since this disparity in topographic involvement reflects ordained biological properties that are found in the scalp of all human beings, there is no scientific justification for regarding baldness as a defect.

The forehead of adult uakaris, stump-tailed macaques, orangutans, and most chimpanzees follow the same pattern of balding as human beings do; as in human beings, only the degree of baldness differs.

Since we are cajoled and harangued daily about the indispensability of head hair, the billion-dollar industries dedicated to hair cosmetics cannot afford to let the truth be known: (1) that baldness is a natural tendency of the scalp follicles to diminish in size as the individual matures; and (2) that the "capillary ornament" commonly known as hair is fated to be replaced by the ornamental phase of the future, total baldness. Because, whether he likes it or not, adult man is becoming progressively more hairless and more bald until in time he may become so entirely.

In searching for evolutionary clues to learn more about such disparate organs as the skin of human and nonhuman primates, one is repeatedly forced to grasp at straws. But if one is willing to temper fancy with what few facts are at hand, the results make for an amusing, if not entirely revealing, story.

References

Carlisle KS, Montagna W (1979) Aging model for unexposed human dermis. J Invest Dermatol 73: 54–58

Lauritzen C, Keep PA van (1978) Proven beneficial effects of estrogen substitution in the postmenopause – a review. Front Horm Res 5:1–25

Leakey MD, Hay RL (1979) Pliocene footprints in the Laetolil beds at Laetoli, northern Tanzania. Nature (London) 278:317, 323

Milligan D, Drife JO, Short RV (1975) Changes in breast volume during normal menstrual cycle and after oral contraceptives. Br Med J 4:494–496

Montagna W, Parakkal PF (1974) The structure and function of skin, 3rd edn. Academic Press, New York

Parker F (1974) Skin and hormones. Textbook of endocrinology, Ch 23, 5th edn, pp 977–981

Robinson JE, Short RV (1977) Changes in breast sensitivity at puberty, during the menstrual cycle, and at parturition. Br Med J 1:1188–1191

The Importance of Theory for Reconstructing the Evolution of Language and Intelligence in Hominids

S.T. PARKER [1] and K.R. GIBSON [2]

Introduction

Although we anthropologists are academically licensed mythologists charged with explaining how "the man got his skin," etc., our stories tend to suffer in plausibility from an absence of theory. In this paper we argue for the importance of theory for reconstructing the evolutionary history and adaptive significance of intelligence and language in our family.

In this paper we will set forth the following problems: the identification of taxonomically relevant traits for comparison of closely related species; the identification of functionally relevant traits for comparison of distantly related species; the identification of adaptive strategies among animals and the various proximal mechanisms subserving them; and the identification of an appropriate animal model for reconstructing the traits of the common ancestor of a group of living species. We will try to demonstrate the heuristic value of Piaget's developmental theory and ethological, ecological, and sociobiological theory for solving these problems.

Identification of Intelligence as a Taxonomically Relevant Trait for Comparative Study of Primate Species

The first problem in evolutionary reconstruction is the identification of taxonomically relevant traits for comparison of individuals from closely related species. This is particularly difficult in the case of behavioral systems which leave no fossil remains. Ethologists approach this problem by observing and describing species-specific behavioral repertoires (ethograms), and comparing the form and function of particular traits in a series of closely related species. When appropriate, they arrange the functionally equivalent traits in a series from the least to the most complex and reconstruct a series of evolutionary stages on the basis of a plausible series of transformations. This procedure implies that the traits of living forms represent different grade-levels, the most

1   Department of Anthropology, Sonoma State University, Rohnert Park, CA 94928, USA
2   Department of Anatomy-Dental Branch, University of Texas, Houston, Texas, USA

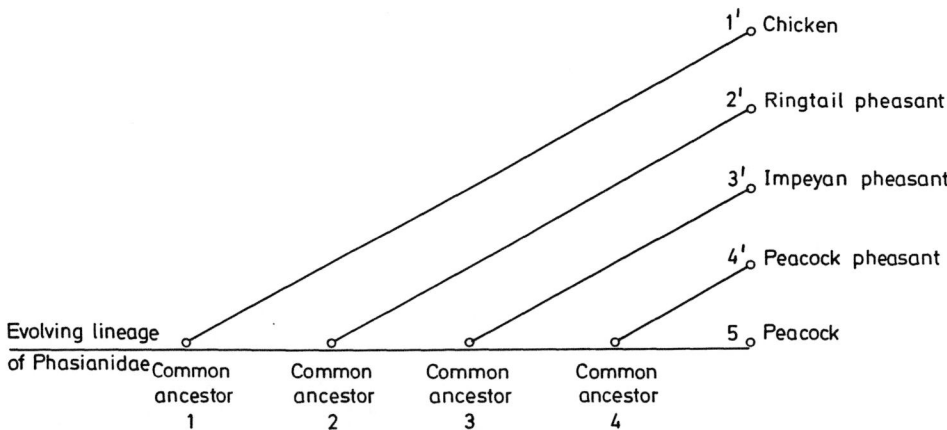

Fig. 1. Cladogram of the family Phasianidae. *Species 1'* (Chicken) most like common ancestor 1 (of whole family). *Species 2'* (Ringtail pheasant) most like common ancestor 2. *Species 3'* (Impeyan pheasant) most like common ancestor 3. *Species 4'* (Peacock pheasant) most like common ancestor 4

primitive traits resembling those of the common ancestor of the next group of descendants, etc. (Fig. 1) (Eibl-Eibesfeldt 1975).

Judging the complexity of forms in a series is based on the notion that earlier forms were prerequisite for the evolution of later forms. Complexity can sometimes be measured in terms of the number of steps necessary to derive one form from another. This is particularly true when new traits are added on top of old traits by terminal addition (Gould 1977).

In principle, naturalistic observation and description of intelligence and its comparative study in closely related primate species is no different from the comparative study of other traits such as expressive movements and courtship displays. In fact, however, comparative study of intelligence presents special problems. First of all, the definition of intelligence as a taxonomically relevant trait is not trivial. Secondly, the methodology of comparative studies is not trivial.

Piaget's model of cognitive development in human children provides taxonomically relevant definitions and descriptions of intelligence which can be used for comparative studies of intellectual development in closely related primate species (Jolly 1972, Parker 1977). Piaget defines intelligence as a (phenotypic) adaptation through active assimilation of reality and accomodation to reality in the service of discovering new means to new and old ends.

Intellectual development in human children occurs in three major domains: (1) the physical domain, including object concepts (permanence, identity, and quantity), space, time, and causality; (2) the interpersonal domain, including imitation, the symbolic functions of drawing, language, and symbolic play, and moral judgment; (3) the intrapersonal domain, including imagery, memory, consciousness, and dreams. The earliest period of intellectual development is called the *sensorimotor period,* spanning from birth to 18 or 24 months of age. During this period the human infant achieves the ability to remember the spatial location of a hidden object, to retrieve it, and finally to search for an invisibly displaced object in a series of locations (object permanence). He achieves the ability to place objects inside, outside, before, behind, underneath, and on top of each other, and to understand simple means-end relationships as revealed, for example, by using a stick as a tool to rake in an out-of-reach object. He also achieves the ability to imitate novel actions long after he has seen them, and to mentally represent actions and images. The achievements of this period can be divided into six sequential stages occurring in six series: sensorimotor intelligence, space, time, causality, imitation, and object concept (Piaget 1954, 1963, 1962).

The subsequent period of intellectual development is the *preoperations period*, spanning from 18 to 24 months to 6 or 7 years of age. During this period children extend their new symbolic capacities in language, drawing, and make-believe play, constructing preconcepts (interiorized actions) concerning object relations and causal relations between events (Piaget and Inhelder 1967, Inhelder and Piaget 1964). During the early part of this period (the symbolic subperiod) they are preoccupied with simple topological relations between objects (such as proximity and enclosure). During the later part (the intuitive subperiod) they are emancipated from this preoccupation and begin to construct simple Euclidean spatial notions such as angularity and straightness. They also begin to construct simple classes of objects based on a single criterion. In subsequent periods they develop true concepts based on reversible mental operations, and finally they develop hypothetical-deductive reasoning (Table 1).

Table 1. Piaget's model of cognitive development

Periods of development	Types of logic	Domains of cognition		
		Physical	Interpersonal	Intrapersonal
Sensorimotor period (birth to 2 yr.)	Sensorimotor trial-and-error; experimentation; discovery of new means	Object permanence; externalized time, space, and causality	Deferred imitation of novel schemes; sensorimotor games	First evoked images
Preoperations period Symbolic subperiod (2 to 4 yr.)	Nonreversible interiorized action schemes, i.e., preconcepts with transductive reasoning	Object identity; topological space; graphic collections	Make-believe games; language	Static evoked images
Intuitive subperiod (4 to 7 yr.)		Incipient projective and Euclidean non-graphic collections		
Concrete operations period (7 to 12 yr.)	Reversible interiorized action schemes, i.e., true concepts with deductive reasoning about concrete phenomenon	Object quantity; true classification with inclusion	Games with rules	Dynamic evoked images
Formal operations period (12 yr. on)	Abstract reasoning	True measurement; systematic hypothesis formation and testing of causality	Universal rules	

Intellectual development occurs through the differentiation and coordination of actions (and interiorized mental representations of actions) on objects: Intelligence arises from action rather than from perception (Piaget and Inhelder 1967). These coordinations create classes of objects (classification) and relations between objects (seriation), revealing properties (quantity) that did not exist before; they also reveal the nature of physical causality (gravity, inertia, equal and opposite forces, transmission of forces, etc.). The coordinations create feedback, which the agent tries to assimilate to his sensorimotor and mental "schemes" (repeatable action patterns). When the feedback does not fit his schemes, he accomodates his schemes to the phenomena as best he can. Mismatches between his schemes and the world create disequilibration and give rise to attempts to reequilibrate on a higher level. Feedback from other people in the form of disagreement also plays an important role in creating disequilibration (Piaget 1978).

Piaget's model is useful for comparative studies of intelligence because it focuses on the metabehavioral level of differentiation, coordination, reinforcement, and application of schemes. This approach is necessary for the identification of taxonomically relevant traits in a behavioral system characterized by unstereotyped goal-directed behavior. A Piagetian approach transcends the limitations of traditional ethological analysis which focuses on the description of fixed action patterns. It transcends the limitations of psychometric approaches to the study of intelligence which impose standardized tasks on the animal and make it impossible to identify and compare spontaneous species-specific behavior (Parker 1977). Most important, it counters the tendency to focus on an arbitrary subset of parameters [e.g., amount of object manipulation (McGrew 1979), persistence, purposiveness, trial-and-error variation between performance of a single scheme (Beck 1980)], while ignoring other diagnostically relevant parameters such as the complexity of the coordinations and the range of applications across situations, which have characterized comparative studies of tool use and intelligence in animals. An ad hoc selection of parameters cannot reveal significant patterns because it lacks a unifying explanatory mechanism.

The use of Piaget's theory for comparative studies has been criticized on the grounds that the theory is wrong. Although the theory is being modified in certain respects through more detailed research, its core concepts are remarkably robust. Moreover, Piagetian theory is orienting virtually all research on cognitive development being done today. Indeed, it is the only developmental theory (Bower, personal communication). Many of the putative disproofs of Piaget's theory are based on simplified training experiments which remove the very parameters Piaget was studying (Sinclair, personal communication, Parker and Gibson 1979).

Comparative Studies of Primate Intelligence

Data from comparative studies of primate intellectual development from a Piagetian perspective and data from other studies suggest that prosimians, Old World monkeys, great apes, and man constitute a series of grade-levels of terminal intellectual abilities corresponding to the sequence of stages of intellectual development in human infants and children (Parker and Gibson 1977):

First, *the prosimian stage: stage 1 and 2 sensorimotor period* reflex grasping with coordination of hand and mouth, apparently no object permanence (Jolly 1972);

Second, *the Old-World monkey stage: stages 3 and 4 of sensorimotor period* schemes of hand-eye coordination without "secondary circular reactions." No 5th stage "tertiary circular reactions and discovery of new means" with objects, no 5th stage imitation of novel vocal, gestural, or object manipulation schemes, but stage 5 object permanence (Parker 1977; Antinucci et al. 1980).

Third, *the great ape stage: stages 5 and 6 sensorimotor period* schemes of "tertiary circular reactions and discovery of new means," stage 5 and 6 causality and spatial schemes, stage 5 imitation of novel gestural and object manipulation schemes, stage 6 deferred imitation; *1st preoperational subperiod* schemes of symbolic play, symbol use, and drawing (Parker 1976, Chevalier-Skolnikoff 1977; Mathieu 1978, Redshaw 1978; Savage-Rumbaugh et al. 1977, Plooij 1977).

Fourth, *the human stage: 2nd preoperations subperiod, concrete and formal operations.*

Although human infants and children display much higher frequencies, longer durations, and wider ranges of application of sensorimotor and preoperational schemes than monkeys and apes do, the fact that the levels of terminal achievement in a series of related species with progressively more recent common ancestor correspond to the sequence of stage of intellectual development in children, strongly suggests ontogenetic recapitulation of evolutionary stages (Fig. 2).

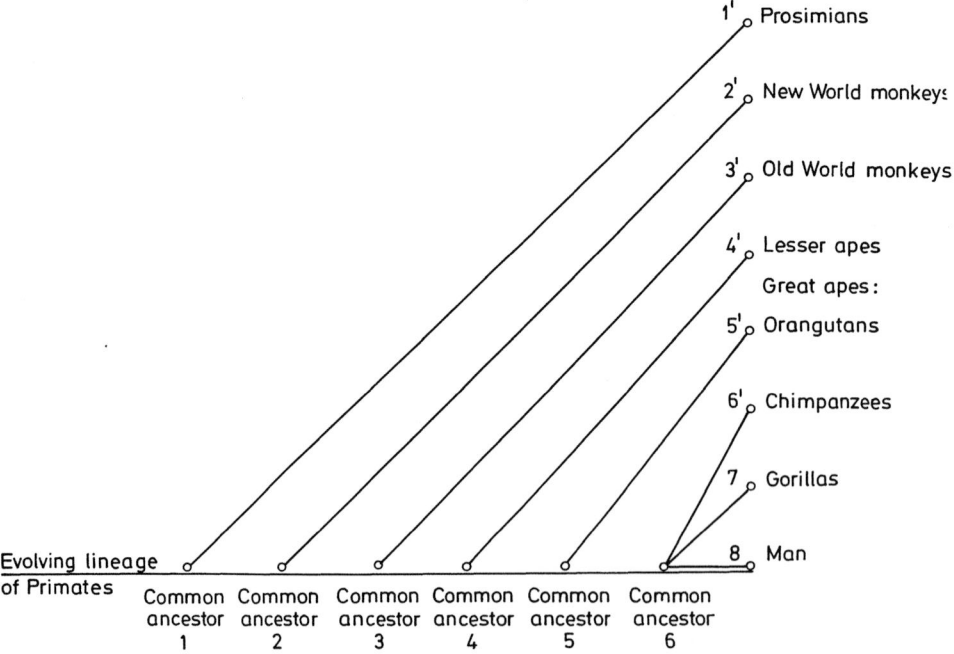

Fig. 2. Cladogram of the order primates. *Group 1'* (Prosimians) most like common ancestor 1 (of whole order). *Group 2'* (New World monkeys) most like common ancestor 2. *Group 3'* (Old World monkeys) most like common ancestor 3. *Group 4'* (Lesser apes) most like common ancestor 4. *Group 5'* (Orangutans) most like common ancestor 5. *Group 6'* (Chimpanzees) most like common ancestor 6

Behavioral recapitulation during development is not unknown in other taxa; it has been observed, for example, in the development of the male courtship display in peacocks. As he matures, the peacock recapitulates the adult male courthsip displays of some of his close relatives, the chicken, the ring-necked pheasant, the impeyan pheasant, and the peacock pheasant, each of whom shows a more ritualized form of courtship derived from the primitive food-enticing pattern of the rooster (Schenkel, cited by Eibl-Eibesfeldt 1975). Ontogenetic recapitulation apparently occurs as a result of a series of terminal additions of new forms one on top of another (Gould 1977). It is especially likely to occur when new forms involve addition to a prior form.

The recapitulation model and the graded series carry useful implications for reconstructing hominid evolution. The first implication is that the common ancestor of chimpanzees, gorillas, and man displayed 5th and 6th stage sensorimotor and symbolic intelligence which are common to all its descendents. This inference is strengthened by the fact that the slightly more distantly related orangutan displays the same pattern. The second implication is that the earliest hominids displayed more elaborated versions of this level of intelligence and probably slightly higher levels. The third implication is that subsequent grades of hominids displayed levels of intelligence intermediate between symbolic and high concrete or formal operational. Specifically, it seems likely that early *Homo* displayed late preoperations; *Homo erectus* perhaps low concrete operations (Wynn 1979); early *Homo sapiens* displayed high concrete and formal operations (Parker and Gibson 1979)[1].

Language

As with intelligence, we argue that stages of prelanguage and language development in human children recapitulate the stages of evolution of language in our ancestors.

Language development begins virtually at birth in human infants and continues for several years. Language is the product of both cognitive capacities for arbitrary representation and recombination of units symbolizing actions, agents, objects and their locations, and of affective capacities for cooperative reciprocal exchanges and joint regulation of actions and attention (Bruner 1975, Stern 1978, Trevarthen 1980).

All languages (whether spoken, gestural, or written) are based on grammatical categories of agent, action, object of action, recipient of action, location, and possession. Indeed, language is "an instrument for regulating joint activity and joint attention"; it is "a specialized and conventionalized extension of joint action" (Bruner 1975, p.2). This joint activity and attention is based on the assumption of reciprocal roles of agent and recipient of action and on the construction of routines of joint interference. These reciprocal role structures emerge through games such as object exchanges and peekaboo (Bruner 1975).

---

1   This general linear trend in the evolution of hominid intelligence coexists with a lateral trend toward retrospective elaboration, during later stages of evolution, of abilities characteristic of earlier stages. Human infants, for example, display a much richer and more elaborated version of 5th and 6th stage sensorimotor and symbolic intelligence than do great apes (Parker and Gibson 1979)

In human children language emerges during the period from 9 to 24 months, between the 4th and 6th stages of the sensorimotor period. The language of this period is called prelanguage or protolanguage, because it precedes the mastery of grammar and entry into the adult language system (Halliday 1975). Protolanguage has content and expression. The content is the meaning within a given social and material context. By 18 months (the 6th stage of the sensorimotor period) these meanings serve the following functions: The *instrumental function* of getting goods and services (the "I want" function); the *regulatory function* of controlling the behavior of others (the "do as I tell you" function); the *interactional function* of initiating and responding to interactions (the "you and me" function); the *personal function* of expressing emotional states (the "here I come" function); the *heuristic function* of gaining information about the environment (the "tell me why" function); and the *imaginative function* of creating make-believe (the "let's pretend" function). The last function, the *informative function* (the "I have got something to tell you" function), appears later.

An "expression" is the particular form that meanings take. While the meanings of protolanguage and their functions remain relatively constant (with the gradual addition of new functions), the expressions of these meanings change radically from phase I to phase II of protolanguage development. Phase I, from 9 to 16 months of age, is characterized by instrumental, regulatory, personal, and imaginative functions. The meanings subserving these functions are expressed in idiosyncratic personal utterances (which are usually not imitations of adult words) and by ritualized referential gestures. During phase I the meaning of an utterance is synonymous with its use in the immediate situation. In other words, each utterance specifies both meaning and contexts and has only one function.

Meanings in phase II of protolanguage, from 16 months to 24 months of age, are expressed in words of the adult language. These utterances have more generalized, less context-specific meanings and can hence refer to object and events outside the immediate situation. During phase II children's utterances begin to display the heuristic function (tell me why) and to differentiate into pragmatic and descriptive functions. Pragmatic utterances have a rising intonation, indicating that they require a response, while descriptive utterances have a falling intonation, indicating that they do not. Children in phase II engage in dialogues (which require the ability to adopt, assign, and reverse social roles), asking and answering yes, no, and "wh'?" (who, where, when, why) questions, as well as lying, joking, rhyming, and analogizing (Halliday 1975).

Protolanguage involves gestures that express all or part of a particular meaning. In fact, the emergence of a "gestural complex" between 9 and 13 months predicts and precedes the emergence of the first words (Bates et al. 1977). This gestural complex is comprised of referential pointing, object showing, object giving, and a gestural request for objects (the "gimme" gesture of rapidly opening and closing the fists, with hands extended: Bates, personal communication). The emergence of this gestural complex correlates with achievement of the 4th and 5th stages of the causality series (involving the realization that other people can act on objects, as revealed by nonverbal requests to reach objects and wind up toys, etc.) and with the achievement of the 5th stage in the imitation series (involving the ability to imitate novel schemes), but not with the stages of the object concept series (Bates et al. 1977).

The recapitulation argument is somewhat less compelling than in the case of intelligence because the stages of language development past 2 or 3 years of age are less well defined than the stages of cognitive development, because the comparative data are weaker, and because human language development today involves the coincident emergence of aspects of language development such as prosodic and phone production which must have evolved much later than the gestural complex and the basic instrumental functions of action and attention coordination.

Comparative data on symbol use by great apes suggests that they can learn gestural and other nonvocal symbols and use them referentially in a manner roughly comparable to that of 2- to 4-year old human children (Parker and Gibson 1979). Recent criticisms of the methodology of sign language studies (Terrace et al. 1979) do not constitute a convincing counter claim; indeed, controlled studies without human interlocutors vindicate the claim that apes generalize symbol meanings and apply them in new contexts and use them for instrumental, regulatory, and interpersonal functions such as requesting food and tools (Savage-Rumbaugh et al.,1977). More compelling even than these controlled experiments on symbol learning are studies of spontaneous referential gestural use by captive pygmy chimpanzees (Sage et al. 1977), and wild chimpanzees (Plooij 1977).

Altogether these comparative data support the theory that the first form of hominid referential communication was gestural (Hewes 1973), and imply that our common ancestor with the great apes and hence the earliest hominids displayed a referential gestural complex similar to that of young children (Bates et al. 1977) and great apes (Parker and Gibson 1977).

We believe that these models of cognitive development and language development provide a framework for distinguishing taxonomically relevant traits, grade-levels of achievement, and evolutionary sequences within Old World monkeys, apes, and man, and that they suggest an appropriate level of analysis of comparative data on spontaneous activities of other primate species. In other words, we believe that developmental models provide the materials for a new ethology of unstereotyped behavior (Parker 1977, Parker and Gibson 1977).

The heuristic power of this new developmental ethology is indicated by the testable predictions it generates concerning the level of intellectual and referential communication displayed by our hominid ancestors, and the range of adaptations it implies.

The recapitulation model implies that formal operations came very late in human evolution and that concrete operations preceded it, and that preoperations preceded concrete operations and was characteristic of the earliest hominids. Moreover, the existence of parallel stages or levels of intellectual functioning in many different domains (i.e., time, space, physical causality, number, classification, seriation) provides some guidelines for cross-checking the plausibility of highly inferential models. A model which postulated formal operational reasoning in the earliest hominids, for example, would be highly suspect, just as a model which postulated formal operational reasoning in one domain and sensorimotor reasoning as the highest achievement in all other domains in a given hominid would be highly suspect [2].

2 Because adults of our species display operations characteristic of all levels of intellectual achievement, it is impossible to diagnose terminal or highest levels of intellectual functioning of our ancestors simply by reconstructing the level of intelligence required for a single activity. Rather,

Animal Strategies and Their Proximate Mechanisms

Attempts to reconstruct the adaptive significance of a trait should be based on a comprehensive taxonomy of resource categories, correlated resource manipulation strategies, and the various proximal mechanisms mediating these strategies in different animal species. In other words, it should be based on a minimal but exhaustive set of categories of animal adaptations. Although we know of no such taxonomy, it is not difficult to produce a preliminary set of categories based on the competitive gene model of animal behavior (Borgia 1970, Dawkins 1978, Emlen and Oring 1977) (Tables 2 and 3).

It is generally recognized that animal adaptations center on a few basic functions including maintenance, defense, and reproduction; maintenance and defense simply being necessary for reproductive success. These basic activities are conditioned by the nature, dispersion, and density of resources (food, water, safe shelter, and potential mates). The success of each animal is determined by his ability to compete for these scarce resources; this in turn is determined by his ability to manipulate animate and inanimate objects in his environment to his advantage. The animate objects include his prey, his predators, his competitors, and resource purveyors of the same and different species.

Social object manipulation (Dawkins 1978) apparently occurs (1) to the extent that other animals are competitors for scarce resources (male/male competition for territories and/or mates is a good example of this); (2) to the extent that other animals are sources of scarce resources (the competition for females as sources of parental investment in gestation and lactation in mammals is a good example of this); (3) to the extent that other animals are potential sources of labor for exploiting scarce resources (domestication and food sharing are good examples of this). Social manipulation of other animals as competitors and as sources of scarce resources is widespread in insects and vertebrates; manipulation of other animals as labor resources is relatively rare, being highly elaborated in social insects and man, but present to some degree in all food-sharing species (e.g., birds and carnivores). It is significant that social insects and man display the highest degree of cooperation, referential communication, and technology in the animal kingdom. Both taxa use other animals as labor sources in cooperative food location, transport, storage, and allocation, and in shelter construction. The proximate mechanisms subserving these strategies are of course radically different. The particular proximate mechanisms seem to depend upon the sensorimotor organization of the species and on the details of resource utilization.

One value in looking at animal adaptations from the broadest comparative perspective is that it allows us to see interrelations betwen adaptive strategies such as animate

---

Footnote 2 (continued)

it is necessary to reconstruct all the major activities and see the highest level necessary to accomplish the most difficult task. If, for example, *Homo erectus* needed 5th stage sensorimotor abilities to grind ochre, we cannot conclude from this that *Homo erectus* displayed no higher levels of intelligence, or that grinding ochre was the primary function of this ability (Marshack 1970). In fact, they apparently display at least concrete operational intelligence in production of Acheulean tools (Wynn 1979)

Table 2. Animate object manipulation

Resource categories	Correlated resource exploitation strategies	Communicative manipulation strategies	Information transfer type
I. *Prey object manipulation*	A. Hunting strategies types: Stalking Sentinel Driving	Cryptic displays Coordination displays	Nonreferential in nonhuman species
Many insects Vertebrates	B. Gathering strategies Types: Selective harvesting Unselective harvesting	Coordination displays Cryptic storage	
II. *Predator object manipulation*	A. Protective strategies: Armour Camouflage Warning	Cryptic displays	Nonreferential in nonhuman species
Insects Vertebrates	B. Defensive strategies: counterattack	Threat displays Aposematic	
III. *Social object manipulation* 1. *Conspecifics as competitors for scarce resources* Vertebrates	Displacement and defensive strategies: Territoriality Dominance	Priority displays Status displays	Nonreferential in nonhuman species
2. *Conspecifics as sources of scarce resources* Insects Vertebrates	Acess and control strategies: Confidence of paternity devices	Male competition displays Epigamic displays Female control displays	Nonreferential in nonhuman species
3. *Conspecifics as labor sources* Social insects Man	Access and control strategies Scarce resource privisioners: Locators Transportators Preparers Storers Sharers	(Cooperation) Information inciting displays Preparation inciting displays Begging displays	Nonreferential and referential (intelligent or stereotyped)

and inanimate object manipulation, and niche factors such as resource type, distribution, and density, which might otherwise escape our notice and/or invite false dichotomies or isolated treatment of factors as, for example, looking at social behavior out of its ecological and technological context.

Table 3. Inanimate object manipulation

Object manipulation categories	Resource exploitation strategies:	Proximate mechanisms
I. *Simple object manipulation* (involving hands, teeth or other anatomical manipulators)	Food-getting strategies: Plucking Grabbing Opening/extracting Excavating	Stereotyped or intelligent
Birds, mammals, some reptiles Some insects	Food preparation strategies: Crushing Grinding Tearing/shearing Cleaning	
II. *Tool use* a) Ordinary tool use Some insects Some birds Sea otters Cebus monkeys Great apes Humans	Food-getting strategies: Extracting Hunting Gathering Food preparation: Crushing Grinding Cooking Butchering Cleaning Grooming, wound care	Stereotyped or intelligent
b) *Special tool use* (manipulating another animal's behavior through tools) Birds Primates Humans	Defense: Missile throwing Clubbing Stabbing Hunting: Driving Baiting Courtship displays: Nuptial giftgiving	Stereotyped or intelligent
III. *Simple manufacture* (involving shape modification of a single object) Insects Birds Chimpanzees Humans	Tool-making strategies and material-making strategies: Addition Subtraction Transformation	Stereotyped or intelligent
IV. *Manufacture by construction* (joining two or more objects together) Insects Birds Rodents Prosimians Apes Humans	Shelter manufacturing strategies, trap or net manufacturing strategies (humans only), engineering strategies: Selection Transport Materials manufacture Joining techniques: Interlocking, weaving, mixing, cementing	Stereotyped or intelligent

Table 3 (continued)

V.	*Engineering*	Hydraulic regulation:	
	Social insects	Dam building	
	Birds	Temperature regulation:	Stereotyped or
	Rodents	Chemical combustion	intelligent
	Humans	Air circulation	
		Solar capture	
		Extraction:	
		Mining	
		Smelting	

Choosing an Animal Model for the Common Ancestor of Great Apes and Man and Reconstructing the Evolution of Language and Intelligence

After we have identified taxonomically relevant aspects of intelligence ahd have developed a taxonomy of adaptive strategies and their proximal mechanisms, we are faced with the task of reconstructing the specific phyletic history of language and intelligence in our lineage. The first maxim in evolutionary reconstructions should be the identification of the necessary conditions for the origin of particular traits and not the identification of the advantages these traits might have once they have evolved. Once they exist, traits may take on many new functions for which they were not originally selected. Conversely, given traits might have been useful proximal mechanisms for similar functions in other taxa lacking them, had they already existed. Once an adaptation exists, sensorimotor and/or symbolic intelligence, for example, is a useful mechanism for a wide range of adaptive strategies, but the point is that these strategies may be equally well or better served by different mechanisms in other taxa depending upon their preadaptations and on the details of their resource utilization. For this reason it is fallacious to conclude that because an adaptive strategy is subserved by one type of prominate mechanism in one species and by a different type of mechanism in another, that adaptive strategy could not have favored both proximate mechanisms under different niche and phyletic conditions. The problem, then, is to reconstruct the phyletic precursors and preadaptations in each lineage, and to reconstruct the specific resource utilization patterns in each lineage.

An inadequately appreciated problem in this area of evolutionary reconstruction is the importance of selecting the most appropriate living model for the common ancestor of the descendant species group being compared. In looking for a living model for the common ancestor of the great apes and man to use as a starting point for our reconstruction, we have our choice of three genera and four species.

Methods for Reconstructing the Adaptive Significance of the Manifestations of Language and Intelligence

The fact that all great apes display essentially the same level of intelligence in captivity might tempt us to conclude that they are all equally good candidates for studies of the adaptive significance of sensorimotor and symbolic intelligence in the wild. This would be wrong.

First of all, the descendants of the common ancestor have diverged in many ways during their adaptive radiation. Only the most conservative descendant species, i.e., the species most like the common ancestor, is a good candidate for reconstructing the primary adaptive functions of the trait which is shared by all descendant species. The other species may display the same trait in a new context or may display it as nonfunctional rudiment. Studying these species would mislead us into identifying secondary adaptive functions as primary ones or into inferring nonfunctionality, as many have in the case of gorilla and orangutan intelligence (Beck 1980).

Our selection of the most conservative great ape species is based on two principles: the centrality of feeding strategies, and dietary divergence in adaptive radiations. The types of food eaten, and hence the spatial and temporal dispersion of food (and of other resources such as water and safe shelter), determine the dispersion of females and young, and hence the potential for female control, resources control, and labor contributions by competing males (Borgia 1970). These factors determine the advantages of particular mating systems, parental strategies (including food sharing), and life history strategies (including dependency period) (Trivers 1972, Emlen and Oring 1977, Gould 1977, Clutton-Brock and Harvey 1978). They also determine the advantages of certain technological strategies such as tool use and shelter construction (Parker and Gibson 1977, 1979, Gibson and Parker 1979).

Dietary needs are determined in part by phylogenetic inertia and in part by body size. Larger animals have a greater ratio of volume to surface area than smaller animals do. Because they tend to conserve body heat they can live on less nutrious foods such as mature leaves and pith, while their smaller relatives with a greater surface area to volume ratio lose body heat more rapidly and require more nutritious foods such as insects, new buds and leaves, and meat. For this reason larger animals within an adaptive array of species descendant from a common ancestor tend to be more omnivorous while the larger animals tend to be herbivorous.

This pattern can be seen in the baboon radiation in *Papio, Mandrillus,* and *Theropithecus* (Jolly 1970), the *Sivapithecid* radiation into *Ramapithecus, Sivapithecus,* and *Gigantopithecus* (Pilbeam et al. 1977), the Pongid radiation into *Pan paniscus, P. troglogytes,* and *Gorilla,* and in the Australopithecine radiation into *Australopithecus africanus, A. robustus, A. boisei* (Pilbeam and Gould 1974). That the largest species in each array has or had an herbivorous diet is suggested by direct observation of *Theropithecus* and *Gorilla* and by structural analogies in *Gigantopithecus* and *A. boisei* (Jolly 1970, Pilbeam and Gould 1974). That the smallest species in each array had an omnivorous diet is suggested by observations of chimpanzees and by paleoecological evidence on the savanna woodland habitat of *Ramapithecus.*

There is some reason to suspect that in each case the common ancestor was a small omnivorous species similar to the smallest descendants in the adaptive array. The theoretical reason for the assertion is that omnivorous adaptations offer greater potential for evolutionary radiation into new niches. These adaptations constitute an important component of generalist strategies which are less subject to extinction with changing environments. Generalists are not limited by exclusive investment in structures and behaviors suitable for a single habitat and food source as living herbivores with the T-complex such as gorillas and pandas appear to be.

If this logic is correct, we can infer that the common ancestor or the great-ape hominid radiation (whether that was *Ramapithecus* or *Australopithecus afarensis*) was a small opportunistic omnivore and not a large herbivore and/or tough-object feeder. This argument and an increasing body of data suggest that chimpanzees are the best living model for this ancestor.

Chimpanzees living in woodland savanna habitats similar to those reconstructed for the first hominid *(Ramapithecus* or *Australopithecus afarensis)* display a broad dietary adaptation embracing fruits, nuts, leaves, buds, bark of many trees, honey, many species of ants and termites, insect grubs, animal eggs, birds, and many species of small mammals and infants of other species of mammals. The availability of each type of food, excepting some animal prey, is highly seasonal. Termiting and anting seem to occur primarily during food shortages in the dry season. Some sexual dimorphism in feeding occurs in termiting and hunting, with females specializing in the former and males in the latter (van Lawick Goodall 1968, Suzuki 1969, Teleki 1973, McGrew 1979).

Chimpanzee tool use occurs primarily in the context of extractive foraging for enclosed foods, i.e., termite fishing, ant dipping, honey dipping, hard-shell opening, and fluid sponging. It does not occur in hunting or in agonistic interactions. This circumstances, combined with comparative data on tool using in other taxa, lead us to propose that tool use arose in the common ancestor of great apes and hominids as an adaptation for extractive foraging on a variety of seasonally and locally variable high-energy enclosed food sources which were unavailable to non-tool-using competitors in the same habitat (Parker and Gibson 1977).

By extension we propose that the earliest hominids were conservative great apes who began to depend primarily on the extractive foraging (supplemented by small-game hunting) as they moved into mosaic riverine and lakeside woodland savanna habitats subject to strong seasonal drying and fluctuations in food availability (Isaac 1976). These creatures began to expand the range of their tool use to include excavating water, roots, and tubers, and small fossorial animals, smashing open turtle shells and scavenging long bones or large mammalian prey. Eventually they extended tool use to include butchery of large animals (extractive foraging of flesh encased in thick skin) and aimed missile driving or stunning competitors and prey. They extended the concept of enclosure so familiar to them in their foraging to the enclosure of collected foods.

Efficient extractive foraging and hunting with tools requires a prolonged apprenticeship to allow observational learning, imitation, and practice. To judge from the data on the development of termite fishing and hunting in chimpanzees, this apprenticeship was probably of the order of 4 or 5 years. One important implication of this reconstruction is that this subsistence strategy would have favored post-weaning food sharing with offspring who could not efficiently procure their own high-energy foods. Post-weaning food sharing is more efficient energetically than extended lactation (Silk 1978). Food sharing between adult males and females, close kin or mates, would be favored by increasing specialization in subsistence activities and by the advantages to both sexes of having access to the reciprocal resource: extracted and gathered foods being more reliable, and hunted foods being more delicious and nutritious.

Food sharing has several aspects: location, transport, and preparation. Extractive foraging on hidden enclosed food sources would favor referential communication of the nature and location of hidden and/or distant food sources, and referential communication of requests for help in preparation of encased foods.

Reconstruction of the function of homologous traits, in this case advanced sensori-motor and symbolic intelligence, through ecological studies of the most conservative species in the adaptive array, gives us a good idea of the ultimate causality of the trait, but it is hardly definitive because it is based on a single case. Our idea should be tested in a broader arena of species displaying analogous traits.

Anthropologists generally limit their comparative studies to the Primates order, or the Anthropoidea suborder, with occasional studies of carnivores. This limitation is apparently based on the fact that no other species display hominid traits such as language and intelligence, and on the idea that analogous traits are irrelevant to evolutionary modeling because they are based on different proximal mechanisms.

This is an advantage, however, because analogous traits in distantly related taxa offer us an opportunity to test hypotheses concerning ultimate causation of traits independent of their proximal mechanisms. Moreover, analogies are the only source of comparative data in the case of some traits such as shelter building which are unique to our order (Gibson and Parker 1979).

The problem comes in identifying traits for comparison across distantly related taxa. In fact, sensorimotor and symbolic intelligence is one possible proximate mechanism mediating a range of inanimate object manipulations including tool use and manufacture, resource division and allocation, shelter construction, hunting, and gathering, as well as a range of animate object manipulations (Dawkins 1978, Kummer 1980); particularly, we suspect those involving exploitation of other animals as labor sources (e.g., for cooperative tool use and food sharing) [3]. It is these activities which we define as functionally relevant for comparative studies of adaptive significance in distantly related taxa. In other words, a different definition of traits is relevant for studies of analogies than for studies of homologies.

Our aim in studying analogies is to establish the broadest possible range for comparison spanning all taxa displaying the trait. Only by studying ecological and populational correlates of all species displaying a given trait can we derive a valid hypothesis concerning the adaptive significance of an activity.

Our extractive foraging hypothesis of the adaptive significance of tool use, for example, is apparently confirmed by analogy: comparative data on tool use in birds and mammals suggests that it virtually always occurs in the context of extracting enclosed foods (Beck 1980).

Tool use apparently arises as an energetically efficient equivalent to organic extraction in species such as Galapagos woodpecker finches, sea otters, and other animals who display feeding niches uncharacteristic of their phyletic group, or are too small to display the structural preadaptations for feeding on certain foods (Alcock 1972, 1975, Parker and Gibson 1977). Comparative data on tool use by wild cebus monkeys reveal a similar pattern (Parker and Gibson 1977).

---

3 Several investigators (Beck 1980) have argued that primate intelligence arose as an adaptation for social strategies. The first problem with this proposal lies in its vagueness, particularly in its failure to specify types and levels and functions of intelligence as a social adaptation. A second problem with this proposal lies in its tendency to dichotomize animate and inanimate object manipulation strategies and its failure to relate social manipulation strategies to resource utilization. Beyond this lies the problem of identifying the types and levels of intelligence involved in various social strategies. Clearly, this is a very important subject for research from a Piagetian perspective

Reconstructing the Adaptive Signifiance of Language and Intelligence Itself

Now that we have outlined a broad comparative approach to reconstructing the adaptive significance of tool use, shelter construction, referential communication, and other manifestations of intelligence, we can reconstruct the form and function of specific intellectual and language abilities mediating these activities. Evolutionary models of form and function must provide a continuous pathway connecting the forms and functions characteristic of a putative common ancestor and those of the living descendants. A model for the evolution of hominid language and intelligence, for example, must reconstruct a series of intermediate forms connecting the communication and intelligence of a chimpanzee-like common ancestor with the language and intelligence of modern humans. This model must also explain how the intermediate forms are both products of preceding forms and precursors of succeeding forms, and what selection pressures mediated the transitions from one form to another. These reconstructions must, of course, be consistent with paleontological, paleoecological, and archeological data.

The last step in our model building, then, involves an analysis of sensorimotor and preoperational intellectual abilities and prelanguage gestural references as proximal mechanisms mediating technological activities such as tool use, shelter construction, aimed throwing, and food sharing in the earliest hominids.

Intelligent tool use as a means to solve particular problems involves trial and error experimentation with object-object relationships and a practical understanding of simple physical causality and its temporal sequence, spatial relations of proximity and enclosure (on top of, underneath, behind, inside), which are characteristic of the 5th and 6th stages of sensorimotor intelligence. Imitation of tool use is also characteristic of this level of intellectual achievement in great apes and man. This conjunction of 5th and 6th stage sensorimotor intellectual achievements and the demands of tool use in extractive foraging is quite remarkable. It seems unlikely that the occurrence of these abilities in the only tool-using extractive foragers among the primates is fortuitous.

We propose, then, that tool use in extractive foraging is mediated by 5th and 6th stage sensorimotor intelligence, and that this level of intellectual achievement was selected in the common ancestor of great apes (and in cebus monkeys) because of the reproductive advantage it conferred on individuals competing for a variety of seasonally and locally variable enclosed food sources.

Stereotyped tool use, by contrast, involves releasing and orienting a few action patterns in response to a key stimuli. This type of tool use is apparently adequate for context-specific tool use for opening a single type of enclosed food. Trial and error imitation and invention characteristic of intelligent tool use are only favored in situations where animals feed on a wide variety of encased foods requiring a variety of tool use techniques. Intelligent tool use is also distinguished from sterotyped tool use by the long period of apprenticeship it requires and by the occurrence of local tool use traditions (Parker and Gibson 1977).

Locating new sources of enclosed foods and communicating their location and planning appropriate extractive tool-using techniques is enhanced by a focus on topological spatial relationships of proximity and enclosure, and symbolic representation of these relationships and techniques which is characteristic of stage 6 of the sensorimotor period and the symbolic subperiod of the preoperations period of intellectual development which follows the sensorimotor period[4].

Aimed throwing of missiles to drive away competing scavengers or to separate a weak or young animal from the herd involves line-of-sight aiming which is characteristic of the intuitive subperiod of preoperational intellectual development following the symbolic subperiod. The production of sharp points and sectioning of solids which is necessary for extracting and butchering meat from large animal prey is also characteristic of this level of intellectual achievement.

Shelter construction involves object-object manipulation of relationships of proximity (placing objects next to, on top of, etc.), and manipulation of partial enclosure relationships of intertwinning in the creation of an enclosure. It also involves formation of simple collections and some seriation of elements. These activities are all characteristic of symbolic and intuitive levels of intelligence.

We propose, then, that discovery of new embedded foods, aimed throwing of missiles, simple stone tool working, and shelter construction are all enhanced by symbolic and intuitive level intelligence, and that this level of intelligence was favored in early hominids because of the reproductive advantages it conferred on individuals who were able to engage in these activities which increased efficiency in hunting and reduced vulnerability to predators, inclement weather, and hostile environments.

Comparative data on shelter construction in other animals suggest that shelters function to protect immature young and adults, to maintain thermal equilibrium, to store and share foods. Shelter construction occurs in species who are exposed to predators and elements without recourse to appropriate natural shelters. This exposure is usually a consequence of open area feeding strategies. Shelter construction is often associated with long-term parental investment and long-term association with extended kin. It also often involves cooperative labor by mates and/or close kin.

It is reasonable to infer from the comparative data that early hominids built shelters to protect themselves and their dependent offspring and to store and share foods with close kin. Shelter construction would have been increasingly important as hominids spent more time in the savanna and as infants lost their ability to cling. As with tool use, however, the proximal mechanisms involved in hominid shelter construction differ from those of most other species in being intelligent rather than stereotyped (Gibson and Parker 1979) (Table 4).

If language is a symbolic structure for regulating joint action and joint attention (Bruner 1975) then we must ask ourselves what resource utilization patterns would have favored these regulations or, to put it another way, these manipulations of the labor potential of conspecifics.

---

4   South American cebus monkeys who, like chimpanzees, are omnivorous extractive foragers, display a least some aspects of 5th and 6th stage sensorimotor intelligence (Parker and Gibson 1977). Currently there is no evidence that they display imitation of object manipulation schemes, gestures, facial expressions or vocalizations, though detailed studies may reveal some of these abilities. There is also little evidence of long apprenticeship in local tool-using traditions in this genus, though it too may occur. These animals may discover new food sources through trial-and-error manipulation of objects rather than imagery of enclosure relationships. Apprenticeship may occur through attention focusing on food items rather then through imitation of movement patterns

Table 4. Primary adaptive functions of primate intelligence, by grade-levels

Kind of intelligence	Prosimian	Old World monkey	Great ape	Early hominid
Sensorimotor intelligence				
Stages 1 and 2				
Simple prehension, hand-mouth coordination	Manual prey catching, branch-climbing by grasping			
Stage 3				
Hand-eye coordination Secondary circular reactions [a]		Manual foraging	Object play for tool use	Same
Stage 4				
Coordination and application of manual schemes on single objects		Manual food preparation and cleaning, manual grooming		
Stage 5				
Object permanence		Food location, memory (?)	Same	Same
Object-object coordinations, trial-and-error investigation of object prop. (tertiary circular reactions), discovery of new means (tool use)			Trial-and-error discovery of tool use for extractive foraging on embedded foods	Same
Stage 6				
Deferred imitation of novel schemes			Imitative learning of tool-use traditions, search for new embedded foods, insightful tool-use	Same
Preoperational intelligence				
Symbolic subperiod Topological preconcepts of enclosure and proximity			Search for rare embedded foods	Same, plus shelter-construction
Make-believe games.				Practice of subsistence roles
Intuitive subperiod Euclidean and projective preconcepts of straight line and angle				Tool manufacture, tool use in butchery, shelter construction
1 to 1 correspondence				Food division
Construction games				Practice in shelter construction and tool manufacture
Aimed-throwing games				Practice for aimed-throwing in hunting and defense

[a] Not present in macaques — arose as retrospective elaboration in great apes

We suggest that sharing with close kin via referential communication of food location and via labor contributions involved in extractive tool use were favored in early hominids because they depended on scarce high-energy hidden food sources which could be more efficiently extracted and allocated through cooperative referential communication.

The "gestural complex" and the content of protolanguage in phase I children are admirably suited for communicating the nature and location of distand and/or hidden foods and for communicating requests for aid in extracting embedded boods; i.e., "regulating joint activity and joint attention", in Bruner's terms, and for manipulating others as sources of labor in Dawkin's terms. Because of this, and on the basis of comparative data on symbol use by great apes, we propose that gestural protolanguage arose as an adaptation for food sharing among close kin in our omnivorous extractive tool using ancestors.

If we take referential communication as the functionally relevant trait we find that it is associated with food sharing with close kin in the only other taxa displaying this, i.e., bees. As in the preceding cases, the proximal mechanism mediating referential communication is different.

The model for the evolution of language and intelligence briefly summarized here, has the advantage of providing a continuous series of forms and functions linking great ape levels of achievement with subsequent hominid levels of achievement. It has the particular advantage of explaining the transition from hunting without tools to hunting with tools.

Conclusion

In this paper we have argued for the importance of deriving the models for the evolution of language and intelligence from basic postulates of Piagetian theory. Specifically, we have proposed the following structures:

1. the importance of selecting taxonomically relevant traits for comparative studies of the form and level of intelligence in closely related species, and the utility of Piaget's model for this purpose;
2. the importance of selecting functionally relevant traits for comparative studies of the adaptive functions in a wide variety of distantly related taxa;
3. the necessity for deducing adaptive function (ultimate causation) by analogy through exhaustive comparative studies of all animal taxa displaying a given trait;
4. the importance of recognizing the effects of phylogenetic inertia in the persistence of certain traits of a common ancestor as rudiments or remnants in some living descendants;
5. the importance of searching for the *necessary* conditions for the origin of a particular trait, as opposed to searching for the numerous advantages it might have conferred once it existed;
6. the importance of judicious selection of a living model for the common ancestor in ecological studies for the purpose of reconstructing primary functions of a trait;

7. the importance of developing a minimal but exhaustive taxonomy of adaptive strategies (and their proximate mechanisms) in relation to competition for scarce resources as a basis for comparative studies of adaptive function of traits.
8. the importance of making a model with continuous transitions between the form and function of traits in the common ancestor and those of the living descendents;
9. the importance of recognizing the primacy of preferred resource distribution and density as selective factors shaping feeding and sheltering strategies, and hence indirectly, tool use, shelter construction, food sharing, referential communication, etc.

These concepts are useful for evaluating alternative models for the evolution and adaptive significance of intelligence in apes and hominids. There are at least two alternative models for the evolution of monkey, ape, and human evolution: the "fig-finding" model or orangutan intelligence, and the "social strategy" model of monkey, ape, and human intelligence.

According to the "fig-finding" model, orangutans' intelligence is an adaptation for locating figs irregularly fruiting in a large home range in the tropical rain forest (Rodman, cited by Webster 1979). First of all, this model violates the concept of distinguishing primary from secondary adaptations by judicious selection of a living model for the common ancestor, taking into account the adaptive radiation into divergent niches. Orangutans are more specialized in their diet than chimpanzees; they are much larger than the putative common ancestor, and are able to engage in extractive foraging without the aid of tools. Even if these animals do use their intelligence to find figs, this does not mean that fig-finding was the primary function of their intelligence. Secondly, this model violates the concept of deducing adaptive function (ultimate causation) by analogy through exhaustive, comprehensive comparison of all animal taxa displaying a given functional trait (in this case, fig-finding). Many animal species, including many monkeys and birds, feed on irregularly fruiting figs in the tropical rain forest. While it is true that some of these species, like howler monkeys, have small home ranges, others like spider monkeys do not. Neither intelligence nor tool use is characteristic of other fig-finders. Another weakness in this model lies in its failure to specify types and levels of intelligence and their observable manifestations.

According to the "social strategy" model, anthropoid intelligence is an adaptation for success in social groups (Humphrey 1976). This model, like the fig-finding model, fails to specify types and levels of intelligence and their observable manifestations. It fails to distinguish different levels of intelligence, apparently confounding the sensorimotor with the formal operational levels by attributing inference to monkeys, for example. It also fails to survey the intellectual correlates of social life in other animal taxa. Of all the social animals, cebus monkeys, great apes, and man are the only species known to display advanced sensorimotor and/or symbolic intelligence. (It is not unlikely, however, that a few other species such as elephants and dolphins do also.) This model violates the concept of distinguishing primary and secondary functions: once evolved, intelligence will take on secondary functions: no doubt including strategizing, but this does not reveal its primary adaptive significance[5].

---

5 This is not to say that selection did not favor intelligence for social manipulation at some stage of hominid evolution. We believe that it did. The point is that this was not the source of intelligence in monkey and apes, and that the model is vague in failing to specify which level and type of intelligence has this adaptive significance

A third possible model, a "nest-building" model of great ape intelligence, looks reasonable on the surface because all the great apes display this pattern (cebus monkeys do not, however). The problem with this model lies in its failure to explain why a simple stereotyped pattern performed in a single context would require an understanding of object-object and spatial relations of interlocking and trial and error experimentation with new means. The implausibility of this argument is revealed by comparison with other shelter- and nest-building species, none of whom display high intelligence. Only when shelter building requires significant trial and error adjustment to a wide range of circumstances and materials, as it did in hominids, does it require advanced sensorimotor or higher levels of intelligence (Gibson and Parker 1979).

We cannot overemphasize the importance of theory in hominid reconstructions. The minimal use of theory involves the avoidance of obvious errors of inconsistency with accepted theory (which is more important than inconsistence with data because the definition of data are contingent on theoretical formulations). The maximal use of theory involves prediction of causal relationships between phenomena which otherwise appear unconnected or insignificant. The value of theories lie in their heuristic power, in their ability to generate coherent and consistent testable hypotheses which would never have emerged in their absence. The broader the range of phenomena explained by a single model, the more compelling it is.

References

Alcock J (1972) The evolution of use of tools by feeding animals. Evolution 26:464–473

Alcock J (1975) Animal behavior: An evolutionary approach. Sinauer Assoc Inc, Sunderland Mass

Antinucci F, Spinozzi G, Visalberghi V, Volterra V (1980) Cognitive development in a Japanese macaque. Presented at VIII IPS Congress, Florence

Bates E, Benigni L, Bretherton L, Camioni, Volterra V (1977) Cognition and communication. From 9–13 months: A correlational study. Program on cognitive and perceptual factors. In: Hum Dev Rep No 12. Institute for the Study of Intellectual Behavior. Univ Colorado

Beck BB (1980) Animal tool behavior. Garland STPM Press, New York

Borgia G (1970) Sexual selection and the evolution of mating systems. In: Blum MS, Blum NA (eds) Sexual selection and reproductive competition in insects. Academic Press, London New York

Bruner J (1975) The ontogenesis of speech acts. J Child Lang 2:1–19

Chevalier-Skolnikoff S (1977) A Piagetian model for describing and comparing socialization in monkey, ape, and human infants. In: Chevalier-Skolnikoff, Poirier F (eds) Primate biosocial development. Garland Publ Inc, New York

Clutton-Brock T, Harvey P (1978) Mammals, resources, and reproductive strategies. Nature (London) 273:191–195

Dawkins R (1978) Animal signals: Information of manipulation? In: Krebs JR, Davies NB (eds) Behavioral ecology: An evolutionary approach. Blackwell Scientific Publications, Oxford

Eibl-Eibesfeldt (1975) Ethology: The biology of behavior, 2nd edn. Holt, Rinehart, Winston, New York

Emlen S, Oring L (1977) Ecology, sexual selection, and the evolution of mating systems. Science 197:215–223

Gibson KR, Parker ST (1979) Extraction, construction, and agonistic missile use: A model of the evolution of early hominid technology. Manuscript

Gould SJ (1977) Ontogeny and phylogeny. Harvard Univ Press, Cambridge

Halliday MAK (1975) Learning how to mean: Explorations in the development of language. Edward Arnold, London

Hewes G (1973) Primate communication and the gestural origin of language. Curr Anthropol 14: 5–24

Humphrey (1976) The social function of intellect. In: Bateson PP, Hinde R (eds) Growing points in Ethology. Cambridge University Press

Inhelder B, Piaget J (1964) The early growth of logic in the child: Classification and seriation. WW Norton Inc, New York

Isaac G (1976) East Africa as a source of fossil evidence for human evolution. In: Isaac G, McCown E (eds) Human origins. Perspectives on human evolution, vol III. WA Benjamin Press, Menlo Park

Jolly A (1972) The evolution of primate behavior. MacMillan Publishing Co, New York

Jolly C (1970) Large African monkeys as an adaptive array. In: Napier J, Napier P (eds) Old World monkeys evolution, systematics and behavior. Academic Press, London New York

Kummer H (1980) on the value of social relationships to nonhuman primates; a heuristic scheme. In: Cranach M von, Foppa K, Lenenies W, Ploog D (eds) Human ethology. Univ Press, Cambridge

Lawick-Goodall J van (1968) The behavior of free-ranging chimpanzees in the Gombe Stream Reserve. Anim Behav Monogr 161–311

Leakey R, Lewin R (1978) People of the lake. Avon, New York

Marshack A (1970) Data for a theory of language origins. Commentary on a developmental model for the evolution of language and intelligence in early hominids. Behav Brain Sci 2:394–396

Mathieu M (1978) Piagetian assessment of cognitive development in primates. Annu Meet Am Anthropol Assoc, Los Angeles, California

McGrew WC (1978) Evolutionary implications of the sex differences in chimpanzee predation and tool use. In: Hamburg DA, McCown ER (eds) Perspectives on human evolution, vol IV. The great apes. Benjamin-Cummings, Menlo Park

McGrew WC (1979) Habitat and the adaptiveness of primate intelligence. Behav Brain Sci 2:393

Parker ST (1973) Piaget's sensorimotor series in an infant macaque: the organization of non-stereotyped behavior in the evolution of intelligence. Ph D thesis, Univ California, Berkeley (Ann Arbor, University Microfilms)

Parker ST (1976) A comparative longitudinal study of the sensorimotor development in a macaque, a gorilla, and a human infant from a Piagetian perspective. Paper presented at the Anim Behav Soc Conf, Boulder, Colorado

Parker ST (1977) Piaget's sensorimotor period series in an infant macaque: a model for comparing unstereotyped behavior and intelligence in human and nonhuman primates. In: Chevalier-Skolnikoff S, Poirier FE (eds) Primate biosocial development. Garland Press, New York

Parker ST (1978) Preoperational intelligence and symbolic communication in protohominids. Paper presented at the 47th Annu Meet Am Anthropol Assoc, Los Angeles

Parker ST, Gibson KR (1977) Object manipulation, tool use, and sensorimotor intelligence as feeding adaptations in cebus monkeys and great apes. J Hum Evol 6:623–641

Parker ST, Gibson KR (1979) A developmental model for the evolution of language and intelligence in early hominids. Behav Brain Sci 2:2

Piaget J (1954) The construction of reality in the child. Ballantine Books, New York

Piaget J (1962) Play, dreams and imitation of childhood. WW Norton, New York

Piaget J (1963) The origins of intelligence in children. WW Norton, New York

Piaget J (1978) The development of thought. Viking, New York

Piaget J, Inhelder B (1967) The child's conception of space. WW Norton, New York

Pilbeam D, Gould SJ (1974) Size and scaling in human evolution. Science 186:892–901

Pilbeam D, Meyer G, Badgley C, Rose MD, Pickford MHL, Behrensmeyer AD, Ibrahim Shah SM (1977) New hominoid primates from the Sawaliks of Pakistan and their bearing on hominid evolution. Nature (London) 270:689

Plooij FX (1977) Some basic traits of language in wild chimpanzees? Intern Rep no 77 ONO5. Vakgroep Ontwikkeling Psyxhologie. Psychologisch Laboratorium. Katholieke Univ, Nijmegen, The Netherlands

Redshaw M (1978) Cognitive development in human and gorilla infants. J Hum Evol 7:133–141

Rumbaugh D, Savage-Rumbaugh S (1981) A response to Herb Terrace: linguistic apes. Science (in press)

Savage-Rumbaugh ES, Wilkerson BJ, Bakeman R (1977) Spontaneous gestural communication among Conspecifics in the pygmy chimpanzee *(Pan paniscus)*. Bourne G (ed) Progress in ape research. Acadmic Press, London New York

Silk J (1978) Food sharing among mother and infant chimpanzees at the Gombe Stram National Park in Tanzania. Folia Primatol 29:129—141

Stern D (1978) The first relationship. Harvard Univ Press, Cambridge

Suzuki A (1969) An ecological study of chimpanzees in savanna woodland. Primates 10:105—106

Teleki G (1973) The predatory behavior of wild chimpanzees. Buckness Univ Press, Lewiston, Pa

Terrace H, Petitto LA, Sanders RJ, Bevers TG (1979) Can an ape create a sentence? Science 206: 891—902

Trevarthen C (1980) Instincts for human understanding and for cultural cooperation: their development in infancy. In: Cranach M von, Foppa K, Lepenies W, Ploog D (edsO Human ethology. Univ Press, Cambridge

Trivers R (1972) Parental investment and sexual selection. In: Campbell B (ed) Sexual selection and the descent of man. Academic Press, London New York

Webster C (1979) Vital triangle: Wasps, figs, clever apes. The New York Times, Dec 18:C1—C2

Wynn T (1979) The intelligence of later Acheulean hominids. Man 14:371—391

Primatology and Sociobiology

J. WIND [1]

Introduction

Primatology is a rather broad concept encompassing various scientific approaches. Best known among them are, of course, morphology, physiology, ethology, conservationism, biomedicine, biochemistry and evolutionary biology. And strictly speaking primatology also includes studying human primates. In this chapter I will review the relations with and the possible incorporation into such a broadly defined primatology of a recently emerged discipline called sociobiology. Because it is now some 5 years ago that Wilson's (1975) impressive volume marked sociobiology's entrance into the scientific — not to say the public — arena, such a discussion may have some use for primatologists by reviewing the topic on the basis of the growing body of literature and the numerous, and often heated, debates that it has triggered during its birth and infancy. Since the most fierce arguments have been about its possible application to human behavior, I will here include some discussion on human primates. For more literature than provided here see Barash (1977), Caplan (1978), Gregory et al. (1978), Parker (1978), Chagnon and Irons (1979), Christen (1979), or Alexander (1980).

For many years many biologists — including ethologists and other primatologists — as well as many social scientists and psychologists have been busy studying primate behavior (human and nonhuman) often focusing on social behavior. So what, then, is new about sociobiology? It is its attempt at merging some sciences that until recently had existed more or less independently from one another, and that individually have been developing rather rapidly during the latter decades. Among these sciences are evolutionary biology, molecular biology, ethology, genetics and ecology. Three of sociobiology's basic paradigms will be mentioned here.

1. Though already suggested some 50 years ago by mathematically oriented biologists like Haldane (1932) and Fisher (1930), it has been realized that the basic unit of natural selection is *not* the species, the group, the individual, or even the chromosome: it is a much smaller fraction of the latter, the formerly hypothetical and still invisible *gene*. In the latter years is has become clear that genes consist of desoxynucleic acid (DNA), and biochemistry and molecular biology has clarified much of its properties. The most important and interesting one is its tendency to make replicas of itself.

1 Institute of Human Genetics, Free University, P.O. Box 7161, Amsterdam, The Netherlands

The chemical and genetical views, now, have lately increasingly come to overlap, and so have led to the so-called Selfish Gene Theory (SGT): as Dawkins (1976) in his anthropomorphical metaphor so clearly puts it, each gene "selfishly" attempts to spread as many copies of itself as possible. In this parlance, then, the individual organisms are merely throw-away survival machines for those selfish genes.

The SGT, however, leaves some semantic (Stent 1978) and operational problems unsolved, e.g., a pitfall consists in the question what exactly is considered as a unit, i.e., what length of DNA? Possibly linked genes? Or maybe, the DNA's individual nucleotides, i.e., the base pairs that during crossing-over of the chromosomes may, occasionally as a consequence of the DNA's splitting, become separated? Assuming that these questions will become more clearly delineated and be solved by the rapidly developing molecular biology, the SGT can be accepted as an interesting working hypothesis having a firm chemical basis. Finally, is has also a general biological basis: units like the species, the group, or the individual are much shorter lived than the DNA's, and hence cannot be selected for except via the genes.

2. The second paradigm on which sociobiology is based and which follows from the Selfish Gene Theory (SGT) is that of the Evolutionarily Stable Strategy (EES) (Maynard Smith and Price 1973). Stated very simply, this indicates that after a new combination of genes or a change in environment has occurred, the genes have to adopt a new "strategy" which maximizes their replication in that particular environment. The concept of ESS largely overlaps the traditional one of adaptation, and it is based on a mathematical approach following from games theory as applied to behavior maximizing the spreading of genes' copies. The word strategy should be considered here as an anthropomorphical and metaphorical shorthand illustrating the seemingly prospective capacities of the genes. In reality, of course, this copies-maximizing behavior is the result of natural selection, genetic variation, and chemical properties of the DNA. For a popularized account of the concept of ESS, see Dawkins (1976), for a more sophisticated one Dawkins (1980).

3. The third basic idea of sociobiology (and the one which triggered its origin) is the paradoxical behavior which (again somewhat anthropomorphically) is called altruism: why should an individual organism decrease its fitness or even sacrifice itself for another individual? This question emerged from observations of social insects many of whom do not reproduce and even sacrifice themselves for their conspecifics. The answer is that their genetic relationship is such that decreasing their own fitness, or even reducing it to zero, may contribute to the survival of the copies of the individual's genes present in other individuals. This is the most apparent in its kin, and this type of se'ection is therefore called kin selection; the fitness of the individual and that of its relatives sharing the same genes is called inclusive fitness (Hamilton 1964).

Use of Sociobiology

The relevance of sociobiology seems greatest for behavioral scientists, especially ethologists. The latter usually occupy themselves with describing and analyzing different types of behavior in animals, and in the latter years there has been an increasing interest in nonhuman primates' behavior and its evolution. However, an annoying problem is that, necessarily, so far nonhuman behavior has to be rendered in human terms or at least in terms that have a heavy anthropomorphic load, like hierarchy, dominance, aggression, territoriality, altruism, slavery, etc. Apparently, when describing, e.g., a particular chimpanzee's behavior as aggression we recognize similarities in its and human behavior (like, for that matter, in many other classifications of behavior). Though sociobiology may have added to causing confusion by metaphorically reshuffling such concepts between human and nonhuman categories (Solomon et al. 1978, Lewontin 1979) it, fortunately, offers also some possibilities of reducing the typological and anthropomorphic load of these classifications.

While in many cases the recognition of obvious analogies or even similarities of human and animal behavior may be useful (Leyhausen 1979, p. 262), one may well ask whether aggression in a chimpanzee is really similar to aggression in humans. In answering such questions there are some problems, e.g., what is the definition of aggression? This is difficult to give, for it is essentially a vague concept, and there is both intraindividually and intraspecifically a large variation of body movements that are classified as such. Second, the observed behavior is often confused with the supposed underlying emotions or intentions (Masters 1979, p. 273, Stent 1977, Solomon et al. 1978, Schneewind 1978, p. 235, Searle 1978, p. 174). The problem may be reduced somewhat if we should have a possibility of translating these typologically rendered phenomena in physico-chemical terms. This is of course quite difficult, and complete avoidance of anthropomorphisms and typology is epistemologically impossible. This is a central problem in ethology and an Achilles heel of sociobiology eagerly attacked by humanistic and other scientists (e.g., Allen et al. 1975, Sociobiology Study Group 1976, Alper 1978, Beach 1978). Yet, some reduction of the problem can be achieved by sociobiology. Let us perform some etiological exploration.

First, aggression in a particular nonhuman primate individual is, in fact, part of a vaguely circumscribed cluster of behavior modes which, due to our pattern recognition capacities, are usually classified as aggression. It is therefore a statistical concept following from a deeper level of behavior modes that are recognized in the species as a whole. These are probably the consequence of an Evolutionarily Stable Strategy particular to that species. The described behavior, therefore, occurs on the phenotypic level, whereas the second, cluster-like, phenomena have to be localized on a deeper, i.e., genotypic, level as determined by its selective value in many previous generations.

The two behaviors are — on both levels — the results of two different sets of organ systems like the nervous, muscular, cardiovascular, etc. (of course shaped via feedback forces concomitant with the behavior both during phylogeny and ontogeny). The organs, however, in two different species resulted, in their turn, from different proteins, e.g., enzymes, as well as other molecules; these from — partly — different RNA's, and these from — partly — different genes. Finally, the gene's action is the result of its selfishness, the common chemically determined property of all the DNA's. This is the basic, common cause of most appearances of aggression.

In this way sociobiology reduces both behaviors to and translates them into common and comparable terms. The etiological chain can, of course, be continued more deeply, i.e., to prebiotic molecules that existed on the primitive earth; the origin of the solar system and that of our galaxy; and ultimately the Big Bang. And a more

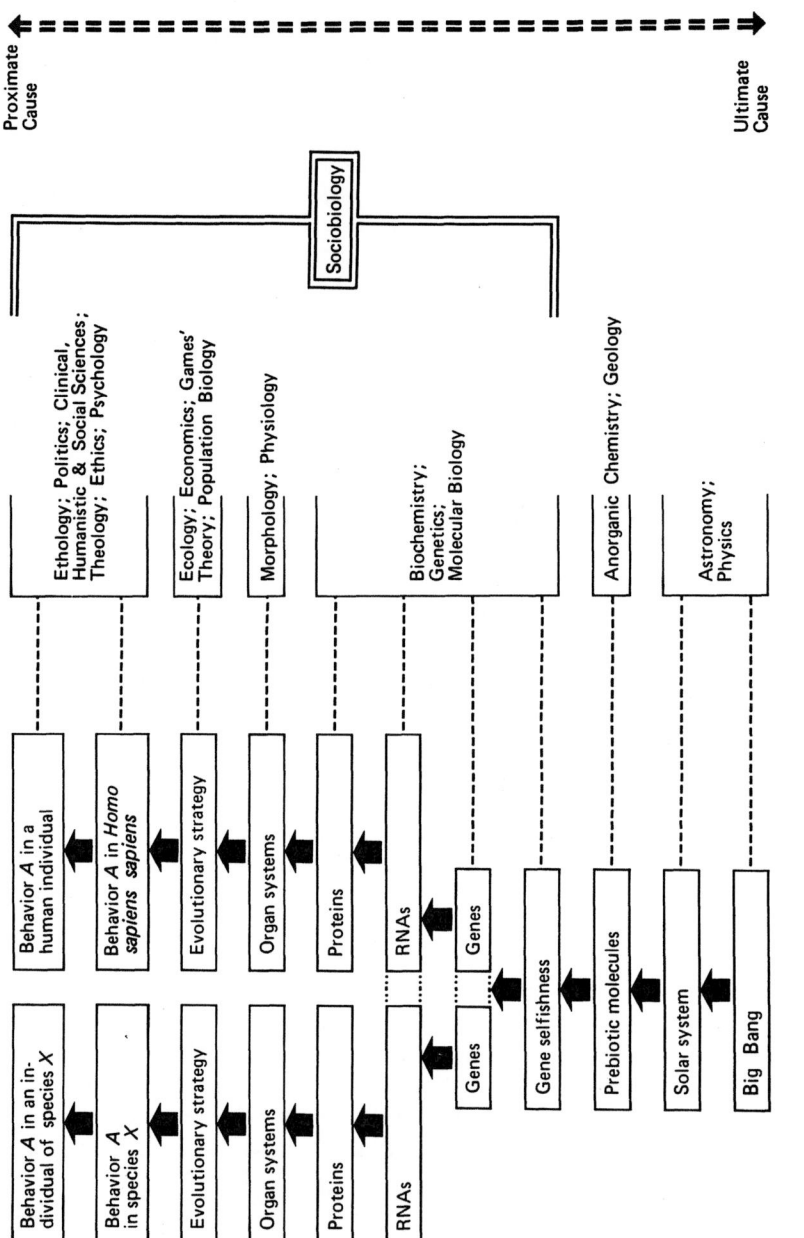

Fig. 1. A schematized attempt at rendering an etiological analysis of the usual ethological observations as well as the place of various sciences relative to the various levels of causality, to each other and to sociobiology

finely grained picture of the chain could be obtained by taking into account its constituting, elementary steps, the ultimate of which are quantum physics' processes. For the present purpose, however, it is sufficient to recognize a crude picture of the chain's first part as rendered in Fig. 1. When we now look at the sciences usually occupying themselves with the various levels of causality concerned, we see that sociobiology encompasses more levels than the other sciences do. It overlaps and glues together various disciplines, recognizing more ultimate causes than ethology, etc. could do, thus providing a more holistic view.

For other primatologists such as conservationists, anatomists, psychologists, geneticists, etc., the relevance of sociobiology is less obvious. And biomedical research appears nicely to be in accordance with the SGT — the human genes exploiting non-human ones. However, sociobiologically the behavior of conservationists itself is rather paradoxical (Mattern 1978, Wind 1980): why should the human genes protect others? This will be discussed below. Morphologists, physiologists, etc. usually work in third level sciences (Fig. 1) and though their studies are less anthropomorphic and more objectified than the second and upper level sciences, they usually remain on the descriptive, phenomenological — sometimes even typological — level. To be sure, some of these scholars render their studies in an adaptive, evolutionary framework; however, sociobiology puts their findings in a wider framework, i.e., it views their findings as being the logical result of the underlying controlling forces of the selfish genes and of the EES's. Such a viewpoint stresses a deeper level than the traditional ones, which imply that anatomy serves function, and function serves the individual's survival.

In conclusion, the use of sociobiology for primatologists and other biologists is that it has some explanatory and even some predictive power, especially as far as behavior is concerned. The latter is to many people more intriguing than morphology and physiology, etc. because it is more malleable and variable, both intraindividually and intraspecifically (being determined less genotypically), and hence, so far, more difficult to predict. Second, understanding the real world, especially life's phenomena, in a wider, more holistic framework of causality may well prove to be useful for other, biological and nonbiological, sciences. Hence, sociobiology helps to understand ourselves better (Fuller 1978). Third, sociobiology suggests that, theoretically, behavior can be expressed as a result of and causing changes in gene frequencies, thus adding to the possibility of quantifying and hence objectifying behavior, including human behavior. Though such an implementation for the time being is only partly possible (see Sect. 3), we can at least attempt to express interspecific differences in behavior (and morphology) in quantitative rather than qualitative terms, because the latter are of smaller value in interspecific comparisons; sociobiology has added to such an approach.

The Limitations of Sociobiology

Like any branch of science, sociobiology has its limitations.

1. It is just one way of looking at and describing life as we perceive it in this real world. Life can as legitimately be described in religious, economic, psychological, political, etc. terms.

2. Scientists using sociobiological theory are not always able to master all the sciences of the various causality levels mentioned in Fig. 1, and hence they often probe less deeply into these sciences than those specialized in each of them are able to do. This lead to misunderstandings and incorrect conclusions.

3. In analyzing the upper level's — i.e., daily life — cause/result relationships and daily-life decision-making, knowledge of sociobiology is rather irrelevant (and that of

quantum physics and cosmogony even more so). The distance between the ultimate cause and the actual result is too great. Its knowledge will, for the time being, hardly or not at all influence our daily-life decisions and strategies. Stated in dialectical rather than in epistemological terms, the irrelevance is caused by the statistical, stochastic nature of the relation between the genes' selfishness and these decisions, and by the relatively great time lag involved: the time scales with which evolution works — and more specifically gene frequency change, such as results in shifts of behavior — are to common human standards usually quite large, i.e., many generations.

4. The predictive power of sociobiology is limited by: (1) The substrate being dynamic rather than static, because of the extremely complicated fabric of continuously changing gene frequencies and environmental interaction. Maynard Smith's concept of ESS, though being of great heuristic value, should therefore rather be called Evolutionarily Less Unstable Strategy (ELUS) (Wind 1980, see also Dawkins 1980, Oster and Wilson 1978); (2) the large number of different genes involved as present in most survival machines; (3) the nonlinear relationships between gene frequencies and behavior characteristics; (4) the forces directing and determining gene selection being statistical in nature and implying time-inertia. This means that behavior with no positive selective value, both on the individual and gene level, does occur, though, of course, less frequently than that having a negative selective value (the lowering of the frequency of a certain gene is usually environmentally determined, i.e., by the action of other, competing, genes or by inorganic causes); (5) epistemologically, complete objectivity being impossible because of the handling of concepts that are necessarily subjective, being reflections of processes in our own mind (Solomon et al. 1978, and many earlier authors).

In conclusion, while the basic paradigm of sociobiology (the selfish gene concept) is quite simple, its practical value and its application to higher vertebrates' — including human — behavior are quite complicated. Hence, for the time being, the value of sociobiology is heuristic rather than practical, especially as far as the higher animals are concerned. Thus, though one of the ultimate causes of a special behavior may be known (i.e., the selfishness of the genes) and its proximate cause can often be assessed — be it sometimes only after the behavior has been performed — daily-life behavior, at least of higher organisms, still largely retains its stochastic (not necessarily "coincidental") character. This applies *a fortiori* to human behavior, thus including political and ethical decisions. From this it follows that the upper level sciences remain completely valid and valuable — if only by providing the raw data to be processed by sociobiology. Hence, they will, for the time being, not be incorporated into sociobiology as has been suggested by Wilson (1975, p. 6). By the same token, sociobiology could be incorporated into other sciences, like theology, etc. The ideal, ultimate, epistemological goal, i.e., that all sciences should coincide and form one great body of knowledge rather than contradicting compartments, will only asymptotically be reached.

Issues Triggered by Sociobiology

Sociobiology has triggered much criticism, especially because of its possible application to human behavior. Critics so far have mainly been found among nonbiologists, but

ave certainly not been restricted to them. However, though the basis of some of
)ciobiology's practitioners — usually called sociobiologists by their critics — may be
1aky, the one of sociobiology itself seems to be quite strong because the "selfishness"
f the genes — while being an anthropomorphism — has a firm chemical ground, i.e,
1e DNA's tendency to replicate. For the rest, there are hardly critics who attack that
asis. Their main points are: (1) sociobiology propagates biological and genetical deter-
1inism which does not apply to man (Alper et al. 1976, Allen et al. 1975, Sociobiology
tudy Group 1976); (2) because of our free will we have taken over the hegemony of
ur genes; (3) human behavior is extremely complicated and variable, and it is strongly
etermined by culture.

   I think that all these criticisms can largely be refuted. A detailed discussion seems
) be somewhat out of place here since it would largely center around purely human
sues. These include sociobiologically seemingly neutral behavior, such as producing
r enjoying "culture", i.e., arts (poetry or painting) and religious activities; and socio-
iologically seemingly odd behaviors like celibacy, infanticide, suicide, adopting chil-
ren (even from overseas), self-sacrifice in a modern war or during overseas missionary
:tivities, wild animal protection, extensive funeral ceremonies, etc. A purely biological
pproach to these questions seems to be able to fit all these human behaviors into
)ciobiological theory (Wind 1980).

   Now, I am unaware of such fierce criticism from students of nonhuman primates.
1 order to anticipate them, however, I will render here some examples of recorded
ehavior of nonhuman primates that seem to falsify the sociobiological theory, like
1e above human examples do. Subsequently, I will attempt to adduce arguments indi-
1ting that these examples do not necessarily imply such a falsification.

   But first, how is it that fierce criticisms have come from students of human rather
1an from those of nonhuman behavior? The most important reason seems to me that
:aditionally many ethologists — mainly focusing on animals — have assumed that all
ehaviors observed by them are adaptive. In contrast, human behavior is not usually
ut into such a preconceived biological framework and thus triggers much more criti-
ism. Yet, well before the era of sociobiology Kummer (1971, p. 90) already warned
1at not all primate behavior is adaptive. Another reason is that human behavior is
mply much better known — even if not necessarily better understood — than non-
uman behavior.

   Bearing in mind the SGT, the usual nonhuman primate behaviors, of course, in fact
em to be in accordance with sociobiological theory. They include feeding, sex,
1aternal care, playing, social behaviors maintaining the group (probably mainly an
ntipredator adaptation; Alexander 1978), including hierarchies, etc. All these are
kely to promote spreading the actors' gene copies; this is confirmed by the more
cent, sociobiologically based ethological reports like those of Kurland (1977) and
iltmann (1980). However, there may be experienced ethologists who have observed
rimate behavior that does not seem to add to spreading gene copies. I, for one, cer-
ainly do not have much of that experience, and it took me some trouble to find
xamples in the literature — though, as an arm-chair scientist, it took me much less than
) actual field workers. What caused these trouble? First, such behavior is undoubtedly
1uch less frequent, but this may not be the only reason why it is less likely to be
;corded and reported; early primatologists may have omitted odd behavior from their

reports as it did not fit into the first, undoubtedly somewhat typological, classifications they attempted to make. This is not to blame them, for this phenomenon is unavoidable in such pioneering work. Primate ethology during the latter decades, however, has clearly outgrown its infancy, and it has provided not only much clearer classifications of common behavior, it has recorded almost all behaviors including anecdotal, odd ones. Now, generally, mentioning anecdotal behavior has a low scientific status. Yet, showing why this is the case may raise that very status, and this may be of some use in the present discussion.

Some Paradoxical Behaviors

Lowering Conspecifics' Fitness

Decreasing a conspecific's fitness by killing may, to be sure, accord with the former "ferocious animal" belief; and from the traditional group selectionist's and the adaptionist's as well as from the sociobiologist's viewpoint it is understandable in males fighting for dominance, such as has been reported by Lindburg (1971, p. 74) in the Rhesus monkey. But often, diminishing conspecifics' fitness seems (socio)biologically odd, e.g., cannibalism of infants occurs in tupaias (Sorenson 1970, Brandt and Mitchell 1970), prosimians (Martin et al. 1976, Mitchell 1979), and chimpanzees (Suzuki 1971, Bygott 1972). Active infanticide in captivity occurs in tupaias (Autrum and von Holst 1968) and the gorilla (van den Berghe 1959); and in the wild in colobines (Poirier 1974), langurs (Hrdy 1979), and macaques (Carpenter 1974, Mitchell 1970). To be sure, langur infanticide mainly occurs by a new male taking over the dominant position. Thus, he promotes spreading his genes by destroying his predecessor's ones (Hrdy 1979). Yet, the langur observations do not always agree with sociobiological predictions (Vogel 1980) and need further testing (Hausfater et al. 1980). In monkeys accidental infanticide occurs due to aunting (Jolly 1972, p. 226, Hrdy 1976); passive infanticide, i.e, by the absence of adequate maternal care (van Lawick-Goodall 1971), has been reported in captive (Yerkes 1943, p. 276) and wild (Sugiyama and Koman 1979, p. 327) chimpanzees, in olive baboons (Ransom and Ransom 1971), and in langurs (P. Jay, quoted by Wilson 1975, p. 350). Finally, there are less impressive examples, i.e., behavior which seems only slightly to decrease spreading genes; Bishop (1979, p. 268) reports that in one group of Himalayan langurs mothers during cold periods do not huddle or keep their arms around their infants. Yet, from a physiological, metabolical point of view this would appear much more economical, as seems to be confirmed by the other individuals of the same troop and by several other langur and macaque troops.

Intraspecific Altruism

Admittedly, most of intraspecific altruistic behavior in primates can readily be explained by sociobiological theory, i.e., in terms of kin selection and reciprocity, and they include, of course, parental care and assistance to sibs. There are, however, cases that do not, at first sight, fit the theory, e.g., cross-group food sharing in chimpanzees (Reynolds and Reynolds 1965, Sugiyama 1969), alloparental care (Wilson 1975, p. 349 ff.) such as aunting which is not necessarily shown by aunts (Lindburg 1971, Rowell 1972, Kummer 1971, p. 80, Poirier 1970, Sackett 1970, p. 136), and chimpanzee non-sib infant adoption (F. de Waal, personal communication).

Cross-Specific Behavior

The majority of interspecific relations is characterized by competition, neutral atti-tudes, or symbiotic mutual benefit, all of which makes sociobiological sense. Feral primates, however, may show behaviors that seem to benefit other species without any obvious reciprocity; e.g., chimpanzee females showing maternal behavior toward a young mangabey monkey (Kortlandt 1967), a macaque inducing allogrooming in two orangutans (Rijksen 1978, p. 113), and oral-sexual contact of an orangutan with a long-tailed macaque (Rijksen 1978, p. 264). Jolly (1972, p. 82 ff) summarizes a num-ber of examples of cross-specific group adoption some of which seem to benefit the stranger but not the "hosts". Van Lawick-Goodall (1971, p. 209) reports a year-long friendly chimpanzee-baboon relation ending in killing by fellow chimps of the baboon's infant. As mentioned above, the most remarkable cross-specific primate altruism seems to me conservationism.

Other Odd Properties

Every experienced observer of animals may adduce odd, enigmatic examples of behav-ior. Here, I will mention only some I happened to find in the primatological literature. Why do males of *Tupaia gracilis* so often mount females of *T. longipes* (Sorenson 1970)? Why are feral gorillas and orangutans so little adapted to their forest habitat that they, rather frequently, fall from trees resulting in fractures and paraplegia (Malbrant and MacLatchy 1949; Schaller 1963, p. 77, Rijksen 1978, p. 140)? Why do two male gelada baboons sometimes engage in fighting likely to kill both parties (Kummer 1971, p. 82)? Why did some pottos leave food for a dead conspecific (Cowgill 1976, p. 266)? Why do feral chimp and gorilla mothers cling for several days to their dead babies (van Lawick-Goodall 1971, p. 221, Schaller 1963, p. 269, 273) thus wasting energy and raising the risk of infection spreading? Why were the Gombe Stream Reserve chimps so badly adapted to infection with the poliomyelitis virus that many of them died from it? And why was the young adult male chimp Flint so attached to his mother that he died within a few weeks after she did (van Lawick-Goodall 1971)? Why do macaques frequently show stillbirths and abortions, and — sometimes lethal — congeni-tal deformities (Brandt and Mitchell 1970)? Why was a neonate gorilla left suffocating — locked in its amniotic sac — by its mother (ibid.)? Why did almost half of a group of wild mountain gorillas (observed by Schaller 1963, p. 101) — in whom the parents had invested so much — die before they had reproduced? And what about homosexual relations that are so frequently shown by primates?

Can Sociobiology Explain Odd Primate Behaviors?

I do not pretend to have exhaustedly treated all (socio)biologically odd behaviors in primates; undoubtedly, experienced primate ethologists will be able to add to the ones mentioned. My examples, however, may be sufficient to provide us here with the tools enabling further analysis. The question now is: can all these paradoxes be solved and sociobiological theory be saved? I think it can, i.e., by using the usual tools of evolutionary biology.

1. It should be realized that not all behavior and other − morphological or physiological − properties are necessarily adaptive (Kummer 1971, p. 90, Wind 1976, 1978, Lewontin 1979, Gould and Lewontin 1979) or contributory to spreading gene copies (Wind 1979). The individual can be considered as a compromise of many different, competing − but necessarily cooperating − genes and hence organs or organ systems. This compromise does not necessarily ensure maximal spreading of these genes. Some change in the composition of the genes might well benefit the spreading of more of them than the actual organism is able to spread. But why does this not happen? The crucial answer is that this change takes time. Adaptation is the change of gene frequencies as a result of new (re)combinations and of changing ecological pressures resulting in a new ESS. Such a change implies time-inertia, i.e., many generations. But even given sufficient time to a species, not every one of its individuals can be expected to be optimally adapted (McFarland 1977); theoretically, only one genotype would be, and absence of genetic variability is nonadaptive.

Hence, properties − behavioral, physiological, or morphological − may exist that do not contribute to fitness (or spreading gene copies) or that even hamper it, though, admittedly, these properties are likely to occur much less frequently than those who do contribute to fitness. So, it is important to remember that some odd behavior may well be in the process of being selected against. Such a process will last longer when that negative selective value is smaller. In some of the mentioned cases of primate intraspecific killing the frequency is quite low so that its negative selective value is probably small. Finally, odd behavior may be the result of recurrent mutations, pleiotropy, linkage and other, formal genetical and other biological, reasons (Lewontin 1979, May and Robertson 1980).

2. When one would be able to assess the number of genes that are in fact alien to the altruists mentioned on pp. 72−73, it may well turn out that the ratio of genes in common (or Rgc) is definitely higher than suggested by the traditionally handled coefficient of relationship r (Wind 1980); the r value is assumed between an individual and his or her offspring, and between sibs to be 1/2, between an individual and his or her grandchildren, uncles or nephews 1/4, etc. However, genetical studies (Darlington 1978), especially concerning polymorphisms (e.g., Harris 1975) and heterozygosity (e.g., Tracey 1979) have indicated at least that two random human individuals have more than half of their genes in common, and probably even 90% (Washburn 1978). In other species the Rgc may range from 50% to 95% (Kimura and Crow 1964, Lewontin and Hubby 1966). Hence, it is likely that also within a nonhuman primate species the Rgc is higher than traditionally assumed. Especially, intragroup altruism seems to fit kin-selection theory (but see the critical discussion concerning humans by Williams 1980). Moreover, aunting and other alloparental care may, in fact, be in the interest of

the altruist's genes, i.e., by providing experience (to would-be parents), a safeguard against eventual conspecifics' attacks, or the seeds for a future more dominant position.

3. Behavior adaptive in one situation may be nonadaptive in another. The low reproductivity of mammals including primates in captivity may be a case in point. Absence of mating as well as stillbirths and infanticide are likely to be the result of a gene-environment interaction which is different from the one which shaped the usual behaviors. These odd properties are likely to have been adaptive in that previous environment, but they are not in captivity (on the contrary); and selection take usually too much time to have brought about a change in the behavior of captive animals (not in domesticated ones).

Hence, a recent environmental change is more likely to be accompanied by odd behaviors than an environment that has been stable for a long time, i.e., many generations.

It can be concluded that the odd behaviors mentioned do not appear to falsify the sociobiological theory. Also the cross-specific altruism as shown by the nonhuman primates mentioned can be explained by the above three points. It can be asked, however, whether the peculiar behavior shown by human primates toward other ones, called conservationism, is sufficiently explained by these points. And because I feel that, in fact, conservationism is an important part of primatology, it warrants a somewhat more detailed discussion.

Conservationism and Sociobiology

Why does man protect nondomesticated species like orangutans and whales? Explaining this by stating that all species have the right to live and that it is our moral and ethical duty to protect them, is not quite satisfactory in this context. For not only do we look here for a biological explanation, but also there are species that we are exterminating without anybody opposing it; in one case a worldwide combined effort by all governments and the cooperation of virtually all mankind has resulted in the extermination of at least one species, i.e., the smallpox virus (Wind 1980). Similar attempts are being made concerning the malaria parasite and its insect host. So the nonbiological reasons are insufficient to explain the above animal-protection movements. Sociobiologically, the above extermination is quite understandable: the survival of *Homo sapiens sapiens* genes, or at least their increase, was seriously threatened by the smallpox ones, and our genes seem to have won the competition.

But what about the orangutans and whales? Does their protection increase human genes? A completely satisfactory answer is difficult to give. There are various possible sociobiological explanations, however.

1. Conservationists' behavior may have a negative selective value, such as spending a lot of labor and money which could have been used for promoting human genes in other ways. But if so, this value is probably small, i.e., in view of the small number of conservationists, the relatively small number of their supporters, the relatively little labor and money spent, and the rarity as well as the recent origin of this particular behavior among all other human behaviors; and it would take biologically speaking

some, and in ordinary parlance a long, time, i.e., many generations, before it will have disappeared.

2. On the other hand, there are reasons for assuming that the behavior has a positive selective value for most human genes. a) The number of genes that mammals have in common may very well accord with some of the 4 billion human survival machines decreasing their individual fitness in order to increase that of whales, as has been shown by some — almost self-sacrificing — Greenpeace members. (I hastily add here that it seems quite possible to me that similar behaviors have been present among primates' conservationists!) Since, however, conservationists' behavior exists only for a few human generations, it is very unlikely to be already part of an ESS, and its selective value must be quite small. b) The protectionists' behavior — either having a negative or a positive selective value — is a neutral spin-off of a general aptitude of humans to protect their own young, i.e., helpless and harmless conspecifics. This aptitude has probably contributed to spreading the altruist's gene copies at the time when our primate ancestors lived in small family groups. c) Not interfering with his environment and keeping sympatric animals alive may have (had) a positive selective value. On the other hand, in spite of profound environmental changes — including animal extermination and decreasing numbers of nonhuman primates — humans and hence their genes have enormously grown in numbers. d) In the case of whale protection there is a clear short-term benefit, i.e., preventing the goose that lays the golden egg from being killed.

In conclusion, as with much other primate and a fortiori human behavior, it is, provisionally, difficult to assess the selective value of animal-protection behavior. For from a sociobiological point of view many of the decisions concerned are really minor phenotypical variations, occurring in the upper level of Fig. 1 and resulting from environmental rather than from genotypical variations. In other words, this behavior is determined by "nurture" rather than by "nature". In the case of animal protection one of the environmental triggers may well have been our Judeo-Christian ethics. And — with many apologies to conservationsists — I cannot help impertinently wondering what our attitude towards, say, orangutans would be, were we convinced that they, like the smallpox or rabies virus, caused to us frequently lethal diseases. While some of animal-protection behavior may seem to oppose sociobiological theory, I think that so far the overall picture of the numbers of individuals and kinds of species killed and exploited by man suggests a parallel primarily with the benefits and secondarily with the taxonomic proximity to *Homo sapiens sapiens*.

Conclusions

The immediate impact of sociobiology on nonhuman primatology will provisionally be small and that on human primatology even smaller. It seems, however, that in the coming years its relevance will increase, especially concerning ethology. It may, in fact, well lead to a better understanding of primate morphology, physiology and behavior. In view of the firm molecular-biological basis of sociobiology, the interpretation of biological phenomena will become facilitated. In addition to the usual, largely descriptive, observations of primate ethologists, the assessment of some other data will become more important. Questions of increasing importance for primate ethology include (1) For how many generations has the present habitat been stable? The lower the number, the more informative is knowledge of the earlier ecology; (2) What is the intra-group genetic variability? E.g., polygyny, outbreeding, and mutagenes should be taken into account; (3) What is the ratio of genes that members of the group under observation have in common? With increasing gene-mapping possibilities such ratios will become available; (4) What is the longevity of the species under observation?

The likelihood of fitting the observations into the sociobiological paradigms and predictions will be positively correlated with the length of the period of stability of the habitat and with the longevity of the species. A simple correlation with genotypic variability seems provisionally unlikely. For the balance or changes of — partly counter-acting — forces like genotypic variation, deleterious inbreeding effects, and ecological changes, will first have to focused upon.

Ultimately, sociobiology may obtain some predictive value; though even in compli-cated organisms like the vertebrates the reliability of predicting morphology, physiol-ogy, biochemistry, and behavior may well increase, there will remain an upper limit to it. This is mainly because of the extremely complicated fabric of — constantly changing and numerous — gene frequencies necessitating very complex, if not impossible, sible, mathematical analyses. The same applies to environmental changes; e.g., for one thing, weather and climate predictions so far have defied the most sophisticated com-puter-based approaches.

The traditional primatological approach, like the other upper level sciences of Fig. 1, will therefore retain its present value. These sciences are unlikely to become incor-porated completely into sociobiology as suggested by Wilson (1975, p. 6). Instead, a mutual benefit is likely, i.e., sociobiology being necessary for putting primatological observations in a more holistic framework, and the upper sciences providing to socio-biology the hardware, the raw material to be processed by it.

Acknowledgment. I thank D.P. Barash, H. Joenje, A. Kortlandt, P.W. Sherman, G.S. Stent, and F. de Waal for their advice.

References

Alexander RD (1978) Natural selection and societal laws. In: Engelhardt HT, Callahan D (eds) Morals, science and sociality. Hastings Center, Hastings-on-Hudson, NY, pp 249–290

Alexander RD (1980) Darwinism and human affairs. Pitman, London

Allen E et al. (1975) Against "sociobiology". The New York Review of Books, Nov 13

Alper JS (1978) Ethical and social implications. In: Gregory MS, Silvers A, Sutch D (eds) Socio-biology and human nature. An interdisciplinary critique and defense. Jossey-Bass, San Fran-cisco, pp 195–212

Alper J et al (1976) The implications of sociobiology. Science 192:424–427

Altmann J (1980) Baboon mothers and infants. Harvard UP, Cambridge Mass

Autrum H, Holst D von (1968) Sozialer „Stress" bei Tupajas *(Tupaia glis)* und seine Wirkung auf Wachstum, Körpergewicht und Fortpflanzung. Z Vergl Phys 58:347–355

Barash DP (1977) Sociobiology and behavior. Elsevier, New York

Beach FA (1978) Sociobiology and interspecific comparisons of behavior. In: Gregory MS, Silvers A, Sutch D (eds) Sociobiology and human nature. An interdisciplinary critique and defense. Jossey-Bass, San Francisco, pp 116–135

Berghe L van den (1959) Naissance d'un gorille de montagne à la station de zoologie expérimentale de Tshibati. Folia Sci Afr Cent 4:81–83

Bishop NH (1979) Himalayan langurs: Temperate colobines. J Hum Evol 8:251–281

Brandt EM, Mitchell G (1970) Parturition in primates: Behavior related to birth. In: Rosenblum LA (ed) Primate behavior, vol I. Academic Press, London New York, pp 177–223

Bygott JD (1972) Cannibalism among wild chimpanzees. Nature (London) 238:410–411

Caplan AL (ed) (1978) The sociobiology debate. Harper & Row, New York London

Carpenter CR (1974) Aggressive behavioral systems. In: Holloway RL (ed) Primate aggression, territoriality and xenophobia. Academic Press, London New York, pp 459–496

Chagnon NA, Irons W (eds) (1979) Evolutionary biology and human social behavior: An anthropological perspective. Duxbury Press, North Scituate Mass

Christen Y (1979) L'heure de la sociobiologie. Albin Michel, Paris

Cowgill UM (1976) Cooperative behavior in *Perodicticus*. In: Martin RD, Doyle GA, Walker AC (eds) Prosimian behaviour. Duckworth, London, pp 261–272

Darlington CD (1978) The little universe of man. Allen & Unwin, London

Dawkins R (1976) The selfish gene. Oxford Univ Press, Oxford New York

Dawkins R (1980) Good strategy or evolutionarily stable strategy? In: Barlow GW, Silverberg J (eds) Sociobiology: Beyond nature/nurture? AAAS Selected Symposium No 35. Westview Press, Boulder Col, pp 331–367

Fisher RA (1930) The genetical theory of natural selection. Oxford Univ Press, Oxford

Fuller JL (1978) Genes, brains and behavior. In: Gregory MS, Silvers A, Sutch D (eds) Sociobiology and human nature. An interdisciplinary critique and defense. Jossey-Bass, San Francisco, pp 98–115

Gould SJ, Lewontin RC (1979) The spandrels of San Marco and the Panglossian paradigm: a critique of the adaptationist programme. Proc R Soc London Ser B 205:581–598

Gregory MS, Silvers A, Sutch D (eds) Sociobiology and human nature. Jossey-Bass, San Francisco

Haldane JBS (1932) The causes of evolution. Longmans and Green, London

Hamilton WD (1964) The genetical evolution of social behaviour. J Theor Biol 7:1–16, 17–52

Harris H (1975) The principles of human biological genetics. North-Holland, Amsterdam Oxford

Hausfater G, Cairns SJ, Aref S (1980) Infanticide: An analytical model. Antropologia Contemp 3:208

Hrdy S (1976) Care and exploitation of nonhuman primate infants by conspecifics other than the mother. In: Rosenblatt J, Hinde R, Beer C, Shaw E (eds) Advances in the study of behavior, vol VI. Academic Press, London New York, pp 101–158

Hrdy SB (1979) Infanticide among animals: A review, classification and examination of the implications for the reproductive strategies of females. Ethol Sociobiol 1:13–40

Jolly A (1972) The evolution of primate behavior. Macmillan, New York London

Kimura M, Crow JF (1964) Number of alleles that can be maintained in finite populations. Genetics 49:725–738

Kortlandt A (1967) Experimentation with chimpanzees in the wild. In: Starck D, Schneider R, Kuhn HJ (eds) Neue Ergebnisse der Primatologie. Progress in primatology. Fischer, Stuttgart, pp 208–224

Kummer H (1971) Primate societies. Group techniques of ecological adaptation. Aldine, Chicago New York

Kurland JA (1977) Kin selection in the Japanese monkey. Karger, Basel New York

Lawick-Goodall J van (1971) In the shadow of man. Collins, London

Lewontin RC (1979) Sociobiology as an adaptationist program. Behav Sci 24:5–14

Lewontin RC, Hubby JL (1966) A molecular approach to the study of genic heterozygosity in natural populations. 2. Amount of variation and degree of heterozygosity in natural populations of *Drosophila pseudoobscura*. Genetics 54:595–609

Leyhausen P (1979) Aggression, fear and attachment: complexities and interdependencies. In: Cranach M von, Foppa K, Lepenies W, Ploog D (eds) Human ethology. Univ Press, Cambridge, pp 253–264

Lindburg DG (1971) The Rhesus monkey in North India: An ecological and behavioral study. In: Rosenblum LA (ed) Primate behavior. Developments in field and laboratory research, vol II. Academic Press, New York London, pp 1–106

Malbrant R, MacLatchy A (1949) Faune de l'équateur Africain Français. II. Mammif`eres. Encycl
  Biol 36:1–323
Martin RD, Doyle GA, Walker AC (eds) (1976) Prosimian behaviour. Duckworth, London
Masters RD (1979) Beyond reductionism: five basic concepts in human ethology. In: Cranach
  M von, Foppa K, Lepenies W, Ploog D (eds) Human ethology. Univ Press, Cambridge, pp 265–
  284
Mattern R (1978) Altruism, ethics, and sociobiology. In: Caplan A (ed) The sociobiology debate.
  Harper & Row, New York, pp 462–475
May RM, Robertson M (1980) Just so stories and cautionary tales. Nature (London) 286:327–
  329
Maynard Smith J, Price GR (1973) The logic of animal conflict. Nature (London) 246:15–18
McFarland DJ (1977) Decision making in animals. Nature (London) 269:15–21
Mitchell G (1970) Abnormal behavior in primates. In: Rosenblum LA (ed) Primate behavior, vol I.
  Academic Press, New York London, pp 195–249
Mitchell G (1979) Behavioral sex differences in nonhuman primates. Van Nostrand-Reinhold,
  New York
Oster GF, Wilson EO (1978) Caste and ecology in the social insects. Univ Press, Princeton
Parker GA (1978) Selfish genes, evolutionary games, and the adaptiveness of behaviour. Nature
  (London) 274:849–855
Poirier FE (1970) The Nilgiri Langur *(Presbytis johnii)* of South India. In: Rosenblum LA (ed)
  Primate behavior, vol I. Academic Press, New York London, pp 251–383
Poirier FE (1974) Colobine aggression: A review. In: Holloway RL (ed) Primate aggression, terri-
  toriality, and xenophobia. Academic Press, New York London, pp 123–157
Ransom TW, Ransom BS (1971) Adult male-infant relations among baboons *(Papio anubis)*. Folia
  Primatol 16:179–195
Reynolds V, Reynolds F (1965) Chimpanzees of the Budongo Forest. In: DeVore I (ed) Primate
  behavior: field studies of monkeys and apes. Holt Rinehart and Winston, New York, pp 368–
  424
Rijksen HD (1978) A fieldstudy on Sumatran orang utans *(Pongo pymaeus abelii* Lesson 1827).
  Ecology, behaviour and conservation. Meded Landbouwhogesch Wageningen 78-2:1–420
Rowell T (1972) The social behaviour of monkeys. Penguin, Harmondsworth UK
Sackett GP (1970) Unlearned responses. Differential rearing experiences, and the development of
  social attachments by Rhesus monkeys. In: Rosenblum LA (ed) Primate behavior, vol I.
  Academic Press, New York London, pp 111–140
Schaller GB (1963) The mountain gorilla. Ecology and behavior. Univ Press, Chicago
Schneewind JB (1978) Sociobiology, social policy, and nirvana. In: Gregory MS, Silvers A, Sutch
  D (eds) Sociobiology and human nature. An interdisciplinary critique and defense. Jossey-Bass,
  San Francisco, pp 225–239
Searle J (1978) Sociobiology and the explanation of behavior. In: Gregory MS, Silvers A, Sutch D
  (eds) Sociobiology and human nature. An interdisciplinary critique and defense. Jossey-Bass,
  San Francisco, pp 164–182
Sociobiology Study Group (1976) Sociobiology – another biological determinism. BioScience 26:
  182, 184–186
Solomon RC et al. (1978) Sociobiology, morality, and culture. Group Report. In: Stent GS (ed)
  Morality as a biological phenomenon. Dahlem Konferenzen, Berlin, pp 283–308
Sorensen MW (1970) Behavior of tree shrews. In: Rosenblum LA (ed) Primate behavior, vol I.
  Academic Press, New York London, pp 141–193
Stent GS (1978) Introduction: The limits of the naturalistic approach to morality. In: Stent G
  (ed) Morality as a biological phenomenon. Dahlem Konferenzen, Berlin, pp 13–22
Stent GS (1977) Book review of "The selfish gene". Hastings Center Rep, December:33–36
Sugiyama Y (1969) Social behavior of chimpanzees in the Budongo Forest, Uganda. Primates 10:
  197–225
Sugiyama Y, Koman J (1979) Social structure and dynamics of wild chimpanzees at Bossou, Gui-
  nea. Primates 20:323–339
Suzuki A (1971) Carnivority and cannibalism observed among forest living chimpanzees. J Anthropol Soc Nippon 79:30–48

Tracey ML (1979) Heterogeneity and evolution (Book Rev). Science 204:759–761
Vogel C (1980) Theoretical concepts of sociobiology tested against new field data on Hanuman
    Langurs *(Presbytis entellus)*. Antropol Contemp 3:286
Washburn SL (1978) Animal behavior and social anthropology. In: Gregory MS, Silvers A, Sutch
    D (eds) Sociobiology and human nature. An interdisciplinary critique and defense. Jossey-Bass,
    San Francisco, pp 53–74
Williams BJ (1980) Kin selection, fitness and cultural evolution. In: Barlow GW, Silverberg J (eds)
    Sociobiology: Beyond nature/nurture? AAAS Selected Symposium No 35. Westview Press,
    Boulder Col, pp 573–587
Wilson EO (1975) Sociobiology. The new synthesis. Harvard Univ Press, Cambridge Mass
Wind J (1976) More on gender differences and the origin of language. Curr Anthropol 17:745–749
Wind J (1978) Abortion, ethics and biology. Perspect Biol Med 21:492–504
Wind J (1979) The selfish human genes. J Hum Evol 8:551–553
Wind J (1980) Man's selfish genes, social behavior and ethics. J Social Biol Struct 3:33–41
Yerkes RM (1943) Chimpanzees. Yale Univ Press, New Haven

Dominance and Subordination: Concepts or Physiological States?

E.B. KEVERNE, R.E. MELLER, and A. EBERHART [1]

Introduction

The concept of dominance and dominance hierarchies has aroused much debate over the last 15 years. Since its adoption by the early primatologists — notably Zuckerman (1932) — from the field of bird social behaviour, dominance came to be considered a fundamental principle underlying all primate behaviour, despite the fact that it was never clearly defined. The subsequent somewhat subjective and careless use of the term to describe some intangible quality possessed by individuals to differing degrees led, perhaps not surprisingly, to a number of criticisms. Thus Gartlan (1964, 1968) and Rowell (1967, 1974) pointed out apparent contradictions in assessments of status when different indices were used, and also argued that the pronounced aggressive hierarchies seen in caged primates must be laboratory artefacts since little aggression was observed in the same species in the wild. This criticism highlights another problem, namely the frequent confusion between dominance and aggressiveness (see Sect. 2).

Some of the controversy surrounding this issue arose, as Hinde (1978) noted, out of the application of the concepts of dominance and subordinacy not only to relationships between pairs of individuals, but also to the pattern of interactions within a social group. It is clearly misleading to think of evolutionary forces having acted at the level of the group to produce "the dominance hierarchy"; likewise it is unhelpful to discuss the function of the hierarchy as has been done in the past. Difficulties also arise when attempting to describe the dominance status of a monkey which is "boss" over one individual but subordinate to two or three others in the group. On the other hand, the concept of dominance can be valid and meaningful if it is restricted to a definition concerning the interactions between any two individuals such that predictions can be made concerning their behavior (see also Keverne et al. 1978a,b). It is commonly the case that in any pair of individuals the behaviour of one is constrained or limited by the presence of the other — or, A bosses B (Hinde 1978). This interaction may be the result of overt or merely potential aggression; in the latter case aggression probably occurred in the past and resulted in the learned relative status. Such relationships maintained with little contact aggression explain the lack of aggression often observed in established, rather than newly formed, social groups (e.g., Richards 1974), making the determination of dominance hierarchies very difficult and leading

---

1   Department of Anatomy, University of Cambridge, Cambridge, CB2 3DY, United Kingdom

to the objections to the concept of dominance mentioned above. However it is not the amount of aggression shown, but its patterning or direction between individuals, that is significant both to the observer and the animals themselves. The highest-ranking animal is often not the most aggressive (see also Meller 1980).

The reality of dominance/subordination relationships and their significance to group members, even in the absence of overt aggression, can be revealed by using measures of behaviour other than attacks or threats. For example, in the talapoin monkey, visual monitoring of one individual by another is a behaviour which, while not having any detectable signalling value, reflects the social structure of the group (Dixson et al. 1975, Keverne et al. 1978b). Furthermore, unlike aggressive interactions, this behaviour is not masked by experience. In addition, there may also be important physiological consequences of status which are not learned or under the control of the individual. In this paper dominance relationships within social groups of talapoin monkeys will be considered with particular reference to the consequences on endocrine status – and hence reproductive ability – of differing social rank.

Dominance and Subordinacy

Dominance rank is defined in terms of the direction of aggression between animals in the social group. In each group, males and females were ranked in order of attacks and threats given to and received from each other, male or female. Although there were some changes in rank during these studies, all such changes occurred during the first 6 months following group formation: after this period dominance orders remained stable over time and across treatments.

Since animals of middle rank dominate some individuals and are in turn subordinate to others, their behavioural interactions are complex. Only the highest ranking male in each group showed sexual behaviour, gave aggression and received little or none, while the lowest ranking male of each group showed no sexual behaviour, received aggression while giving little or none. Hence monkeys at the extremes of the social hierarchy are unique in their behavioural experiences and provide the appropriate contrasts for consideration in this paper.

Materials and Methods

Twelve adult male, and nine adult female, talapoin monkeys *(Miopithecus talapoin)* were used. Two of the males in one group had been castrated and all the females had been ovariectomised at least 3 years previously. The castrated males received testosterone replacement (in the form of subcutaneous implants) throughout the study, while females were given oestradiol implants at intervals during the study to mimic follicular phase oestrogen levels.

Individuals whose behaviour was being observed were housed in groups of four males and four or five females in large group cages ($5' \times 6' \times 11'$). The details of behavioural observations are given elsewhere (Dixson et al. 1975); in brief, sexual, aggressive and social interactions were scored, with all animals being monitored continuously. In Group I, males and females had continuous access to one another, while in Groups II and III the sexes were separated except during the test periods. A minimum of 1000 min of observations was obtained for each treatment. Animals that were not being observed were housed singly in a separate room.

The following experimental manipulations were carried out:

1. The females in the groups were periodically made sexually attractive, by the administration of subcutaneous implants of oestradiol.
2. Each male in the group was given access to oestrogen-treated females in the absence of any other males, while the other males were removed to single cages (in isolation).
3. Oestrogen surges were produced by implanting two additional oestradiol-filled capsules into females bearing a single capsule. This treatment resulted in an elevation of plasma oestradiol which lay within the physiological range for the late follicular phase of the cycle and similar to that found before ovulation. This is the standard technique adopted for inducing an LH surge in primate neuroendocrine studies (Knobil 1974).

Plasma samples were taken twice weekly by femoral venipuncture under ketamine anaesthesia, and testosterone levels measured using radio-immunological technqiues described elsewhere (Keverne et al. 1978a). No behavioral observations were scored on the day following plasma sampling.

Results

Relationship Between Social Rank and Behaviour

Figure 1 (upper section) shows the sexual behaviour and (lower section) the aggressive behaviours given and received by highest and lowest ranking males in each of three social groups of monkeys when females in the group received oestradiol. Aggression received (withdrawals) was, by definition, rank related. In each group, it was the highest ranking male (dominant) which received least aggression, while the lowest ranking male (subordinate) of each group received most aggression. However, no highest ranking male was clearly the most aggressive in his group; most often the highest ranking male ranked with one or more others showing high or moderate levels of aggression. In each group the highest ranking male directed some aggression towards females although significantly less frequently than to other males. The lowest ranking male of each group initiated no aggressive behaviour to other monkeys in the group (Fig. 1). All measures of aggression given were significantly higher for highest ranking males than for subordinates (P > 0.0001), while aggression received was significantly higher for lowest ranking males when compared with dominants (P < 0.0001).

For sexual behaviour the general pattern was for highest ranking males to be sexually active, usually the most active, and for the lowest ranking males to be sexually inactive. For mounts and ejaculations shown in Fig. 1, these sexual behaviours were significantly higher in all cases (P < 0.0001) for highest-(dominants) versus lowest-(subordinates) ranking males.

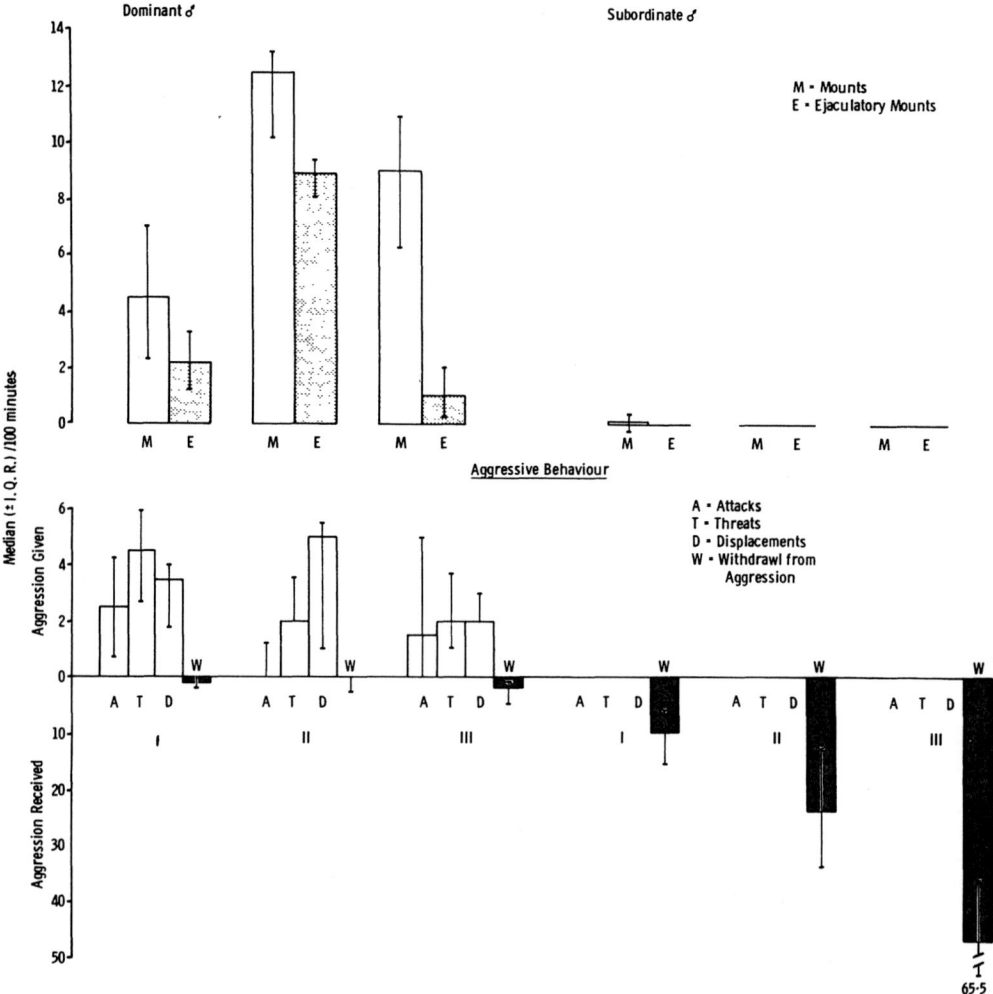

Fig. 1. The sexual *(upper section)* and aggressive *(lower section)*, behaviour of highest *(left section)* and lowest *(right of Fig.)* ranking males in three social groups *(I, II, III)* of talapoin monkeys. These behaviours are all significantly different according to rank (P < 0.0001)

It has been observed that in talapoin monkey social groups individuals repeatedly glance at each other and that in well-established groups low-ranking individuals look at others more frequently than high-ranking monkeys. Even when the levels of overt aggression are low, the direction of visual monitoring is closely related to the dominance hierarchy (Fig. 2). Thus, the highest ranking male in each group monitors others very little, but is himself monitored more than any other male by all subordinates. Even the lowest ranking male of each group who received most aggression from lower ranking males and rarely interacted with the highest ranking male, nevertheless showed highest levels of visual monitoring to the dominant male.

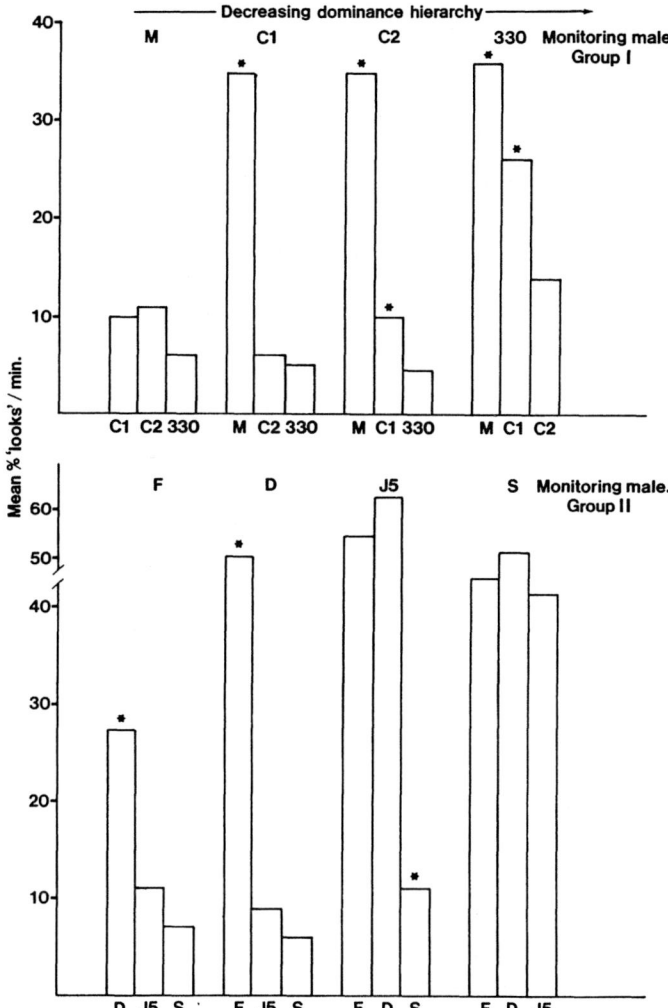

Fig. 2. Proportion of visual monitoring each male directed to other males in the group when females were oestrogenised (Keverne et al. 1978b)

Relationship Between Social Rank and Plasma Testosterone

In all three groups when males had access to oestrogen-treated females, the dominant male tended to have higher plasma testosterone levels (and over weeks or months these were significantly so) than did the subordinate male (Group I t = 2.36, P < 0.05; Group II t = 5.29, P < 0.001; Group III t = 3.05, P < 0.001). On a day-to-day basis, however, there were occasions when other males in the group showed higher plasma testosterone levels than the dominant male. When the males were removed from the mixed-sex social group and caged singly, there was no difference in plasma testosterone levels between dominant and subordinate males, and in no way could plasma testosterone levels in singly-caged males serve as a predictor of dominance (Fig. 3). The most significant finding, therefore, was the *increase* in plasma testosterone which occurred

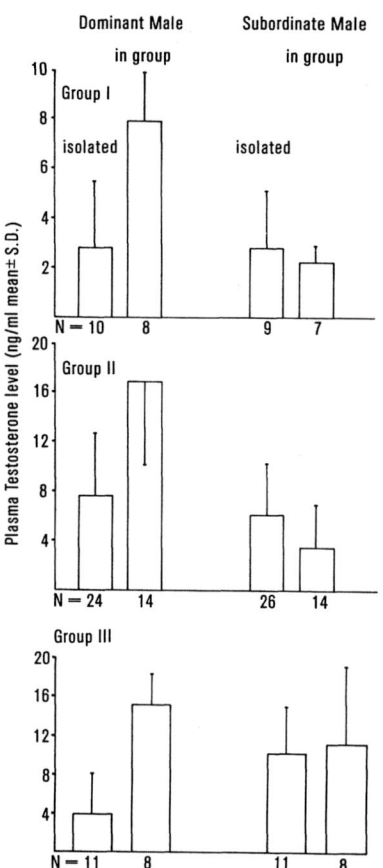

Fig. 3. Differences in plasma testosterone levels between *dominant* and *subordinate* males of three social groups in response to interaction with oestrogen-treated females. Samples were collected twice weekly during a 10–12 week period in isolation and a 7–8 week period in the social group with oestrogen-treated females

in the dominant male with access to attractive females (Group I, t = 4.6, P < 0.001; Group II, t = 3.83, P < 0.001; Group III, t = 6.19, P < 0.001), an increase which failed to occur in the most subordinate males.

The presence of higher ranking males and the intermittent but potentially continuous threat of aggression in the social group may have been instrumental in preventing both sexual behaviour and increased plasma testosterone in the subordinate male. To test this possibility, males were removed from the social group to individual caging in isolation. The dominant and subordinate males were each separated, and then returned to the social group of females, but in the absence of any other males. In this situation both males showed sexual behaviour (Fig. 4) and each showed an increase in plasma testosterone. Neither male was receiving of giving high levels of aggression because no other males were present and aggressive interactions with females remained unchanged and extremely low. Interestingly, the sexual performance of the former subordinate male was considerably lower than that of the former dominant male, requiring more mounts to achieve fewer ejaculations (Fig. 4). Moreover, the sexual behaviour of this "subordinate" was entirely paced by the females, which showed high levels of sexual presentations and initiated all the male's mounts (Fig. 4).

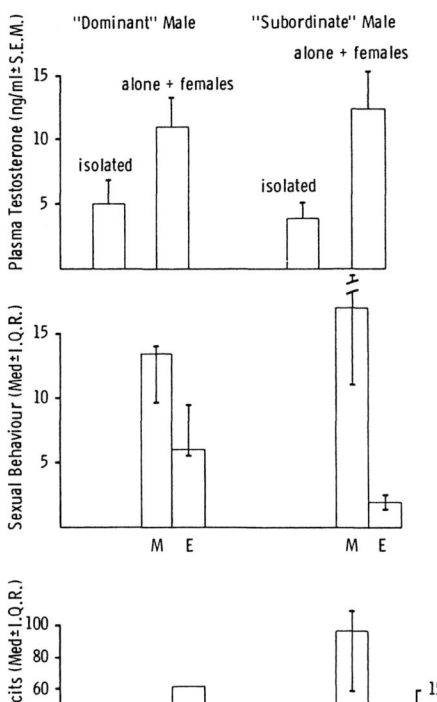

Fig. 4. Increases in plasma testosterone and the sexual interactions of "dominant" and "subordinate" males in moving from isolation into a social group of oestrogen-treated females for 3 weeks in the absence of other males. The former subordinate male receives more female solicitations *(P)*, initiates fewer mounts *(MIM)* and shows a lower sexual performance (mount:ejaculation ratio) than the former dominant male

Effects of Social Rank on the "Stress" Hormones

Although testosterone increased in both the former "subordinate" and "dominant" males when they were each with females in the absence of other males, a totally different picture emerged for the stress hormones cortisol and prolactin. Thus, in the former dominant male moving back to the social group of females was associated with a significant decrease in prolactin and no change in cortisol (Fig. 5). The levels of cortisol and prolactin increased dramatically on moving into the social group, and cortisol remained high even on removal of this male back to isolation (Fig. 5). Hence, although we see no significant difference between the levels of stress hormones in dominant and subordinate males in isolation, moving into the group of females produces markedly different hormonal profiles in these animals. Whereas the stress hormones decrease in the dominant male, they increase significantly in the subordinate, giving the latter considerably higher levels (P < 0.001) than the dominant male. This contrast in the levels of stress hormones between dominant and subordinate male is in no way directly related to the aggressive behaviour. There is no significant difference in the aggressive behaviour shown by females to each male and this is extremely low (less than 45 episodes to each male over 4 weeks of observation). Indeed, this represents a considerable decrease in aggression for the subordinate male when compared with that he received in the

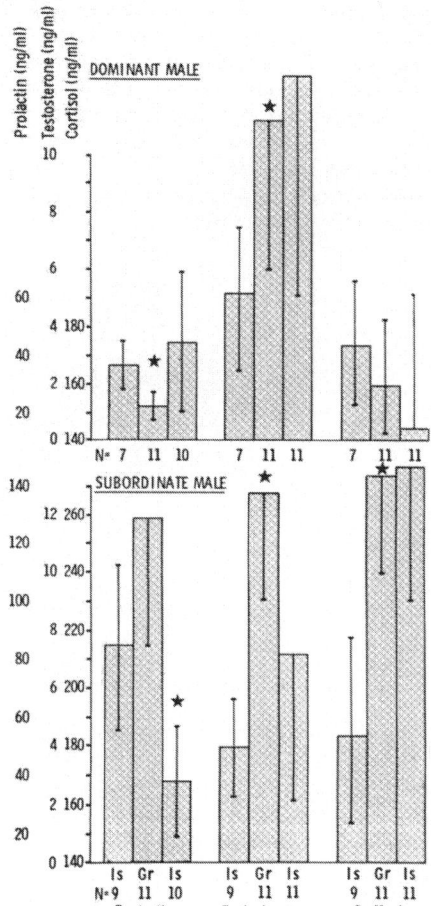

Fig. 5. Changes in prolactin, testosterone, and cortisol in the "dominant" and "subordinate" males on moving from isolation into the oestrogen treated female group for 3 weeks in the absence of other males. The former subordinate male shows a markedly different and higher "stress" response than the former dominant on moving into the group of females even in the absence of overt aggressive behaviour

mixed-sex social group situation (216 episodes in 4 weeks of observation). Nevertheless, his prolactin level was significantly higher when alone with the females than in the presence of the whole group, although cortisol levels in the subordinate male were equally high in both situations. It is of interest to note that this male's cortisol in fact increased significantly over the first year of the group's establishment (Fig. 6), although aggressive interactions had markedly declined over this time.

Dominance/Subordination Among Females; the Consequences for Behavioural and Endocrine Status

Female talapoin monkeys, like the males, also form a social hierarchy which can be assessed by the direction of aggressive encounters. Although aggressive behaviour among females is less frequent than among males, social rank has marked consequences both for other behaviours, and for the endocrine state of the individuals.

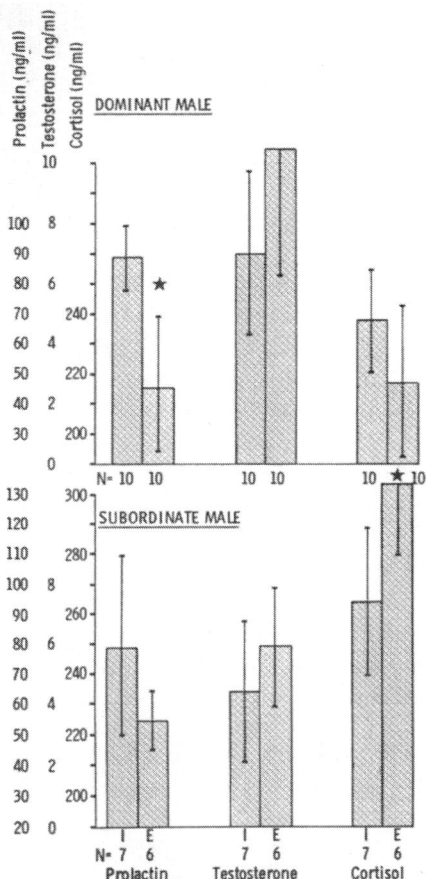

Fig. 6. Changes in plasma prolactin, testosterone, and cortisol in the most dominant and subordinate males of the group in the initial period, *I* (within the first 3 months), and the established period, *E* (12 months later) of group formation. The females in the group were oestrogen-treated in both cases. Prolactin fell in the dominant male, and cortisol rose in the subordinate, from *I* to *E*

Among females, the visual monitoring by each individual of the other females of her group was related to her position in the social hierarchy. Thus the most dominant females received most attention (30%–40% of all visual monitoring) from other females, while the dominant females themselves monitored other females relatively little (5%–10% of all visual monitoring). The converse was found for the subordinate females, which showed high levels (30%–40%) of visual monitoring of other females but received relatively little attention from any other females (less than 10%).

When the extremes of the hierarchy are contrasted, it is seen that the highest ranking female received significantly more sexual attention from males (339 mounts and 139 ejaculations, cf. 64 mounts and 6 ejaculations/100 h observations; P < 0.0001) and less aggressive behaviour (45 attacks, cf. 110 attacks/100 h observations; P < 0.02) than the subordinate female. In addition, the subordinate female received relatively high levels of aggression from other females (55 attacks/100 h observation; P < 0.01) while the highest ranking female, by definition, received none.

Since the group females were all ovariectomised and received oestradiol implants, it is impossible to assess ovarian hormone levels. However, two stress-related hormones, prolactin and cortisol, have been measured and show marked differences between extremes of the hierarchy. Thus, over the course of a year, both prolactin and cortisol were significantly higher ($P < 0.001$) in the subordinate female, which was receiving little sexual attention and the highest levels of aggression. In the human primate, high levels of prolactin are associated with infertility. While we cannot check for pregnancy or ovulation in the ovariectomised talapoin monkeys, we can challenge the hypothalamo-pituitary axis with increased oestrogen feedback. This should induce an LH surge, an indication of potential fertility since the LH surge is the trigger of ovulation.

The outcome of challenging the hypothalamo-pituitary axis in the dominant and subordinate females on a number of occasions can be seen in Fig. 7. Although both females received oestradiol surges, only the dominant female responded with an LH surge. Plasma prolactin was significantly higher in the subordinate than the dominant female at this time (115 ± 35 cf. 50 ± 17 ng/ml; $P < 0.0001$). The subordinate female was subsequently treated with a dopamine agonist, bromocryptine, a procedure which effectively lowers prolactin in humans and can restore ovulation. This, as expected, resulted in the reduction of plasma prolactin levels, and when challenged again with oestrogen feedback, an LH surge occurred in the subordinate which resembled that formerly seen in the dominant female.

Since it was possible that the raised prolactin levels in the subordinate female might have been due to factors other than her social environment she was removed from the group and housed singly. Eight weeks later, her prolactin levels had fallen to those comparable with the higher ranking animals, and the oestrogen challenge now induced an LH surge without the need for bromocryptine treatment.

Discussion

These experiments show that in a social group of talapoin monkeys a number of behaviours are not shown equally by different individuals but are related to the status of that individual in the group's aggressive hierarchy (Eberhart et al. 1981). Thus, high-ranking individuals take part in more sexual interactions, are monitored more by others, receive little aggression and display aggressive behaviour to others. In marked contrast, the lowest extreme of the hierarchy is characterised by monkeys which take no part in sexual activity, show high levels of visual monitoring while being monitored little, receive relatively high levels of aggressive behaviour, but are themselves not aggressive to others. Additional studies have reported spacing (Keverne et al. 1978a), grooming (Simpson 1973, Seyfarth et al. 1978), display behaviour (Nishida 1970) and access to food (Ploog et al. 1963) to be affected by a primate's position in the social hierarchy.

These findings are in line with reports for a number of species of primate in which hierarchies have been observed. With the passage of time (2–3 years) and the increased stability of the hierarchy, there are few changes in the patterning of interactions. Two notable, and on the face of it, contradictory exceptions are recorded: (1) overt aggressive behaviour declines, and (2) the stress hormones become elevated in subordinate individuals. Subordinacy therefore not only provides some indication as to the behavioural propensities of an individual, but, in addition, it appears to lead to physiological

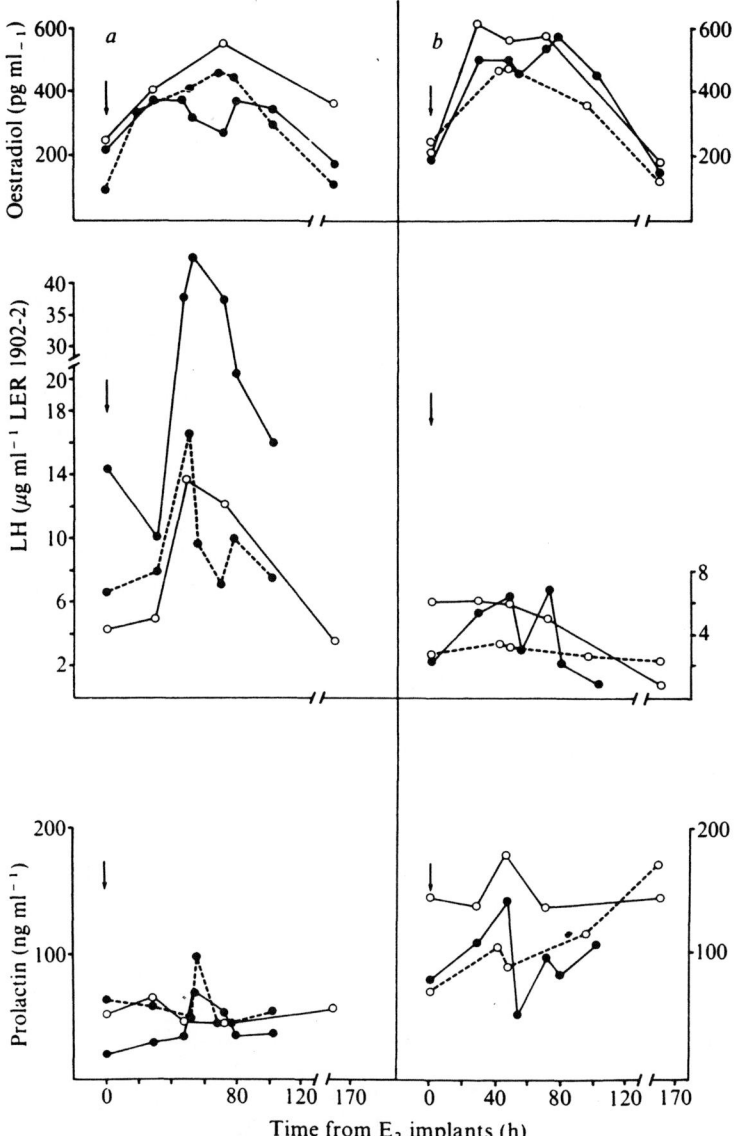

Fig. 7. Plasma LH and prolactin levels in highest-ranking *(a)* and lowest-ranking *(b)* female talapoin monkeys when given extra oestradiol implants (indicated by *arrows*) over a 5-day period. The subordinate female has significantly higher prolactin levels and unlike the dominant female does not respond with an LH surge (Bowman et al. 1978)

changes. Thus, in the social group the subordinate male has elevated levels of cortisol and at times elevated prolactin, but shows no increases in plasma testosterone similar to those seen in the dominant male (Eberhart and Keverne 1979). Among the females, subordinate rank is associated with high prolactin and high cortisol and a failure to show an LH surge in response to an oestrogen challenge (Bowman et al. 1978, Keverne 1979).

Although these neuroendocrine differences between dominant and subordinate monkeys represent quantifiable differences, their gradual appearance suggests that they reflect rather than determine social rank. In doing this, they reinforce the behavioural hierarchy by producing neuroendocrine profiles that are selectively advantageous to the reproductive success of the dominant individuals. Thus, although subordinate females are not entirely sexually excluded, a behaviourally induced neuroendocrine mechanism exists whereby the hypothalamo-pituitary axis is "switched off" to the positive oestrogen feedback normally triggering the ovulatory LH surge. Hence, the relative reproductive success of dominant females is increased by suppressing that of the subordinates (see also Dunbar and Dunbar 1977, Abbott and Hearn 1978).

In dominant males with marked increases in plasma testosterone we observe high levels of sexual performance, and in the social group, when all the males are present, these males receive the highest levels of sexual invitations from the females. It is possible that the high testosterone of dominant males may enhance their attractiveness to females (Keverne 1979), while subordinates would thus remain unattractive, and so reduce the aggression they might otherwise provoke from the dominant males. Such endocrine differences may be part of the subordinate male's adaptation to minimise the cost of staying in a social group while maintaining the potential for reproduction should the opportunity arise.

One may pose the question as to the advantage to the subordinate monkeys in this situation, and why, if there is none, they do not leave the social group. Of course, this possibility is not available to them in captivity, but even in the wild successful mobility between groups seems to be the prerogative of dominant males (Packer 1979). A possible answer to this question emerges when we consider how the subordinate male, in comparison with the dominant, fails to cope with the opportunities even when they are available in new social situations. Hence, when the subordinate male is moved into the group of females in the absence of other males, the females take all the initiative in pacing his sexual interactions, and although we see an increase in testosterone at this time, we also observe a marked increase in the stress hormones, cortisol and prolactin. This occurs in the virtual absence of overt aggressive behaviour since no other males are present, and moreover represents a considerable reduction in the amount of aggression this male was receiving from the mixed-sex social group. Clearly then, the subordinate male, possibly because of his past experiences of chronic subordination, experiences great problems in coping with a changed social environment. A possibly similar heightened responsiveness of the adrenal gland has been reported for subordinate rhesus monkeys (Sassenrath 1970). In relative terms, the subordinate monkey copes better with the low levels of aggression received in the established social group than he does with a new social situation that involves sexual interactions in the absence of aggression. Whether or not given time he could adapt to the new environment cannot be stated here.

In these studies it was not our intention to reproduce the natural habitat of the talapoin monkey, but rather to establish a model for studying the influence of social factors on behavioural and endocrine states. Nevertheless, the physiological constraints which these studies have shown to distinguish animals of differing social rank are of some significance for field workers. One might predict from these neuroendocrine data that high-ranking males are more likely to succeed in a changed social environment because of their ability to cope with the novel situation. Among females on the other hand, it might pay for subordinates to show intergroup mobility because of the cost to reproduction of staying in their natal group.

While this is speculation, these studies clearly show that dominance interactions are of real significance to individuals, having particularly striking consequences for individuals at the extremes of the hierarchy. Moreover, the consequences are very real even when the aggressive behaviours have tended to stabilise at low levels. Field workers observing stabilised social groups have tended to conclude that such low levels of aggression are probably too low to support a dominance hierarchy (Gartlan 1968, Rowell 1974, Dunbar and Dunbar 1975). Clearly, this assertion is misleading, and we strongly support Dunbar (1981) in providing a word of caution on drawing inferences about dominance hierarchies solely on the rates of agonistic encounters.

Acknowledgments. We are grateful to Sue Dilley and Penny Hackett for their excellent technical assistance with radioimmunoassays, and to the help and constructive criticism of our colleague Joe Herbert. The LH for iodination were prepared by Dr. L.E. Reichert and obtained through NIAMDD. The LH antiserum was supplied by Dr. G.D. Niswender. The prolactin standard was obtained from the MRC and the prolactin for iodination from Dr. P.S. Lowry. The prolactin antiserum was a gift from Dr. H. Friesen. The work was supported by a programme grant from the Medical Research Council, a Mental Health Foundation Junior Research Fellowship to Rachel E. Meller and a Marshall Scholarship to Jerry A. Eberhart.

References

Abbott DH, Hearn JP (1978) Physical, hormonal and behavioural aspects of sexual development in the marmoset monkey. J Reprod Fertil 53:155–166

Bowman LA, Dilley SR, Keverne EB (1978) Suppression of oestrogen induced LH surges by social subordination in talapoin monkeys. Nature (London) 275:56–58

Deag JM (1977) Aggression and submissions in monkey societies. Anim Behav 25:465–474

Dixson AF, Scruton D, Herbert J (1975) Behaviour of the Talapoin monkey *(Miopithecus talapoin)* studied in groups in the laboratory. J Zool London 176:177–210

Dunbar RIM (1981) Determinants and evolutionary consequences of dominance among female gelada baboons. Behav Ecol Sociobiol (in press)

Dunbar RIM, Dunbar EP (1975) Social dynamics of gelada baboons. Contrib Primatol 6

Dunbar RIM, Dunbar EP (1977) Dominance and reproductive success among female gelada baboons. Nature (London) 266:351–352

Eberhart JA, Keverne EB (1979) Influences of the dominance hierarchy on LH testosterone and prolactin in male talapoin monkeys. J Endocrinol 83:42P–43P

Eberhart JA, Keverne EB, Meller RE (1981) Social influences on plasma testosterone levels in male talapoin monkeys. Horm Behav (in press)

Gartlan JS (1964) Dominance in East African monkeys. Proc East Afr Acad 2:75–79

Gartlan JS (1968) Structure and function in primate society. Folia Primatol 8:89–120

Hinde RA (1978) Dominance and role – two concepts with dual meanings. J Social Biol Struct 1: 27–38

Keverne EB (1979) Sexual and aggressive behaviour in social groups of talapoin monkeys. Ciba Symp 62. Sex, hormones and behaviour. Excerpta Medica, Amsterdam Oxford London, pp 271–298

Keverne EB, Meller RE, Martinez-Arias A (1978a) Dominance, aggression and sexual behaviour in social groups of Talapoin monkeys. In: Herbert J, Chivers D (eds) Recent advances in primatology, vol I. Academic Press, London New York, pp 533–547

Keverne EB, Leonard RA, Scruton DM, Young SK (1978b) Visual monitoring in social groups of talapoin monkeys *(Miopithecus talapoin)*. Anim Behav 26:933–944

Knobil E (1974) On the control of gonadotropin secretion in the rhesus monkey. Rec Prog Horm Res 30:1–46

Meller RE (1980) Aggression in primate social groups: Hormonal correlates. In: Campbell A, Marsh P (eds) Perspectives on aggression and violence. Blackwell Press, Oxford

Nishida T (1970) Social behaviour and relationship among wild chimpanzees of the Maliali mountains. Primates 11:47–87

Packer C (1979) Male dominance and reproductive activity in *Papio anubis*. Anim Behav 27:37–45

Ploog DW, Blitz J, Ploog F (1963) Studies on social and sexual behaviour of the squirrel monkey *(Simiri sciurens)*. Folia Primatol 1:29–66

Richards SM (1974) The concept of dominance and methods of assessment. Anim Behav 22:914–930

Rowell TE (1967) A quantitative comparison of the behaviour of a wild and caged baboon group. Anim Behav 15:499–509

Rowell TE (1974) The concept of social dominance. Behav Biol 11:131–154

Sassenrath EN (1970) Increased adrenal responsiveness related to social stress in rhesus monkeys. Horm Behav 1:283–298

Seyfarth RM, Cheney DL, Hinde RA (1978) Some principles relating social interactions and social structure among primates. In: Herbert J, Chivers D (eds) Recent advances in primatology, vol I. Academic Press, London New York, pp 39–52

Simpson MJA (1973) The social grooming of male chimpanzees. In: Michael RP, Crook JH (eds) The comparative ecology and behaviour of primates. Academic Press, London New York

Zuckerman S (1932) The social life of monkeys and apes. Kegan Paul Trench and Trubner, London

Sexual Behavior in Aging Male Rhesus Monkeys

C.H. PHOENIX and K.C. CHAMBERS [1]

Introduction

There have been very few experimental studies on sexual behavior in old animals. Several reproductive studies using aging farm animals have been carried out, but the purpose of such studies has been primarily to determine the effects of age on fertility (Biship 1970). Nevertheless, they have provided evidence of a decline in sexual interest and activity with increasing age.

There are a few experimental studies on the sexual behavior of aging male rodents. Larsson (1958) found that in 1-h pairings, 2-year old male laboratory rats had fewer intromissions and ejaculations than 1-year old rats. However, in a later study, when the test duration was extended, the 2-year old males achieved a greater number of ejaculations than the younger males before reaching sexual exhaustion (Larsson and Essberg 1962). Jakubczak (1964) reported that the rates of intromission and ejaculation in 30-month old guinea pig were significantly lower than those in 6-month old animals.

Information obtained in interviews and in clinical practice by Kinsey et al. (1948), and by Masters and Johnson (1966), indicates that sexual activity in men declines steadily from adolescence to old age.

Robinson et al. (1975) reported that rhesus monkeys 20-years old or older (N = 4) had a significantly lower percentage of tests with ejaculations than did middle-aged or young males. The old males ejaculated in 48% of the tests, whereas the middle-aged and young males ejaculated in 81%.

Eaton (1978) has concluded that sexual vigor, i.e., the frequency of mounting and ejaculation, of male Japanese macaques *(Macaca fuscata)* declines in old age. He based his conclusion on observations of a troop of Japanese macaques housed in an outdoor enclosure at the Oregon Regional Primate Research Center. The reason for the decline in ejaculation frequency is not known, but according to Eaton loss of dominance can be excluded.

Most explanations of the decline in sexual behavior with increasing age may be characterized primarily as physiological or psychological. Physiological explanations of the decline in sexual behavior in old age have generally assumed that there is a

1 Department of Primate Behavior, Oregon Regional Primate Research Center, 505 N.W. 185th Avenue, OR 97005, USA

decline in testosterone levels in old age. There has been much theorizing but not much documentation. Several researchers failed to find a difference between testosterone levels in young and old men. More recently, some investigators have found that in adult men mean serum levels of free testosterone begin to decline at age 50 (Stearns et al. 1974), but this finding is not universally accepted (Harman and Tsitouras 1980). Some evidence suggests that luteinizing hormone (LH) and follicle-stimulating hormone (FSH) serum levels show a slight but steady rise after age 40 and a more abrupt rise after 70 years of age. There is also a suggestion that gonadotropin secretion is regulated by free rather than bound testosterone (Rubens et al. 1974).

Psychological explanations for the decline of sexual behavior in men were offered by Kinsey et al. (1948). They suggested that boredom developed from repetition of the same sort of experience, and they referred to the loss of interest as "psychologic fatigue". They claimed that a change in partners increased sexual performance. In the mouse, the rat, and other species, a male that stops copulating with a female with whom it has been paired again becomes sexually active when a different female is introduced (Bermant and Davidson 1974). Kempf (1917) reported that male rhesus monkeys were sexually reinvigorated by new sexual partners. Although all of these experiments were carried out with adult males, they were not old males.

Larsson (1963) has demonstrated that lowered sexual performance in aging rats can be increased by nonspecific sensory stimulation such as that provided by handling. The same kind of handling of young rats, however, does not increase sexual performance. In a similar vein, he has shown that old rats are not likely to mate during the light phase of the diurnal cycle, although they may exhibit high levels of sexual activity during the dark phase of the cycle.

Sexual Behavior in Male Monkeys

In our laboratory, we observed what appeared to be a decline in the sexual performance of male rhesus monkeys during the 6-year period from 1969 to 1975. These males had served as controls for studies involving, among other things, the effects of castration on the sexual behavior of adult males (Phoenix et al. 1973) and the effects of replacement therapy with various androgens (Phoenix 1974, 1976, 1977). The apparent, but by no means proven, decline in sexual behavior of the control males prompted us to conduct a retrospective study on their performance.

The eight male and ten female rhesus monkeys that were the subjects of this study were born in the wild and purchased as adults. The males were at least 10 years old when the studies began; the females were young adults. The animals were kept in individual cages in the same air-conditioned room. Tests of sexual behavior were given in a quiet, light- and temperature-controlled room. Ovariectomized stimulus females received injections of 10 μg of estradiol benzoate for 12 or 13 days before serving as test partners. They served in this capacity for 2 consecutive days each month in which tests were conducted. Each male was tested once or twice a week, and data were analyzed in blocks of eight to ten tests given in consecutive weeks. A few seconds after an estrogen-primed female had been introduced into the test cage, a male was permitted to enter, a timer was started, and behavior was recorded on an inventory sheet. Tests lasted 10 minutes, or until the male ejaculated.

Although the number of behaviors that we observed and recorded were numerous, we will limit our discussion to relatively few and will emphasize those behaviors displayed by the male. The behaviors observed have been described in detail elsewhere (Phoenix 1974). Briefly, proximity response (prox) is defined as approaching and sitting within 1 foot of a seated partner; contact is defined as placing the hands on the lower back or hips of a female when she is not in a present posture; mount is defined as placing the hands on the hips of the partner and clasping the ankles or calves with the feet; intromission is defined as inserting the erect penis into the vagina; thrusts per intromission is defined as the average number of pelvic thrusts accompanying an intromission; and erection is defined as a fully visible glans and shaft of penis without any suggestion of a specific degree of rigidity. Latencies to the first mount, to intromission and to ejaculation (the time elapsed from the beginning of the test to the first occurrence of the particular response) also were measured. In addition to observing these behaviors, we took samples of peripheral vein blood at various intervals beginning in 1970.

The results of our retrospective study revealed that over the 240 weeks there had been a statistically significant decline in the mean rates of proxing, contacting, mounting, and intromitting ($p < 0.05$). The percentage of tests with ejaculations declined significantly, and the mean latency to ejaculation increased ($p < 0.05$). The mean number of thrusts per intromission did not change over time (Phoenix 1977). Thus, ejaculation latency was positively correlated with time, and all the other measures except thrusts per intromission were negatively correlated with time. Correlations varied from $r = -0.67$ for the prox rate to $r = -0.96$ for the intromission rate. All correlations were statistically significant: for prox, $p < 0.05$, and for all other measures, $p < 0.01$.

It was obvious that sexual performance had declined, and in view of the advanced age of the animals the decline might well have been causally related to an aging factor. We wanted to find out whether the level of sexual behavior would continue to decline and to determine the physiological or environmental factors responsible for the decline. We continued periodic testing of the animals and took blood samples for hormone assays.

Table 1 shows the mounting rate and percentage of tests with ejaculations for six of the males in 1969 and in 1979. Two of the eight males died before the 1979 tests. All of the six males showed a decline in the percentage of tests in which they ejaculated, and all but one male showed a decline in mounting rate. The mean decline for the group in the two measures of performance was statistically significant.

We know that in young adult male rhesus monkeys the level of bound testosterone is not correlated with an individual animal's level of sexual performance, under the conditions that we have described, but castration nevertheless leads to a decline in behavior. The rate of decline and the baseline reached over time vary widely among individuals. Treatment with testosterone restores sexual behavior to precastration levels (Phoenix et al. 1973, Phoenix 1977).

Was the decline in performance in our old monkeys associated with a decrease in testosterone levels? Robinson et al. (1975) in their study on old male rhesus monkeys

Table 1. Individual mean mounting rate and percentage of tests with ejaculations
in the two series of ten tests given 10 years apart

Animal no.	Mounting rate in min		% tests with ejaculation	
	Nov. 1969	Sep. 1979	Nov. 1969	Sep. 1979
800	1.82	0.45	80	0
1058	1.54	1.65	80	60
3156	1.56	0.38	70	10
3527	1.73	1.39	70	60
3530	2.39	1.52	100	30
3535	1.67	1.27	80	50
Mean	1.79	1.11	80	35

failed to find a significant difference between the testosterone levels of young and old
males. We also failed to find a significant decline in plasma levels of bound testosterone
in our males (Table 2). We have concluded that a decline in plasma levels of bound
testosterone cannot account for the decline in sexual behavior in old age.

Table 2. Plasma levels of testosterone in eight adult rhesus males over
a 10-year period [a]

Animal no.	Plasma testosterone (ng/ml)			
	Jan. 1970 sample	Apr. 1970 sample	Apr. 1975 sample	Sep. 1979 sample
800	4.83	– [b]	2.10	2.37
1058	2.17	3.44	1.01	1.36
1073	2.96	0.30	3.49	– [c]
3156	11.28	3.38	2.23	3.44
3527	0.79	0.93	2.82	3.92
3530	9.69	10.30	4.95	6.02
3535	0.82	1.15	1.41	3.27
4591	3.45	4.69	2.31	1.54

[a] Blood samples taken in January and April of 1970 were assayed by means
   of gas-liquid chromatography, and other samples were analyzed by radioimmunoassay
[b] S. mple was lost in processing. Mean value for that date was used in the
   analysis of variance
[c] Animal died. Mean value for that date was used in the analysis of variance

   The idea of increasing sexual performance by treatment with testosterone is not
new. The classic case is that of Brown-Séquard, who, when he was 72 years old in
1889, injected an aqueous extract of dog testes into himself in an attempt at rejuvena-
tion (Brown-Séquard 1974). Testosterone is known to increase muscle mass and is
used by some athletes to improve performance, but the problems associated with sexual
activity in old age do not, as far as we know, involve muscular weakness.

Although there were no significant differences in testosterone levels over the 10 years, we felt it was possible that the tissues that mediated the behavior might be less sensitive to testosterone in old age and that by increasing testosterone we might be able to compensate for the reduced sensitivity. Jakubczak (1964) had proposed a similar hypothesis regarding the old guinea pigs that he had studied. He had found that injected testosterone did not increase the sexual performance of the old males. Nevertheless, we decided to test the hypothesis with our rhesus males. From the eight males, we selected the four with the lowest percentages of tests with ejaculations. Into these we injected testosterone propionate. The males received daily injections of 1 mg/kg of body weight for 28 days and were given two tests of sexual behavior each week beginning 1 week after the first injection. Males with the higher rates of intromission and ejaculation were untreated and served as controls.

The results of the study are shown in Table 3. The group of low-level performers differed significantly from the group of high-level performers with respect to four measures of behavior before treatment: contact rates, mounting rates, percentages of tests with ejaculations, and latencies to ejaculation. These results were not surprising since the animals had been assigned to the groups on the basis of ejaculation frequency in previous tests. After treatment the low-level performers differed significantly from the high-level performers with respect to five measures of behavior. The two groups differed in intromission rate as well as in the four measures in which they had differed before the low-performance group had been treated with testosterone. This difference, however, was due to an increase — not statistically significant — in the rate of intromission of the control group. The pretest rate was 0.89 intromissions per minute and the posttest rate was 1.27. This change resulted in a statistically significant difference in the intromission rate between the high and low performers in the posttreatment tests.

Table 3. Mean level of performance for six measures of sexual behavior in low-level performers before and after testosterone treatment and in untreated high-level performers

Group	Behavior					
	Prox	Contact	Mount	Intromission	% ejaculation [a]	LTEJ
Before						
Low group	0.05	0.68	0.60	0.32	25.00	6.50
High group	0.21	1.57 [b]	1.36 [b]	0.89	88.00 [b]	3.01 [b]
After						
Low group	0.02	0.74	0.60	0.40	21.00	8.87
High group	0.12	1.61 [b]	1.51 [b]	1.27 [b]	81.00 [b]	4.30 [b]
Low group						
Before	0.05	0.68	0.60	0.32	25.00	6.50
After	0.02	0.74	0.60	0.40	21.00	8.87 [b]
High group						
Before	0.21	1.57	1.36	0.89	88.00	3.01
After	0.12	1.61	1.51	1.27	81.00	4.30

[a] The % ejaculation = percentage of tests with ejaculations
[b] The $p < 0.05$
Abbreviations: Prox = proximity response; LTEJ = latency to ejaculation

In pretest/posttest comparisons of the low performers, there was only one significant difference in the behaviors observed. The latency to ejaculation increased from 6.50 to 8.87 min. An increase in ejaculation latency can hardly be thought of as improved performance as a result of testosterone treatment. There were no significant differences in the high performers for these pretest/posttest comparisons. We have concluded, therefore, that increasing testosterone levels in animals whose plasma levels fall within the normal range do not increase sexual performance. It is possible, of course, that much higher levels of testosterone would have been effective, but we have no a priori reason for making that assumption. In a previous study 1 mg of testosterone propionate/kg injected into rhesus males castrated a year earlier restored behavior to precastration levels (Phoenix et al. 1973).

There is evidence from radioimmunoassay studies that in men free testosterone (the hormone presumably available for diffusion into target tissues) declines sooner in aging men than bound testosterone (Stearns et al. 1974). However, not all investigators agree (Harman and Tsitouras 1980). We are now working with David Hess of the Oregon Reional Primate Research Center to measure the apparent free testosterone concentration (AFTC) in serum using the method of Harman and Danner (1977). Blood from 20 old males and 10 young males was analyzed for total testosterone levels and AFTCs. So far, AFTC tests have only been completed on the serum of four old animals and three young ones. The old males had significantly greater fractional binding of testosterone than the young males (t [5] = 5.19, p < 0.01). However, we need to complete the AFTC assays on all our males before drawing conclusions about the ratio of bound to free testosterone or about a correlation with performance level. If we find less free serum testosterone in old males and a correlation between the free testosterone level and sexual performance, we will need to reexamine much of our thinking about the role of testosterone in sexual behavior. It is unlikely that hormone levels will provide the whole answer to the observed decline in sexual behavior.

Early studies on primate sexual behavior suggested that when an observer approached and stood before a caged pair of monkeys, they frequently responded by engaging in sexual activity. Bingham (1928) found that when monkeys were unaware of his presence he could watch them without observing sexual activity, but when he made his presence apparent they immediately became sexually active. He concluded that monkeys had a a tendency "to show off sexually in the presence of interested observers" (p. 21). Tinklepaugh (1932) noted that when male monkeys were excited or emotionally disturbed by an asexual stimulus and were prevented by confinement from attacking or fleeing, they were likely to become sexually active. He interpreted this sexual behavior as a substitute response that reduced tension. Male monkeys provoked by animals in adjacent cages repeatedly mounted their female partners according to Maslow (1936). These investigators probably wrote accurate reports of their observations, but they failed to provide quantitative measures of behavior or to introduce control conditions for comparisons.

In another experiment, we sought to increase sexual performance in our males by introducing a source of nonspecific stimulation during tests of sexual behavior. A vigorous young adult male "stranger" was used as a nonspecific stimulus. He was introduced into the room but not into the test cage.

We picked eight young vigorous males that were strangers to our old males. Each old animal was tested with a female in the presence of each of the eight strangers.

Hence they were given eight tests of sexual behavior with strangers present and eight tests without strangers present.

The strangers produced only one statistically significant difference in behavior: mean latency to the first mount increased from 58.2 s to 114.5 s.

Mean latency to the first intromission was 162.34 s without the stranger and 122.11 s with the stranger. This difference was not statistically significant, an indication that the delay in the latency to the first mount had no significant effect on the time of the subsequent intromission. The mean latency to ejaculation differed by only 20.4 s. It was apparent that the animals were not sufficiently distracted or excited to alter their customary level of sexual activity.

Kempf (1917) reported that male rhesus monkeys were sexually reinvigorated by new sexual partners, and Kinsey et al. (1948) suggested that in older men a change in partners increases sexual performance. We decided to test this possibility with our animals.

To examine differences in performance with new sexual partners, we picked ten unfamiliar ovariectomized females of about the same age as our standard stimulus females. This new group received the same estradiol priming as the old familiar females. Each male was tested once with each female for a total of 20 tests per male. The results of the study may be seen in Table 4. The male performance with the familiar partners was not significantly different ($p < 0.05$) from the male performance with the unfamiliar partners. Had our males been tested with a single female partner over the years rather than with ten females, the outcome might have been different. In any event, simply introducing unfamiliar females is not enough to increase performance in old males.

There are reports that older men find young women more arousing sexually than old women. The statement has little scientific validity, but it did suggest the following experiment. We chose ten young cycling females (about 4 1/2 years of age) and another group of ten old cycling animals (about 16 years of age). Both groups were unfamiliar to the old males. Each male was tested once with each female at midcycle for a total

Table 4. Selected measures of performance by ten old rhesus males paired with old ovariectomized females, familiar and unfamiliar

Behavior	Familiar [a]	Unfamiliar
Prox	0.05	0.05
Contact	0.76	0.90
Mount	0.67	0.70
Intromission	0.28	0.32
TPI	7.93	8.12
% Ejaculation [b]	30.00	28.00
LTEJ (min)	4.52	3.35

[a] None of the differences were significant ($p > 0.05$)
[b] The % ejaculation = percentage of tests with ejaculations

Abbreviations: Prox = proximity response; TPI = thrusts per intromission; LTEJ = latency to ejaculation

of 20 tests per male. The male contacted the young females more frequently than they contacted the old females (p < 0.05), and the mean percentage of tests in which they developed erections was also greater in the young-female series (p < 0.01). Other behaviors, however, did not differ significantly (Table 5). Mounting, intromission rates, and the percentage of tests with ejaculation did not differ between the young- and old-female series.

Table 5. Selected measures of performance by ten old rhesus males paired during the periovulatory period with young, cycling, unfamiliar females and old, cycling, unfamiliar females

Behavior	Young	Old
Prox	0.01	0.01
Contact	0.64	0.41 [a]
Mount	0.56	0.34
Intromission	0.26	0.19
TPI	7.33	6.50
% Ejaculation [b]	23.00	20.00
LTEJ (min)	4.48	4.63

[a] Significantly lower (p < 0.05). Other differences were not significant

[b] The % ejaculation = the percentage of tests with ejaculations

Abbreviations: Prox = proximity response; TPI = thrusts per intromission; LTEJ = latency to ejaculation

To clarify the issue, we repeated the experiment with eight different old cycling females. As in the previous experiment, each male was tested with each female. We found no difference in the rate of contact or in the percentage of tests with erections between the young females and this second group of females.

We have assumed that some random error in the first experiment led to differences in two of the many measures and have concluded that the level of sexual performance of old males was as high with old unfamiliar females as it was with young unfamiliar females.

Overview

Our work has demonstrated that sexual performance declines in old male rhesus monkeys. The decline cannot be ascribed to a reduction in plasma levels of bound testosterone. In previous work and in the studies reported here, we found no correlation between the plasma level of testosterone and the level of sexual performance. Administration of testosterone propionate to increase the plasma testosterone level did not increase sexual performance. Preliminary data suggest that old male rhesus monkeys have less free testosterone than young males, but our results are incomplete. Findings in man remain controversial.

Nonspecific sensory stimuli, e.g., the presence of a rhesus male stranger, do not increase sexual performance. Introduction of an unfamiliar female sex partner does not increase sexual performance. Young females are no more effective than old females in stimulating old males to copulate.

Important factors to be considered in old age are increasing physical disability and diminished coordination and dexterity. These limited the capacity of some of our males to perform appropriately. Only six of the original eight males were living at the time of our sexual behavior tests in late 1979. In two of the original eight, severe arthritic conditions limited their ability to pursue and mount the females. In addition, one of the males had serious dental problems and so could not easily eat his customary biscuits; these problems may have had a secondary impact on his total well-being and vigor.

Increasing physical disability, however, cannot account for the decline in sexual performance of all the males. In general, most of the males currently being studied appear capable of responding physically. We are struck by the fact that in youth almost any female is acceptable as a sexual partner, and reluctant females are pursued with considerable vigor. As the males grow old they are less persistent, and a single rejection by a female is often sufficient to end their attempts. They do not seem to try as hard, but we do not know why. We do know that partner preferences that have been demonstrated in young adult males seem to be accentuated in old males. The animals become more discriminating, more finicky. Certain subtle aspects of the behavior of these preferred females may be critical in arousing old males to respond. Studies are now being designed to investigate this possibility.

Acknowledgments. This study is publication No. 1151 of the Oregon Regional Primate Research Center. The study was supported in part by Grants RR-00163 and AG-01608 from the National Institutes of Health, U.S. Public Health Service. We thank Jens N. Jensen for his valuable assistance in observation and computer programing and John A. Resko for his help with all the steroid hormone assays.

References

Bermant G, Davidson JM (1974) Biological bases of sexual behavior. Harper & Row, New York
Bingham HC (1928) Sex development in apes. Comp Psychol Monogr 5:1–165
Bishop MWH (1970) Aging and reproduction in the male. J Reprod Fertil (Suppl) 12:65–87
Brown-Séquard CE (1974) The effects produced on man by subcutaneous injections of a liquid obtained from the testicles of animals. In: Carter CS (ed) Hormones and sexual behavior. Dowden, Hutchinson and Ross, Stroudsbury PA (Reprinted from Lancet, 1889, 2:105–106)
Eaton GG (1978) Longitudinal studies of sexual behavior in the Oregon troop of Japanese macaques. In: McGill TE, Dewsbury DA, Sachs BD (eds) Sex and behavior: status and prospectus. Plenum Press, New York, p 35
Harman S, Danner R (1977) Rapid measurement of an index of testosterone binding to serum binding globulin using ion exchange columns. J Clin Endocrinol Metab 45:953–959

Harman SM, Tsitouras PD (1980) Reproductive hormones in aging man. I. Measurement of sex steroids, basal luteinizing hormone, and Leydig Cell response to human chorionic gonadotropin. J Clin Endocrinol Metab 51:35–40

Jakubczak LF (1964) Effects of testosterone propionate on age differences in mating behavior. J Gerontol 19:458–461

Kempf EJ (1917) The social and sexual behavior of infrahuman primates with some comparable facts in human behavior. Psychoanal Rev 4:127–154

Kinsey AC, Pomeroy WB, Martin CE (1948) Sexual behavior in the human male. W.B. Saunders, Philadelphia

Larsson K (1958) Sexual activity in senile male rats. J Gerontol 13:136–139

Larsson K (1963) Non-specific stimulation and sexual behavior in the male rat. Behaviour 20:110–114

Larsson K, Essberg L (1962) Effect of age on the sexual behavior of the male rat. Gerontologia 6:133–143

Maslow AH (1936) The role of dominance in the social and sexual behavior of infrahuman primates. III. A theory of sexual behavior of infrahuman primates. J Genet Psychol 48:310–338

Masters WH, Johnson VE (1966) Human sexual response. Little & Brown, Boston

Phoenix CH (1974) The effects of dihydrotestosterone propionate on the sexual behavior of castrated male rhesus monkey. Physiol Behav 12:1045–1055

Phoenix CH (1976) Sexual behavior of castrated male rhesus monkeys treated with 19-hydroxy-testosterone. Physiol Behav 16:305–310

Phoenix CH (1977) Factors influencing sexual performance in male rhesus monkeys. J Comp Physiol Psychol 91:697–710

Phoenix CH, Slob AK, Goy RW (1973) Effects of castration and replacement therapy on sexual behavior of adult male rhesuses. J Comp Physiol Psychol 84:472–481

Robinson JA, Scheffler G, Eisele SG, Goy RW (1975) Effects of age and season on sexual behavior and plasma testosterone and dihydrotestosterone concentration of laboratory-housed male rhesus monkeys *(Macaca mulatta)*. Biol Reprod 13:203–210

Rubens R, Dhont M, Vermeulen A (1974) Further studies on Leydig Cell function on old age. J Clin Endocrinol Metab 39:40–45

Stearns EL, MacDonnell JA, Kaufman BJ, Padua R, Lucman TS, Winter JSD, Faiman C (1974) Declining testicular function with age: Hormonal and clinical correlates. Am J Med 57:761–766

Tinklepaugh OL (1932) Sex behavior in infrahuman primates as a substitute response following emotional disturbance. Psychol Bull 29:666

Simian-type Blood Groups of Hamadryas Baboons. Population Study of Captivity-born Animals at the Sukhumi Primate Center – Preliminary Report

B.A. LAPIN [1], W.W. SOCHA [2], and J. MOOR-JANKOWSKI [2]

Introduction

As shown in numerous earlier publications (Moor-Jankowski et al. 1964, 1967, Wiener and Moor-Jankowski 1969, Verbitski et al. 1967), baboons, like many other species of Old World monkeys, display two kinds of the serological polymorphism: one, that is based on individual differences in the content of saliva and other body secretions, as defined by inhibition tests for the human-type A, B, H blood group substances, and confirmed by the serum tests for the presence of anti-A and anti-B isoagglutinins, and the other, based on antigenic properties of the red cells, defined by the so-called simian-type specificities, and detectable by hemagglutination tests, carried out with the antisera purposely produced by iso- or cross-immunization of baboons and other monkeys.

Since our last population study on blood groups of baboons (Socha et al. 1977), considerable progress has been achieved in the production of the simian-type blood grouping sera. Extensive immunization program resulted in the production of new reagents, thus increasing the number of the simian-type specificities detectable on the baboon red cells. Recently, large numbers of baboon blood samples became available for blood grouping tests, and the purpose of the present report is to give the preliminary results of the tests performed with an entire battery of typing sera that included old, as well as the newly introduced reagents.

Materials and Methods

All the blood samples originated from hamadryas baboons maintained in semi-isolate condition (Asanov and Mirvis 1972) at the Sukhumi Primate Center. The blood specimens were collected from anesthetized animals and shipped in batches, by air freight, to the Primate Blood Group Reference Laboratory in New York. In all instances, the specimens were received in good condition and suitable for testing. The blood samples were tested fresh, with batteries of typing reagents. If adequate quantity of plasma could be separated, cross matching tests were performed with 15–20 randomly selected

---

1   Institute of Experimental Pathology and Therapy, P.O. Box 66, Gora Trapetziya, Sukhumi, USSR
2   New York University Medical Center, LEMSIP, 550 First Avenue, New York, NY 10016, USA

red cell samples from the same batch. Unused portions of blood were frozen and stored in liquid nitrogen freezers for future use. Since most of the baboon antisera gave the best reactions by the antiglobulin method, results described in this report refer to those obtained by that hemagglutination technique. The cross matching tests were carried out by saline agglutination, antiglobulin and enzyme treated red cell techniques. During the study, a modification of the agglutination technique was introduced, namely the isotonic saline was replaced by low ionic-strength (LISS) medium (Moore and Mollison 1976). For details of blood grouping techniques used for typing primates, see earlier publications (Socha et al. 1972, Erskine and Socha 1978).

Results and Discussion

The list of antisera used in the present study and shown in Table 1, illustrates the scope of tests routinely performed on all samples. The reagents included antisera produced earlier and described in previous publications (Moor-Jankowski et al. 1967, 1973, 1974, Socha et al. 1977, Wiener et al. 1970) as well as the newly produced antisera (items 12 through 22 in Table 1) never before described nor tested against large numbers of blood samples. In addition, each batch of blood specimens was tested with different sets of antisera produced by isoimmunization of rhesus monkeys. Some of those sera were known to cross-react with the red cells of various species of macaques and were expected to detect the same or similar combining structures also on the red cells of baboons. The results of those experiments will be the subject of another publication.

As can be seen in Table 2, the standard baboon typing sera used in tests with the present series of hamadryas baboons by and large detected individual differences among animals, thus confirming the value of red cell antigens as genetic markers in baboons. Of special value seem to be the blood groups that are defined by monospecific reagents (which constitute the majority of antisera used in the present study), and the statistical analysis of the data obtained is currently carried on in order to detect possible associations among specificities so far defined.

Of the 22 specific antisera used in the study, 6 were used previously for testing Ethiopian baboons, and Table 2 gives comparative data on blood group frequencies in Sukhumi and Ethiopian baboons. Of particular interest are, naturally, comparisons between two populations of hamadryas baboons. As can be seen, in three instances (groups A^p, C^p and G^p) the differences between their frequencies in the hamadryas populations were found highly significant (as judged by the Student's test) and could not be ascribed to chance alone. This is quite understandable in view of the fact that the Sukhumi colony has grown mainly through internal breeding (Asanov and Mirvis 1972). It will be of interest to investigate the level of these differences between particular genealogical groups within the population of Sukhumi baboons.

Blood specimens of all Sukhumi baboons gave positive reactions with the so-called anti-hu reagent which was also found to agglutinate the red cells of all hamadryas baboons previously tested, independently of their geographical origin. The anti-hu reagent was produced by immunizing olive baboons with the red cells of hamadryas,

Table 1. List of standard antisera used for baboon blood grouping

No.	Immunized animal	Donor animal	Speci-ficity	No.	Immunized animal	Donor animal	Speci-ficity
1.	B-2, Paula olive baboon	B-11, Chubby olive baboon	A^P	12.	B-15, Irwin yellow baboon	B-127, Blue hybrid	O^P+W^P
2.	B-13, Jack yellow baboon	B-9, Eugene yellow baboon	A^P+Y^P	13.	B-15, Irwin [c]	B-127 hybrid	O^P
3.	B-9, Eugene yellow baboon	B-13, Jack yellow baboon	B^P+Z^P	14.	B-17, Janny yellow baboon	B-133 hybrid	P^P
4.	B-9, Eugene [a]	B-13, Jack	B^P	15.	B-127, Blue hybrid	B-15, Irwin yellow baboon	A^P+X^P
5.	B-11, Chubby olive baboon	B-2, Paula olive baboon	C^P+Q^P	16.	B-131, Ralph hybrid	B-776, Ilex hamadryas baboon	S^P+T^P
6.	B-11, Chubby [b]	B-2, Paula	C^P	17.	B-131, Ralph [d]	B-776, Ilex	T^P
7.	B-32, Lynn olive baboon	B-60012, Janna olive baboon	G^P	18.	B-133, Roman hybrid	B-17, Janny yellow baboon	U^P
8.	B-268, Grayma olive baboon	B-85, Nosmo chacma baboon	hu	19.	B-143, Dave hybrid	B-230, Vivian yellow baboon	$V^P+V_i^P$
9.	B-176, Camilla hamadryas baboon	B-252, Eberle yellow baboon	ca	20.	B-143, Dave [e]	B-230, Vivian	V_i^P
10.	B-87, Souds chacma baboon	B-553, Benita yellow baboon	$ca+E^P$	21.	B-145, Thorb hybrid	B-774, Yucca hamadryas baboon	L^P
11.	B-60012, Janna Olive baboon	B-32, Lynn olive baboon	N^P	22.	B-776, Ilex [f] hamadryas	B-131, Ralph hybrid	M^P

[a] Absorbed with pooled blood of B-2 & B-11
[b] Absorbed with pooled blood of B-9 & B-13
[c] Absorbed with blood of B-2
[d] Absorbed with blood of B-9
[e] Absorbed with pooled V_2 red cells
[f] Absorbed with blood of B-790

and was supposed to define a "species-specificity" shared by all hamadryas (h) and chacma ("u" for *ursinus*) baboons, while being polymorphic in olive and yellow baboons. The present findings strongly support the notion of species specificity of the anti-hu sera, and — once more — point to a taxonomic value of serological tests in baboons. Not surprisingly, the two other red cell specificities found on the red cells of all Sukhumi baboons tested so far, namely the T^P and L^P, were also detected by antisera produced by cross-immunization using hamadryas red cells (see Table 1). As shown by comparative tests carried out with the red cells of olive and yellow baboons, the specificities defined by the newly produced sera were different from hu, their relationship to hu, however, could not be tested by statistical analysis since the overall number of olive and yellow baboons was still too small. It is noteworthy that, in the present series of immunization, as well as in the previous immunization experiments (Moor-Jankowski et al. 1973, 1974) all olive, yellow, or hybrid baboons injected with the red cells of various hamadryas baboons produced antibodies that recognized structures common to the red cells of all hamadryas baboons. This indicates that this "species-specific" characteristic of hamadryas blood is highly immunogenic for the donors of other baboon species.

Table 2. Frequencies of the blood group specificities in hamadryas baboons from Sukhumi and in feral baboons from Ethiopia

Blood group specificity	Frequency				
	Hamadryas baboons			Olive baboons	Olive/hamadryas
	Sukhumi		Ethiopia	Ethiopia	hybrids
	n = 730	p	n = 172	n = 192	Ethiopia, n = 129
A^p	0.248	< 0.001	0.058	0.635	0.465
$A^p + Y^p$	0.415		n.t.	n.t.	n.t.
B^p	0.851	> 0.05	0.907	0.505	0.457
$B^p + Z^p$	0.922		n.t.	n.t.	n.t.
$C^p + Q^p$	0.904		n.t.	n.t.	n.t.
C^p	0.589	< 0.01	0.703	0.802	0.674
G^p	0.008	< 0.001	0.256	0.167	0.295
hu	1.0		1.0	0.771	0.925
ca	0.433	> 0.05	0.372	0.792	0.698
ca + E^p	0.759		n.t.	n.t.	n.t.
N^p	0.781	> 0.05	0.832	0.974	0.925
$O^p + W^p$	0.902		n.t.	n.t.	n.t.
O^p	0.875		n.t.	n.t.	n.t.
P^p	0.638		n.t.	n.t.	n.t.
$S^p + T^p$ [a]	1.0		n.t.	n.t.	n.t.
T^p [a]	1.0		n.t.	n.t.	n.t.
U^p	0.875		n.t.	n.t.	n.t.
V^p_1	0.798		n.t.	n.t.	n.t.
V^p_2	0.568		n.t.	n.t.	n.t.
L^p [a]	1.0		n.t.	n.t.	n.t.
M^p	0.086		n.t.	n.t.	n.t.
$A^p + X^p$	0.248		n.t.	n.t.	n.t.

[a] Polymorphic in LEMSIP baboon immunized panel; n.t. = not tested

Sera of 347 Sukhumi baboons were tested for the presence of agglutinating antibodies, and of these, 36 were found to contain antibodies reactive with all or some of the red cells tested. In the majority of cases, the antibodies were of low titer (1 to 2 units) and avidity, and detectable by enzyme-treated red cell method only. In a few instances, the antibodies were of autoagglutinin type and reacted best at low temperature. A special attention has been devoted to 12 sera specimens that were found to contain antibodies of relatively higher titers (4 to 10 units by antiglobulin method) and avidity (Table 3). In contrast to the more common, weak antibodies that agglutinated only ficinated red cells, the rare, higher titer agglutinins reacted best by antiglobulin technique, and displayed type-specific properties. As can be seen from the data in Table 3, among 11 sera that were found to contain type-specific antibodies, at least 7 had agglutinins identical, similar to, or somewhat related to the specificity B^p, defined by one of the blood grouping reagents produced at early states of our immunization program (Moor-Jankowski et al. 1967, Wiener et al. 1970). This points to the highly immunogenic properties of the B^p antigen and its special position among baboon red cell antigens, at least in hamadryas baboons. As for the origins of antibodies detected in the sera of the animals, information available at the time of this

Table 3. Agglutinating antibodies in the sera of Sukhumi baboons

No.	Identification of animal	Characteristics of antibodies	Remarks
1.	4737	Medium avidity in (AG[a]/ETC[a]) type-specific, probably new specificity	Female, immunized with baboon blood (68-75), 11 pregnancies, incl. 2 abortions, pregnant (5th month at time of bleeding)
2.	4266	High avidity in S[a]/AG/ETC, contains two kinds of antibodies, both type-specific, one probably B^P-like	b
3.	4263	High avidity (S/AG/ETS) autoantibody	b
4.	7323	Medium avidity (AG/ETC), type-specific, probably B-like	Female, immunized with baboon blood (69-75), 6 pregnancies, incl. 4 stillbirths
5.	7351	Medium avidity (AG only), type-specific, probably B^P-like	b
6.	11099	High avidity (S/AG), type-specific, probably B^P-like	b
7.	12176	Medium avidity (S/AG), type-specific, probably new specificity	Male, not immunized. No activity in subsequent bleeding (10 months later)
8.	12329	High avidity (AG/ETC), type-specific, probably new specificity	Female, not immunized, two normal pregnancies, last 8 months prior to bledding
9.	12486	Medium avidity (AG/ETC), type-specificity not determined	b
10.	12883	Medium avidity (AG only), type-specific, B^P-like	Female, not immunized, one normal pregnancy terminated 2 y prior to bleeding
11.	13075	Medium avidity (AG only), type-specific, probably B^P-like	b
12.	15472	Medium avidity (AG only), type-specific, B^P-like	b

[a] S = saline agglutination; AG = antiglobulin technique; ETC = enzyme-treated red cell technique
[b] Animal's colony records not available at the time of writing

writing indicate, that while agglutinating properties of some of the sera could be related to the earlier immunization by experimental administration of baboon blood, no such immunization could be traced in at least three cases in which, therefore, the detected antibodies must be classified as "natural antibodies", produced by undetermined immunizing factors (Socha et al. 1976, Wiener and Socha 1976). Since in two of such instances, natural antibodies occurred in the sera of female animals with one or more previous pregnancies recorded, there was a possibility of a transplacental immunization due to materno-fetal incompatibility. The third case concerned a nonimmunized male in whose serum medium-strong, type-specific agglutinins were discovered at the time of the initial bleeding. Another serum specimen, however, obtained from the same animal 10 months later did not show any detectable activity. Leaving aside the problem of the mechanism of formation of "natural antibodies", their occurrence in the sera of some of the baboons appears to be of practical importance in husbandry and breeding, and should be taken into consideration in the management of a baboon colony.

References

Asanov SS, Mirvis AB (1972) The breding of hamadryas baboons at the Sukhumi Primate Center. In: Diczfalusy E, Standley CC (eds) The use of non-human primates in research on human reproduction. Karolinska Institute, Stockholm, pp 483–489

Erskine AG, Socha WW (1978) Principles and practice of blood grouping, 2nd edn. CV Mosby, St Louis

Moore HC, Mollison PL (1976) Use of low-ionic-strength medium in manual tests for antibody detection. Transfusion 16:291–296

Moor-Jankowski J, Wiener AS, Gordon EB (1964) Blood groups of apes and monkeys. I. The A-B-0 blood groups in baboons. Transfusion 4:92–100

Moor-Jankowski J, Wiener AS, Gordon EB, Davis JH (1967) Blood groups of baboons demonstrated with isoimmune sera. Nature (London) 214:81

Moor-Jankowski J, Wiener AS, Socha WW, Gordon EB, Davis JH (1973) A new taxonomic tool: serological reactions in cross-immunized baboons. J Med Primatol 2:71–84

Moor-Jankowski J, Wiener AS, Socha WW (1974) A new taxonomic tool. II. Serological differences between baboons and geladas demonstrated by cross-immunization. Folia Primatol 22:59–71

Socha WW, Wiener AS, Moor-Jankowski J (1972) Methodology of primate blood grouping. Transplant Proc 4:106–110

Socha WW, Wiener AS, Moor-Jankowski J, Scheffrahn W, Wolfson SJ Jr (1976) Spontaneously occurring agglutinins in primate sera. Int Arch Allergy Appl Immunol 51:656–670

Socha WW, Wiener AS, Moor-Jankowski J, Jolly CJ (1977) Blood groups of baboons. Population genetics of feral animals. Am J Phys Anthropol 47:435–442

Verbitski MSj, Asanov SS, Andreeva AV, Kiriyam NE (1967) Eritrotsytarnyie isoantigeny pavianov gamadrilov. Vopr Antropol 214:81

Wiener AS, Moor-Jankowski J (1969) The A-B-0 blood groups of baboons. Am J Phys Anthropol 30:117–122

Wiener AS, Socha WW (1976) Spontaneously occurring agglutinins in primate sera. II. Their classification and implications for the mechanisms of antibody formation. Haematologia 10:463–467

Wiener AS, Socha WW, Moor-Jankowski J, Gordon EB (1970) The A^P-B^P blood groups of baboons. Am J Phys Anthropol 33:433–438

Rhesus Macaques: Pertinence for Studies on the Toxicity of Chlorinated Hydrocarbon Environmental Pollutants

W.P. MCNULTY [1]

Environmental Dissemination

Polyhalogenated polyaromatic hydrocarbons (PHPAHs) — polychlorinated biphenyls (PCBs), polybrominated biphenyls, polychlorinated dibenzo-*p*-dioxins, polychlorinated dibenzofurans, and polychlorinated naphthalenes — have been disseminated, knowlingly and by accident, into the environment for the past 40 years. In particular, PCBs have been extensively used as dielectrics in transformers and capacitors and as heat exchanger fluids; they have also been incorporated into paints, inks, sealers, and plastics. Although increasingly strict controls have been imposed on the uses of PHPAHs and on the uses of industrial products in which they are contaminants, they persist in the environment because they are chemically stable and poorly biodegradable. Accidental discharges of PCBs still occur, primarily from electrical equipment, and polychlorinated dibenzofurans and dibenzo-*p*-dioxins continue to be released in industrial accidents and as low-level contaminants in fungicides and herbicides.

Human Poisoning

Some members of each of these classes of compounds are quite toxic, and severe human illness has occurred after ingestion of cooking oil contaminated with PCBs (e.g., the Yusho incident of 1968) (Kuratsune 1976) and after exposure to 2,3,7,8-tetrachlorodibenzo-*p*-dioxin (TCDD) in industrial accidents (Hay 1979, Pocchiari et al. 1979).

Almost nothing is known of the pathogenesis of the human diseases caused by these chemicals. The most constant sign has been an affliction of the skin called chloracne, so named by Herxheimer (1899), who mistakenly believed that elemental chlorine was the cause. Chloracne is characterized by comedones, intradermal cysts, and pustules, mostly on the face, trunk, and axillae (Crow 1970, Taylor 1979). Hyperpigmentation of the skin and nails and deformation of the nails were observed in the Japanese epidemic of PCB poisoning (Urabe and Koda 1976). In addition to these effects in the skin,

1 Laboratory of Pathology, Oregon Regional Primate Research Center, 505 N.W. 185th Avenue, OR 97006, USA

a variety of hepatic, neurological, and gastrointestinal disorders have been suspected but not proven to be caused by PHPAHs, and there has been no pathological documentation of lesions in these systems in human beings.

The diseases caused by polychlorinated naphthalenes, PCBs, and polychlorinated dibenzo-p-dioxins are similar, perhaps identical, although the toxic potencies of individual compounds may be widely different. The onset of symptoms is delayed after exposure, and repeated doses are cumulative in effect (Hayabuchi et al. 1979). Chloracne, as well as the other poorly defined symptoms, can persist for months after exposure has ceased (Urabe et al. 1979). Polychlorinated dibenzofurans and PCBs are stored in liver and adipose tissue for prolonged periods (Nagayama et al. 1977) and are excreted in milk (Yakushiji et al. 1978). Presumably the same is true for polychlorinated dibenzo-p-dioxins, but chemical detection of minuscule amounts of dioxins is difficult and data are lacking. We know nothing about the metabolism, pharmacokinetics, or biochemical actions of PHPAHs in human beings. Clearly, a suitable animal model is desirable.

Pathologic Lesions in Animals

Some features of toxicological responses to PCBs and TCDD in several laboratory species and people are shown in Table 1. The responses to PHPAHs in human beings are in fact responses to complex mixtures. For example, commercial PCB mixtures are known to contain polychlorinated dibenzofurans (Bowes et al. 1975), and the PCB mixture contaminating Japanese cooking oil had undergone considerable modification during its prolonged use as a heat exchanger fluid (Miyata et al. 1977).

Table 1. Lesions of polychlorinated biphenyl and 2,3,7,8-tetrachlorodibenzo-p-dioxin toxicity: Distribution by species

Species	Thymus	Skin	Liver	Stomach	Weight loss
Rat	+	−	+	−	+
Mouse	+	−	+	−	+
Guinea pig	+	−	−	−	+
Rabbit	+	+	+	−	+
Rhesus macaque	+	+	−	+	+
Man	?	+	?	?	+

Thymus

Atrophy of the thymus is common to all species tested. In a prepubertal male rhesus macaque *(Macaca mulatta)* fed a toxic PCB congener, 3,4,3',4'-tetrachlorobiphenyl, at 1 ppm for 1 month, the thymic cortex disappeared, the cellularity of the medulla lessened, and the corpuscles became cystic (Fig. 1). The appearance of the thymus was identical to that in monkeys chronically fed 2,3,7,8-tetrachlorodibenzofuran (TCDF)

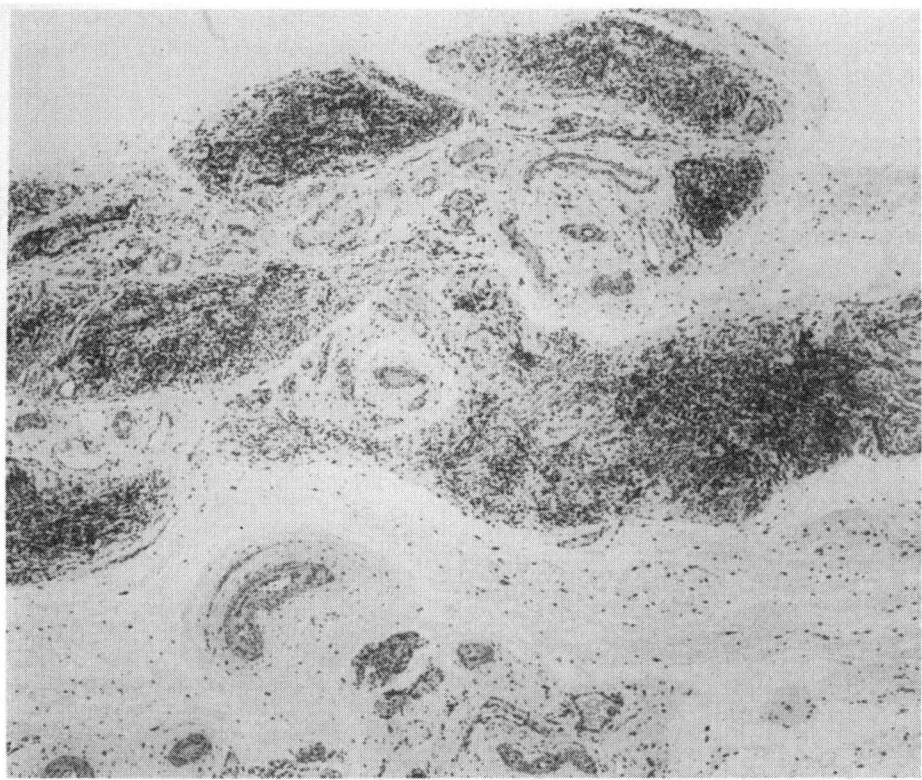

Fig. 1. Thymus gland of an immature rhesus male given 3,4,3′,4′-tetrachlorobiphenyl for several months. The cortices of the lobules have disappeared and the medullas are shrunken. × 50

or TCDD. The thymus glands of rats, mice, and guinea pigs are similarly affected in poisoning with PCBs, TCDD, or TCDF (McConnell and Moore 1979). The mechanism of this involution is not known.

Liver

Although these effects on the thymus have been similar in all species tested, marked species differences have been noted in the responses of liver, stomach, and skin (reviewed by Kimbrough et al. 1978, McConnell and Moore 1979). In rats, mice, and rabbits poisoned with PCBs or TCDD, the liver increases in size and may show fatty change; there is necrosis and giant cell formation among the hepatocytes. After prolonged exposure of rats to PCBs, a structural alteration called adenofibrosis occurs in the liver and ultimately neoplastic nodules appear.

In monkeys poisoned with PCBs, TCDF, or TCDD, the liver usually becomes larger, but the hepatocytes show minimal histological changes. When examined in an electron microscope, these cells show concentric membrane arrays and an increase in smooth endoplasmic reticulum (Allen et al. 1973), but these are probably adaptive changes reflecting the induction of mixed function oxidase enzymes (see below) rather than cell damage.

However, the biliary tree, as distinct from the hepatic parenchyma, is affected in poisoned monkeys. The epithelial cells lining the intrahepatic bile ducts are tall and contain an abnormally large number of large mucous vacuoles (Fig. 2). This hyperplastic epithelium becomes irregularly folded, and fragments of the epithelial lining are sloughed into the lumen. Similar epithelial changes can be seen in the mucosa of the extrahepatic bile ducts and gallbladder. The hepatic and common ducts, but not the gallbladder, become strikingly dilated, although there is no obstruction. No explanation is yet available for these changes in the biliary epithelium. Polychlorinated biphenyls and probably TCDD are largely excreted in the bile, and the epithelial changes may be due to high concentrations of the compounds or their metabolites in the bile stream.

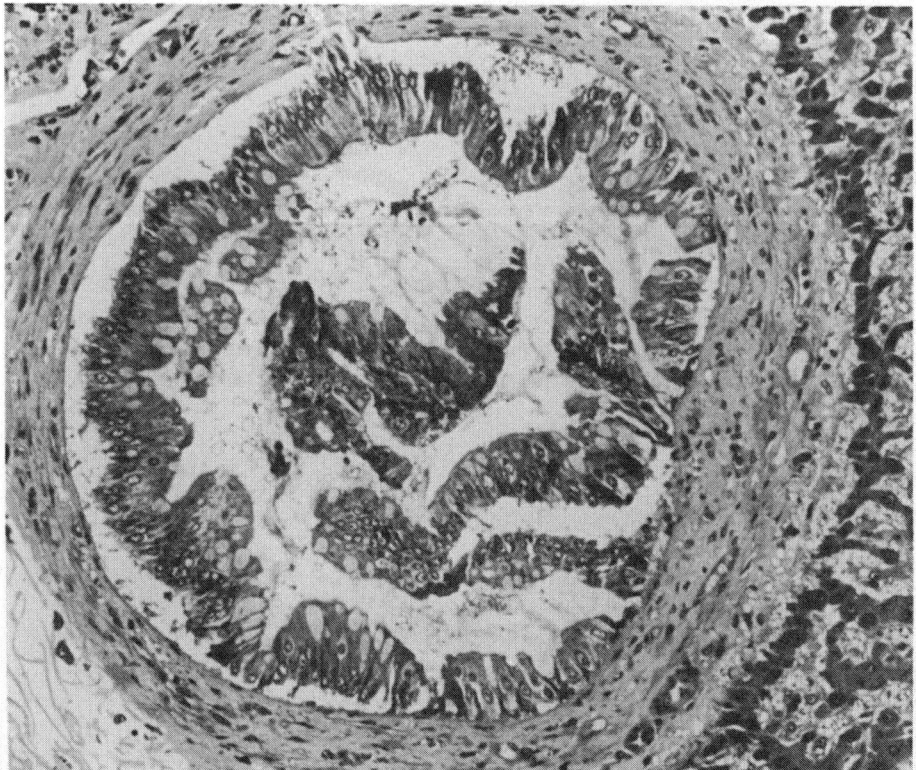

Fig. 2. Intrahepatic bile duct of a young rhesus male given 3,4,3′,4′-tetrachlorobiphenyl for several months. The mucosal cells are tall and contain abnormally large numbers of mucous vacuoles. The mucosa is partly desquamated. × 200

Stomach

In rhesus macaques, chronic exposure to PCBs, TCDD, and TCDF is accompanied by a change in the pattern of growth and cellular differentiation in the mucosa of the body of the stomach. In a normal stomach, gastric epithelial cells of several types are replenished from a generative zone in the midmucosa. Only here do DNA synthesis and mitosis occur. From this zone, mucous cells migrate toward the surface to line the gastric pits; parietal (acid-secreting) cells and zymogenic cells migrate downward to line the gastric glands. In poisoned monkeys, this orderly system is profoundly disturbed (Becker et al. 1979). The production of new parietal and zymogenic cells is interrupted, and all the new cells secrete mucus. The regular architecture of pits and glands is replaced by a disorderly pattern of irregular epithelial clefts and cysts lined with hypersecretory mucous cells (Fig. 3). The synthesis of DNA, as measured by thymidine uptake, occurs throughout the mucosa (Becker, unpublished data), and irregular mucous glands extend down into the submucosa and even into the muscularis. This dysplasia of gastric mucosa is limited to the body of the stomach; the prepyloric

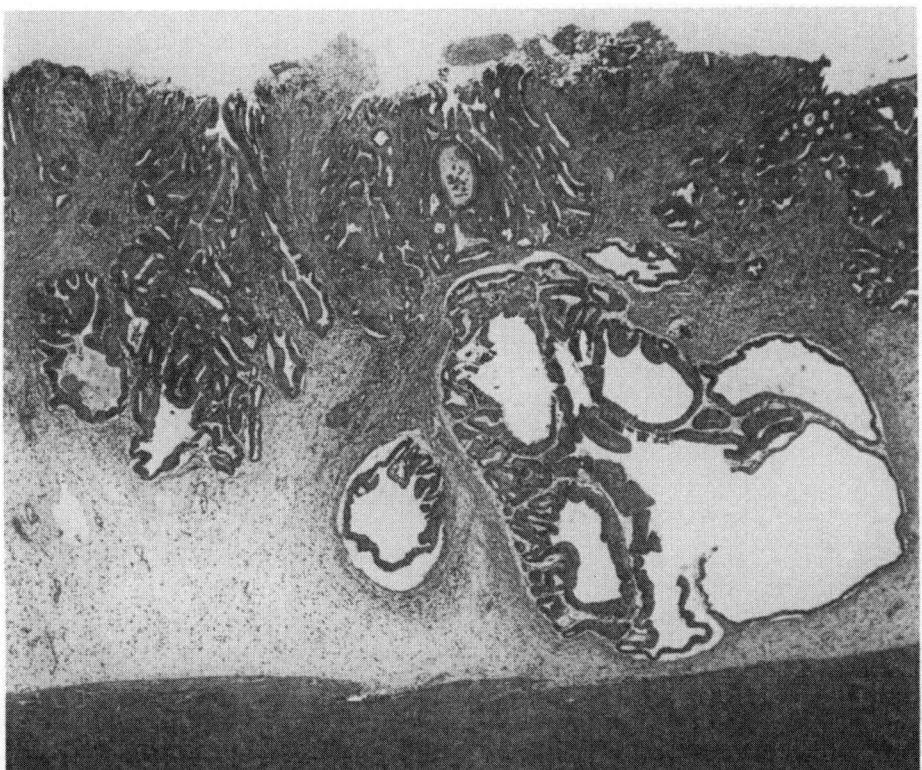

Fig. 3. Mucosa of body of the stomach of an immature rhesus male given 3,4,3′,4′-tetrachlorobiphenyl for several months. The normal pattern of crypts and glands is obliterated by the disorderly growth of mucous cells, which have grown into the submucosa and formed cysts. × 20

mucosa remains histologically normal. The lesion develops when PCBs are given intra-peritoneally as well as when they are ingested, and so it is probably not caused by direct contact of the mucosa with ingested PCB.

Superficial ulcerations have been noted in the stomachs of pigs fed PCBs (Hansen et al. 1975), but such lesions do not resemble the profound growth disturbance regularly seen in rhesus stomachs. Whether similar lesions occur in the stomachs of poisoned people is not known. Gastrointestinal complaints were frequent among Yusho victims. No person is known to have died from PCB or TCDD poisoning, and gastric lesions have not been described in autopsy reports of Yusho victims who died from other causes.

Skin

The first noticeable physical signs of poisoning in rhesus macaques are thickening and reddening of the eyelids and loss of the lashes. These signs reflect a progressive squamous metaplasia of the large racemose sebaceous glands in the tarsal plate — the meibomian glands — and atrophy of the hair follicles (Figs. 4 and 5). The phenomenon is

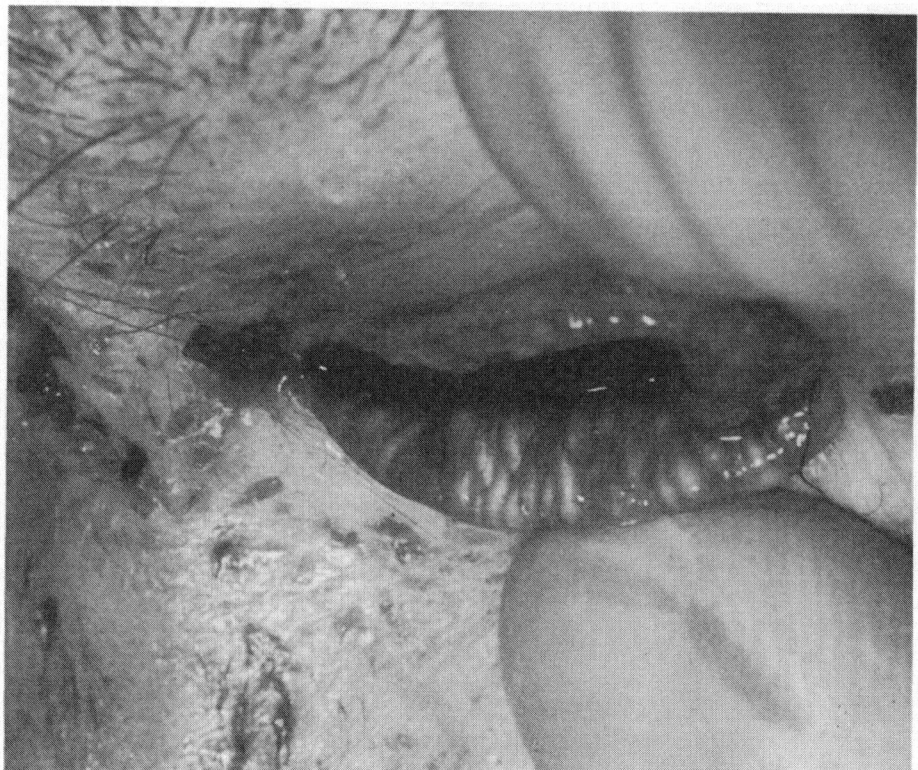

Fig. 4. Everted eyelid of an immature rhesus male given 2,3,7,8-tetrachlorodibenzofuran for several months. The meibomian glands are swollen and tortuous, and the eyelashes have been lost

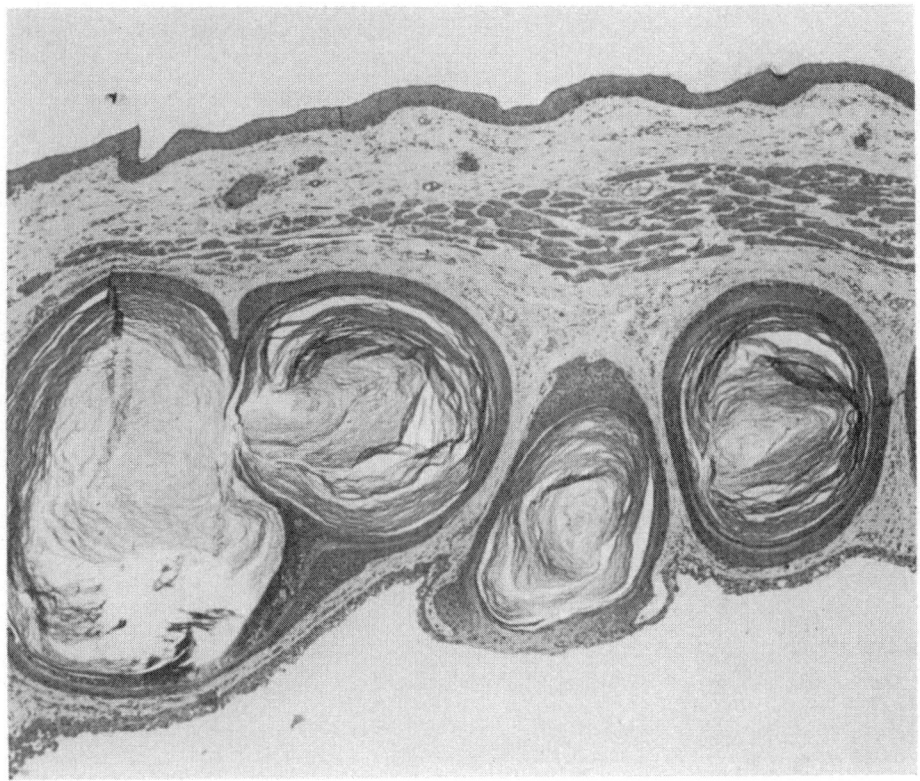

Fig. 5. Cross section of an eyelid of an immature rhesus male given 3,4,3′,4′-tetrachlorobiphenyl for several months. The normally sebaceous meibomian glands have been converted to squamous cysts. × 50

generalized over the whole body; sebaceous glands shrink to small squamous buds and hair follicles enter telogen; eventually the hair shafts are lost and the follicles become solid epithelial strands. In some locations, where sebaceous glands are normally large, such as the eyelids, ear canals, and lips, the sebaceous glands become cysts distended with keratinous material.

Each nail bed, which normally provides only a stratified nonkeratinizing epithelial surface to which the overlying nail plate is attached, changes to keratinizing epidermis, and accumulating keratinous material elevates the nail, which is finally lost (Fig. 6).

These changes are the equivalent of chloracne in man, a disease in which hyperkeratosis of the pilosebaceous apparatus is a prominent feature (Shelly and Ligman 1957). The eyelids and fingernails of PCB-poisoned monkeys resemble those of Yusho victims both clinically and microscopically (Urabe and Koda 1976). These kinds of changes in skin appendages have never been reported for laboratory rodents, but in rabbits "follicular hyperkeratosis" occurs after topical application of TCDD or PCBs (Vos and Beems 1971).

Fig. 6. Section of a fingernail of an immature rhesus male given 3,4,3′,4′-tetrachlorobiphenyl for several months. The nail bed is hyperkeratotic and the nail is lost. × 20

Metabolism

Although there are pronounced species differences in the morphological expression of toxicity, the metabolism of PCBs is much the same in rats, mice, rabbits, cows, and monkeys. No information on the metabolism of TCDD or TCDF is available. The major metabolites of PCBs in all species are conjugated hydroxylated derivatives (Sundstrom et al. 1976); a variety of methoxylated and sulfur-containing metabolites have also been reported. In general, the rate of metabolism is inversely related to the number of chlorine substitutions on the phenyl rings. More specifically, the metabolism is rapid if two adjacent positions are unsubstituted. Thus, the highly chlorinated 2,3,6,2′,3′,6′-hexachlorobiphenyl, in which the adjacent 4 and 5 positions are unchlorinated in both rings, is rapidly excreted in rats. On the other hand, the 2,4,5,2′,4′,5′-isomer, in which no two adjacent positions are open, is very slowly excreted (Kato et al. 1980).

   This dependence on vicinal hydrogens is presumptive evidence that the formation of an arene oxide intermediate (Jerina and Daly 1974) is an important, and perhaps obligatory, step in the metabolism of PCBs. Direct evidence has been found for arene oxide formation during the metabolism of 2,5,2',5'-tetrachlorobiphenyl in rats, rabbits, and rhesus macaques (Hsu et al. 1975, Preston and Allen 1980). Arene oxides are reactive electrophiles that can become covalently bound to macromolecules, in particular DNA.

The toxicities of individual pure chlorobiphenyls differ strongly, and the differences depend more on the distribution of the chlorine atoms than on their number. The key to toxicity seems to be a lack of substitution in the *ortho* positions of the phenyl rings. The compounds 3,4,3',4'-tetrachlorobiphenyl and 3,4,5,3',4',5'-hexachlorobiphenyl have no chlorine atoms in the *ortho* positions (2 and 6) and are very toxic. Their symmetrical isomers 2,5,2',5'-tetrachlorobiphenyl and 2,4,5,2',4',5'-hexachlorobiphenyl, with *ortho* substitutions, have not been found to be toxic at all for rhesus macaques at the highest dietary levels tested (McNulty et al. 1981, and unpublished data).

Toxicity is thus dissociated from metabolism (and from accumulation in body tissues, primarily liver and fat). Both 3,4,3',4'- and 2,5,2',5'-tetrachlorobiphenyls have open vicinal positions and are rapidly metabolized; the former is toxic and the latter is not. On the other hand, 3,4,5,3',4',5'-hexachlorobiphenyl and 2,4,5,2',4',5'-hexachlorobiphenyl, neither of which has adjacent unsubstituted positions, are both slowly metabolized and strongly accumulated in adipose tissue. The former is very toxic, but an animal fed large dietary amounts of the latter had no lesions at autopsy, even though the level of hexachlorobiphenyl in the adipose tissue had reached 1,000 ppm (unpublished data).

The toxicity of commercial PCB mixtures is still unexplained. 3,4,3',4'-tetrachlorobiphenyl is a minor component of Aroclor 1242 (a few parts per thousand) (Albro and Parker 1979), and 3,4,5,3',4',5'-hexachlorobiphenyl is not detectable at all. Asymmetrically chlorinated biphenyls with one *ortho* substitution in one ring and none in the other have been found to be toxic for rats, and one member of this class, 2,3,4,5,3',4'-hexachlorobiphenyl, is a component of Yusho oil and has persisted in the tissues of Yusho victims (Yoshihara et al. 1979).

Enzyme Induction

Both PCBs and TCDD are potent inducers of hepatic microsomal cytochrome P-450-associated monooxygenases. These are enzymes that convert a broad range of lipid-soluble substrates into more polar compounds, which can be more readily excreted, often after conjugation. There are several classes of these enzyme systems, each associated with a particular cytochrome P-450 with specific spectral and immunologic properties. The non-*ortho*-substituted chlorobiphenyls, as well as TCDD, induce cytochrome P_1-450, or P-448, so named because the wavelength of the absorption maximum in the carbon monoxide-difference spectrum of this cytochrome is 2 nm shorter than that of the classic cytochrome P-450, which is induced by phenobarbital and by *ortho*-substituted chlorobiphenyls. The clusters of enzmymatic activities linked with cytochromes P-448 and P-450 are also different. For example, the activity of aryl hydrocarbon hydroxylase increases when cytochrome P-448 is induced, whereas that of aminopyrine-N-demethylase does not; the reverse is true for cytochrome P-450 induction.

Thus, both great toxicity and cytochrome-P-448 induction have been associated with chlorobiphenyls without *ortho* substitution, whereas the *ortho*-substituted congeners are relatively nontoxic and induce cytochrome P-450, if they induce any enzyme at all, in rhesus macaques (unpublished data) and in rats (Goldstein et al. 1977). The highly toxic TCDD is also a P-448 inducer in rodents (Poland and Glover 1974) and rhesus macaques (unpublished data), and the same is true for TCDF in rats (Goldstein et al. 1978). The molecules of TCDD and TCDF are constrained to a planar configuration, and chlorobiphenyls without the steric hindrance of *ortho* substitution can also assume a planar molecular form. 2,3,7,8-tetrachlorodibenzo-*p*-dioxin induces and binds with high affinity to a protein receptor in the cytoplasm of mouse liver cells, and the receptor complex is transported to the nucleus. 3,4,3′,4′-tetrachlorobiphenyl, TCDF, and 3,4,5,3′,4′,5′-hexachlorobiphenyl all compete for binding to this same receptor (Poland et al. 1979).

Not only do asymmetric chlorobiphenyls with an *ortho* substitution in only one ring exhibit toxicity, they also show mixed inductive properties, i.e., their pattern of induced enzyme activities has characteristics of both P-448 and P-450 (Yoshimura et al. 1979, Parkinson et al. 1980).

Thus, in all species thus far tested, among the PHPAHs there has been a strict association between the type of hepatic enzyme induction and toxicity, even though large species differences have been noted in the sensitivities to PHPAHs and the pathologic manifestations of toxicity. No biochemical action of PHPAHs, other than enzyme induction, is known. It is tempting to speculate that these species differences are expressions of quantitative differences among species in the degree of disorder introduced into metabolic pathways. Hepatic monooxygenases not only accept drugs and foreign compounds as substrates, they also metabolize steroid hormones, fatty acids, prostaglandins, and probably many as yet unrecognized endogenous substances. Doubtless many tissues besides liver will be shown to have enzyme-inducing responses to PHPAHs, once this possibility has been explored. Species differences in the inductive capacities of different tissues (e.g., skin and stomach) may account in part for species differences in pathologic lesions.

Fetotoxicity

In rodents, TCDD is fetotoxic and teratogenic; it increases the number of resorptions and causes cleft palate and renal malformations (Sparschu et al. 1971, Courtney and Moore 1971). The pace of development, the nature of placentation, and the maternal endocrine support are much different in rodents and primates, but the anatomy of the reproductive tract and placenta and the physiology of pregnancy are much the same in rhesus macaques and women.

We investigated the effects of small doses of TCDD given orally early in pregnancy, after implantation but before completion of organogenesis (Table 2). The chemical caused a high rate of fetal death, but the few fetuses that did not die in utero were anatomically normal, that is, a fetotoxic but not a teratogenic effect was seen.

Table 2. Fetal loss in rhesus macaques after oral doses of 2,3,7,8-tetrachlorodibenzo-p-dioxin during weeks 4 through 6 of pregnancy

Group	Total dose (μg/kg)	Schedule	Fetal losses	Gestational age of lost fetus (days)	Maternal toxicity
I	5	9 divided doses, 3 times a week	2/2	47, 50	2/2
	1	Same	3/4	50, 57, ?	1/4
	0.2	Same	1/4	?	0/4
II	1	1 dose, day 25	3/3	48, ?, ?	2/3
	1	1 dose, day 30	3/3	50, 51, 55	3/3
	1	1 dose, day 35	2/3	53, 108	1/3
	1	1 dose, day 40	2/3	~ 100, ~ 100	0/3
III	0	9 divided doses, 3 times a week	3/12	118, ?, ?	0/12

The cause of the fetal deaths was not determined. The TCDD was given after (earliest date 20 days after fertilization) the rhesus fetal-placental units had become independent of maternal hormonal support (Goodman and Hodgen 1979). It is thus unlikely that the fetuses were lost because of toxic effects on the maternal ovaries or uterus. Some, but not all, of the mothers did show toxicity from TCDD, and 2 of the 16 mothers given 1 μg/kg died. However, toxicity did not appear until 70 to 100 days after administration of the poison and most of the abortions occurred earlier, within 10 to 30 days, when the mothers were still clinically well.

Few of the dead fetuses were recovered, and these were so badly autolyzed that anatomical study was impossible; possibly the dead fetuses did have developmental anomalies. Most of the placentas, or parts of them, were recovered, but these showed no lesions other than autolysis.

Conclusions

The use of nonhuman primates in toxicological research has limitations. The determination of 50% effective doses and dose-response curves requires large numbers of uniform animals for statistical validity, but monkeys are expensive, relatively scarce, and outbred — and therefore likely to show a wide range of individual variation in response. Individual variation is not wholly a disadvantage, however; in environmental medicine, we are particularly interested in the level of exposure that can affect a minority of a highly variable human population. The chief justification for the use of primates as models for human beings in toxicological studies is evidence (1) that considerable interspecies variation in toxicological response exists and (2) that the response in primates resembles that in people.

Rhesus macaques have other practical advantages for research with PHPAHs. Their size makes it practical to perform repeated blood samplings and biopsies of skin, fat, and liver in longitudinal studies on the development of and recovery from toxicity, on enzyme induction, and on pharmacokinetics and metabolism. I have found that individual animals differ widely in susceptibility to poisoning and suspect, but have not yet shown, that such differences may reflect differences, possibly genetically determined, in metabolism or enzyme induction. There is murine evidence; inbred mouse strains show great differences in the affinity of the hepatic receptor for TCDD and the strength of enzyme induction, and mice with a low-affinity receptor are less susceptible to the toxic action of TCDD. Furthermore, the presence of the high-affinity receptor of TCDD to the receptor necessarily accompanies toxicity (Poland and Glover 1980).

Acknowledgments. Publication No. 1107 of the Oregon Regional Primate Research Center. Supported by Grants ES-01522, RR-00163, and RR-05694 of the National Institutes of Health.

References

Albro PW, Parker CE (1979) Comparison of the compositions of Aroclor 1242 and Aroclor 1016. J Chromatogr 169:161–166

Allen JR, Abrahamson LJ, Norback DH (1973) Biological effects of polychlorinated biphenyls and triphenyls on the subhuman primate. Environ Res 6:344–354

Becker GM, McNulty WP, Bell M (1979) Polychlorinated biphenyl-induced morphologic changes in the gastric mucosa of the rhesus monkey. Lab Invest 40:373–383

Bowes GW, Mulvihill MJ, Simoneit BRT, Burlingame AL, Risebrough RW (1975) Identification of chlorinated dibenzofurans in American polychlorinated biphenyls. Nature (London) 256:305–307

Crow KD (1970) Chloracne, a critical review, including a comparison of two series of acne from chloronaphthalene and pitch fumes. Trans St Johns Hosp Dermatol Soc 56:79–99

Courtney KD, Moore JA (1971) Teratology studies with 2,4,5-trichlorophenoxyacetic acid and 2,3,7,8-tetrachlorodibenzo-*p*-dioxin. Toxicol Appl Pharmacol 20:396–403

Goldstein JA, Hickman P, Bergman H, McKinney JD, Walker MP (1977) Separation of pure polychlorinated biphenyl isomers into two types of inducers on the basis of induction of cytochrome P-450 or P-448. Chem Biol Interact 17:69–87

Goldstein JA, Hass JR, Linko P, Harvan DJ (1978) 2,3,7,8-tetrachlorodibenzofuran in a commercially available 99% pure polychlorinated biphenyl isomer identified as the inducer of hepatic cytochrome P-448 and aryl hydrocarbon hydroxylase in the rat. Drug Metab Dispos 6:258–264

Goodman AL, Hodgen GD (1979) Corpus luteum-conceptus-follicle relationships during the fertile cycle in rhesus monkeys: Pregnancy maintenance despite early luteal removal. J Clin Endocrinol Metab 49:469–471

Hansen LG, Byerly CS, Metcalf RL, Bevill RF (1975) Effect of a polychlorinated biphenyl mixture on swine reproduction and tissue residues. Am J Vet Res 36:23–26

Hay A (1979) Accidents in trichlorophenol plants: A need for realistic surveys to ascertain risks to health. Ann NY Acad Sci 320:321–324

Hayabuchi H, Yoshimura T, Kuratsune M (1979) Consumption of toxic rice oil by 'Yusho' patients and its relation to the clinical response and latent period. Food Cosmet Toxicol 17:455–461

Herxheimer K (1899) Über Chlorakne. Muench Med Wochenschr 46:278

Hsu IC, Miller JP van, Seymour JL, Allen JR (1975) Urinary metabolites of 2,5,2′,5′-tetrachlorobiphenyl in the nonhuman primate. Proc Soc Exp Biol Med 150:185–188

Jerina DM, Daly JW (1974) Arene oxides: A new aspect of drug metabolism. Science 185:573–582

Kato S, McKinney JD, Matthews HB (1980) Metabolism of symmetrical hexachlorobiphenyl isomers in the rat. Toxicol Appl Pharmacol 53:389–398

Kimbrough R, Buckley J, Fishbein L, Flamm G, Kasza W, Marcus W, Shibko S, Teske R (1978) Animal toxicology. Environ Health Perspect 24:173–184

Kuratsune M (1976) Epidemiologic studies on Yusho. In: Higuchi W (ed) PCB poisoning and pollution. Academic Press, London New York, pp 9–23

McConnell EE, Moore JA (1979) Toxicopathology characteristics of the halogenated aromatics. Ann NY Acad Sci 320:138–150

McNulty WP, Becker GM, Cory HT (1980) Chronic toxicity of 3,4,3′,4′- and 2,5,2′,5′-tetrachlorobiphenyls in rhesus macaques. Toxicol Appl Pharmacol 56:182–190

Miyata H, Kashimoto T, Kunita N (1977) Detection and determination of polychlorodibenzofurans in normal human tissues and Kanemi rice oils caused "Kanemi Yusho". J Food Hyg Soc Jpn 18:260–265

Nagayama J, Masuda Y, Kuratsune M (1977) Determination of polychlorinated dibenzofurans in tissues of patients with 'Yusho'. Food Cosmet Toxicol 15:195–198

Parkinson A, Cockerline R, Safe S (1980) Polychlorinated biphenyl isomers and congeners as inducers of both 2-methylcholanthrene- and phenobartione-type microsomal activity. Chem Biol Interact 29:277–289

Pocchiari F, Silano V, Zampieri A (1979) Human health effects from accidental release of tetrachlorodibenzo-p-dioxin (TCDD) at Seveso, Italy. Ann NY Acad Sci 320:311–320

Poland A, Glover E (1974) Comparison of 2,3,7,8-tetrachlorodibenzo-p-dioxin, a potent inducer of aryl hydrocarbon hydroxylase, with 3-methylcholanthrene. Mol Pharmacol 10:349–359

Poland A, Glover E (1980) 2,3,7,8-Tetrachlorodibenzo-p-dioxin: Segregation of toxicity with the Ah locus. Mol Pharmacol 17:86–94

Poland A, Greenlee WF, Kende AS (1979) Studies on the mechanism of action of the chlorinated dibenzo-p-dioxins and related compounds. Ann NY Acad Sci 321:214–230

Preston BD, Allen JR (1980) 2,2′,5,5′-Tetrachlorobiphenyl: Isolation and identification of metabolites generated by rat liver microsomes. Drug Metab Dispos 8:197–204

Shelly WB, Kligman AM (1957) Experimental production of acne by penta and hexachloronaphthalenes. Arch Dermatol 75:689–695

Sparschu GL, Dunn FL, Rowe VK (1971) Study of the teratogenicity of 2,3,7,8-tetrachlorodibenzo-p-dioxin in the rat. Food Cosmet Toxicol 9:405–412

Sundstrom G, Hutzinger O, Safe S (1976) The metabolism of chlorobiphenyls – a review. Chemosphere 5:267–298

Taylor JS (1979) Environmental chloracne: Update and overview. Ann NY Acad Sci 320:295–307

Urabe H, Koda H (1976) The dermal symptomatology of Yusho. In: Higuchi K (ed) PCB poisoning and pollution. Academic Press, London New York, pp 105–123

Urabe H, Koda H, Asahi M (1979) Present state of the Yusho patients. Ann NY Acad Sci 320: 273–276

Vos JG, Beems RB (1971) Dermal toxicity studies of technical polychlorinated biphenys and fractions thereof in rabbits. Toxicol Appl Pharmacol 19:617–633

Yakushiji T, Watanabe I, Kuwabara K, Yoshida S, Koyama K, Hara I, Kunita N (1978) Long-term studies of the excretion of polychlorinated biphenyls (PCBs) through the mother's milk of an occupationally exposed worker. Arch Environ Contam Toxicol 7:493–504

Yoshihara S, Kawano K, Yoshimura H, Kuroki H, Masuda Y (1979) Toxicological assessment of highly chlorinated biphenyl congeners retained in the Yusho patients. Chemosphere 8:531–538

Yoshimura H, Yoshihara S, Ozawa N, Miki M (1979) Possible correlation between induction modes of hepatic enzymes by PCBs and their toxicity in rats. Ann NY Acad Sci 320:179–192

The Role of a Kenyan Primate Center in Conservation

J.G. ELSE [1]

Introduction

As with all countries with indigenous nonhuman primate populations there is a need in Kenya for a defined comprehensive programme in primate conservation, field management and utilization. Such an endeavour must be extremely broad and integrate all aspects, from conservation of an endangered species to commercial trapping/export and domestic production.

Implementation of such a programme does not require direct control of all activities. What is required is a concerted effort by those involved to define a realistic approach and then design, implement and support relevant studies. This will lead to the establishment of sound policies which meet the needs of Kenya as well as other countries. The Institute of Primate Research (IPR) field research programme is designed along this line. It is hoped that this will provide the nucleus for the establishment of a broad government primate scheme as part of an overall wildlife programme for Kenya.

Such a research endeavour, by design, necessitates close collaboration with various agencies of the Kenya Government as well as universities, international organizations, research institutions, and individuals. In addition, utilization of the existing government infrastructure combined with an educational programme for Kenyans and the participation of the large number of primatology research scientists coming to Kenya must be included.

National Primate Census Programme

A mandate, before any sensible programme can be implemented, is accurate knowledge of existing primate populations and trends. This information is not presently known for Kenya. A primate census programme is a major endeavour in which the primate species are systematically recorded and numbers estimated throughout Kenya. This must be continuous study to take into account long-term population trends.

1 Institute or Primate Research, National Museum of Kenya, P.O. Box 114, Limuru, Kenya

Such census studies are notoriously complex and expensive. Until proper support is obtained the IPR is pursuing other approaches which will later be integrated into the overall census.

Mail Questionnaire Census Survey: Regional personnel of the Ministry of Natural Resources and Environment (game wardens and rangers) and Ministry of Education (science teachers) will be requested to fill out a questionnaire concerning primate species and numbers in their area. This questionnaire will be suppplemented with basic descriptions and colour photographs of each species.

Field Primatology Research Scientists: Researchers coming to Kenya to conduct primate field studies will be required to provide basic ecological and census information on primate species in their study area. There are about 20 such researchers coming to Kenya each year.

Primate Pest Problems

As with some other wildlife species, there are severe conflicts between certain primate populations and the people of Kenya. Although primarily an agricultural problem, forest reserves, tourist hotels, picnic sites, and suburban areas are also affected. There have been sporadic attempts to alleviate these problems, some probably marginally successful. However, there has been no long-term evaluation on such attempts. The IPR begun several projects in this area, the approach to each project has been the same, i.e., define and document the problem, design a research approach and undertake pilot studies.

Primate Utilization and Production

The continued and critical need for primates in biomedical research and pharmacological testing comes at a time of drastic reductions in feral primate populations throughout the world. The latter has been brought about by a variety of reasons, the major one being human development of natural primate habitats.

As mentioned, there are severe conflicts between people and wildlife in many areas of Kenya. This is especially true in agricultural areas, a prime example being devastation of maize (corn) fields by baboons. Common primate species which are causing such problems in land designated for agricultural use should be removed and utilized (locally or exported). The use of primates for biomedical research should not be considered in conflict with conservation, but rather a source of revenue to support it. The rarer species (e.g., DeBrazza monkey), also causing agricultural destruction, require special consideration.

Commercial Trapping and Export: The trapping and export of primates is a viable and justified source of revenue that could be used to partially support and sustain an overall conservation programme. However, supply of feral primates is only a temporary solution for biomedical needs. As rural development in Kenya continues and primates are utilized, there will be a point where the practice must be discontinued. Hopefully, by this time, alternate methods of primate supply, such as domestic production, will have been developed. An annual quota for export should be established for each species. Actual trapping must be closely monitored and restricted to areas where the need for removal has been verified. At the same time the effectiveness of procedures in alleviating the pressure in problem areas must be evaluated.

Minimum standards of care for primates held in the field after trapping, at centralized animal holding areas and in transit, have been established by other countries and international agencies. These need to be slightly modified for Kenya and implemented. Support for such a programme could be obtained by imposing a government levy on each animal exported from Kenya as well as trapping fees.

Domestic Production: The only long-term solution for a continued and reliable supply of primates is mass domestic production. Domestic production of macaques has been very successful in the United States but is still quite limited in scope. The unavailability of macaques has led to a search for alternate primate models. African species such as the vervet monkey, baboon, and bushbaby are rapidly increasing in popularity.

The IPR has been involved in limited domestic breeding of vervets, sykes, baboons and galagoes for the past 4 years. The basic husbandry techniques and expertise required for the establishment of large-scale domestic production have been established for each of these species. Proposals to establish such a programme are being prepared.

Field Cropping: The feasibility of setting aside large areas naturally inhabited by primates and establishing a cropping scheme for primate supply has been explored in other countries. This should also be considered for Kenya. Advantages include a low overhead leading to a substantial profit yield for the investment and an excellent opportunity for applied field studies. Disadvantages that must be considered include the need to develop practical methods for trapping (most primates quickly become trap-shy) and the scantier information (e.g., ages, medical history) available compared with domestic production.

Permanent Primate Reserves and Study Sites

Information obtained from those undertaking studies at primate study sites and reserves should be made available to Kenya. Projects most directly applicable to the needs of the country should be given priority. Applicants to conduct research at these sites will be required to include the relevancy of their study in the overall goals of wildlife research in Kenya. Primate field studies can be divided into two broad categories, each with specialized requirements.

Non-Interventive Studies: (Requiring no manipulations). These may be undertaken in National Parks. The number of researchers must be limited to what each park can support. In cases where number restrictions are imposed, those participating in long-term projects and those undertaking studies most relevant to Kenya are given priority.

Interventive Studies: (Trapping, marking, etc.). Interventive studies are becoming increasingly important in areas such as sociobiology, genetics, reproduction, and infectious diseases. Permanent study sites with laboratory facilities need to be made available as such studies cannot be undertaken in National Parks.

Endangered and Threatened Species

Special efforts must be made to protect those primate species which are endangered or rare in Kenya. Several approaches can be taken.

Reference Collection: A reference collection of information and literature on each species needs to be prepared. This will be made available to those wishing to undertake studies on these species. Likewise, basic information plus progress and final reports will be required for each new study.

Primate Reserves: Areas containing rare primate species should be set aside whenever possible. An example of this is the Tana River Primate Reserve where the Tana River Mangabey and Kirk's Colobus are found. However, means of long-term monitoring of primate populations in such reserves still needs to be established.

Further Declines in Rhesus Populations of India

C.H. SOUTHWICK [1], M.F. SIDDIQI [2], J.A. COHEN [3], J.R. OPPENHEIMER [4], J. KHAN [5], and S.W. ASHRAF [5]

Introduction

Previous studies of rhesus monkey populations in northern India throughout the 1960s showed considerable declines in these populations due to a variety of ecological, economic, and social factors (Southwick and Siddiqi 1966, 1968, Southwick et al. 1961a,b, 1965, 1970). Since 1970 there have been indications in Aligarh District of western Uttar Pradesh that this decline was leveling, and some improvement in rhesus population status was occurring (Southwick and Siddiqi 1977, Southwick et al. 1980). From 1962 to 1970, the Aligarh rhesus population declined 60% (from 403 to 163 individuals), but from 1970 to 1978 this population recovered part of this loss with a 44.8% increase (from 163 to 236). Seth and Seth (1980) have also observed population increases in a number of rhesus groups in western and northern U.P. and eastern Rajasthan since 1975.

The purpose of the present research has been to survey a wider area of northern India to determine the more general status of rhesus populations in agricultural areas, and to see if rhesus populations throughout a broader region of northern India were also improving. In other words, the purpose of the present field work was to determine if changes in the Aligarh population were representative of rhesus populations in other parts of northern India.

Methods

The present study consisted in repeating population surveys along roadside, canal bank, and village and town habitats that were first surveyed in 1959 and 1960. The rationale and justification for the use of roadsides and canal banks as survey routes,

1 Department of EPO Biology, University of Colorado, Boulder, Campus Box B. 334, CO 80309, USA
2 Department of Geography, Aligarh University
3 Department of Zoology, University of Florida
4 College of Staten Island, 50 Bay Street, Staten Island, NY 10301, USA
5 Department of Zoology, Aligarh University

and villages and towns as survey units in agricultural areas have been presented in the publications cited above. In brief, the environments immediately surrounding roadsides, canals, villages, and towns provide the best primate habitat in agricultural areas of northern India where so much of the total environment is open space without trees or water. In the first few years of our field studies, we found that the roadside, canal bank, village, and town habitats contained an estimated 85% of the total rhesus population of nothern India.

The localities, field methods, seasons, times of day, and field personnel involved in the population surveys were all replicated as closely as possible to provide an accurate comparison over a 20 year span. All surveys were done with systematic and consistent field methods to minimize sampling error. These methods have been described in previous publications (Southwick et al. 1961a, 1965).

Results

Roadside surveys throughout Uttar Pradesh and adjacent portions of neighboring provinces (Rajasthan, Haryana, Punjab, and Madhya Pradesh), showed a striking decline in the abundance of roadside rhesus groups (Table 1). In a survey route of 7018 km, the numbers of rhesus groups declined from 399 in 1959—60 to 296 in 1964—65 to only 94 in 1978—79. This represents a total decline in the numbers of rhesus groups from an average of 5.68 groups/100 km to 1.27 groups/100 km, a decline of 77.6%.

Table 1. Roadside surveys of rhesus monkeys in India, 1959—1979

Years of survey	Total distance	No. of rhesus groups seen	Groups per 100 km	Percent change 1959—60
1959—60	7018 km	399	5.68	
1964—65	7018 km	296	4.22	− 25.8%
1978—79	7408 km	94	1.27	− 77.6%

There were regional differences in the extent of these declines. The least regional decline (60.8%) occurred in western U.P. (including part of Rajasthan), whereas the greatest regional declines (91.6% to 92.2%) occurred in central, eastern and southern U.P. (Table 2).

Table 2. Change in roadside rhesus populations

Region	Number of groups/100 km		% change
	1959—60	1978—79	
Western U.P.	4.47	2.05	− 60.8%
Central U.P.	9.12	0.71	− 92.2%
Eastern and southern U.P.	7.15	0.60	− 91.6%
Northern U.P.	6.54	0.76	− 88.4%
	5.68	1.27	− 77.6%

Particularly astounding losses of rhesus groups in roadside habitats occurred in certain specific localities. For example, in northern and central U.P. along the Nepal border and including the districts of Pilibhit, Tanakpur, Bilaspur, Bareilly, Shajahanpur, Sitapur and Lucknow, there was a 100% loss of roadside rhesus in the survey routes (Table 3). Whereas 58 rhesus groups had been found in roadside habitats in these districts in 1959—60, none was found in 1978—79. Even in one of the areas of northern India most sacred to Hindus, the area around Ajodyha and Faizabad, where rhesus had been formerly very abundant and 43 rhesus groups were known to have existed in 1959—60, now only 3 groups could be found, a decline of 93% (Table 3).

Table 3. Some specific locality comparisons of roadside rhesus groups in Uttar Pradesh, 1959—60 to 1978—79

General	Specific locality	Observed rhesus groups 1959—60	Groups observed 1977—78	Percent change
Northern U.P. (along Nepal border)	Pilibhit Tanakpur Bilaspur	33	0	− 100%
Central U.P.	Bareilly Shahjahanpur Sitapur Lucknow	25	0	− 100%
Central to eastern U.P.	Lucknow Faizabad Ajodhya Gorakpur	45	3	− 93%
Eastern and southern U.P.	Gorakpur Azamgarh Benares Allahabad	28	3	− 89%
Eastern to central U.P.	Jaunpur Rae Bareili Lucknow	17	3	− 71%
Western U.P.	Aligarh Bareilly Moradabad Agra Delhi	44	14	− 68%

Not only have the numbers of rhesus groups declined, but in three regions out of four, the average sizes of rhesus groups also declined. In central, eastern, and southern U.P., average group sizes declined from 14.2 to 12.4 respectively to 10.4 and 9.2 respectively, declines in both cases of approximately 25% (Table 4). In northern U.P., a lesser and nonsignificant decline of 7% occurred, and in western U.P. average group size actually increased from 16.2 to 20.6, an increase of 27%. In other words, although there were fewer groups in western U.P., the average size of the remaining groups was larger. This phenomenon had previously been observed in our studies of village and town groups in the 1960s (Southwick and Siddiqi 1966, 1968).

Table 4. Changes in roadside rhesus – group sizes

Region	Average group sizes (No. of groups in sample)		% Change in group size
	1959–60	1978–79	
Western U.P.	16.2 ± 1.3 (65)	20.6 ± 2.9 (17)	+ 27.2%
Central U.P.	14.2 ± 1.1 (63)	10.4 ± 2.5 (8)	– 26.8%
Eastern and southern U.P.	12.4 ± 0.7 (62)	9.2 ± 1.7 (4)	– 25.8
Northern U.P.	19.3 ± 1.9 (36)	17.0 ± 3.8 (11)	– 7.2
Totals	15.1 ± 0.6	16.7 ± 1.8	+ 10.6

Total population/100 km
1959–60   15.1 × 5.68 = 85.77 monkeys/100 km
1978–79   16.7 × 1.27 = 21.21 monkeys/100 km

Difference   64.56
% Decline   75.3

Combining all data on the numbers of rhesus groups and average group sizes, the total rhesus population in roadside habitats on northern India declined 75.3% from 1959 to 1979 (Table 4).

Canal banks afford virtually ideal habitat for rhesus monkeys, with an abundance of trees, water, surrounding agricultural fields, and absence of vehicular traffic. Canal bank surveys showed comparable declines, however, in rhesus populations. A survey of 322 km on canal banks of western and central U.P. showed a decline from 11.5 groups/100 km to 2.8 groups/100 km, a decline of 75.6%. Average group size declined from 19.0 to 15.8, creating a total population decline of 79.8% (Table 5).

Table 5. Canal bank survey, 1959–1979

	1959–60	1978–79
Kilometers surveyed	322	322
Groups seen	37	9
Groups seen/100 km	11.5	2.8
Average group size	19.0 ± 2.0	15.8 ± 3.7
% Change, No. of groups		– 75.6%
% Change, aver. group size		– 16.8%
% Change in total population		– 79.8%

In population surveys throughout 1959–1960, census data showed that villages and towns contained, at that time, the majority of rhesus in India, estimated to be 63% of the total population. There were thought to be several reasons for this, primarily: (1) rhesus are highly commensal primates, and seem to be attracted to human settlements and activities; (2) villages and towns afford all the essentials of rhesus habitat, including trees, other forms of shelter, water, and food; and (3) in the agricultural areas of northern India, much of the inter-village space, except for roadsides, canal banks and mango groves (from which rhesus are usually expelled) is open space, fields and pastures, with few or no trees. Villages and town actually contain clusters of trees and suitable vegetation for rhesus. This is especially evident from the air, though photographs to show this cannot be taken from aircraft in India.

In 1959—60, 280 villages and 30 towns were surveyed in four different regions: 108 villages and 16 towns in Aligarh district (western U.P.), 53 villages in Dehra Dun district (norhtern U.P.), 69 villages and 6 towns in Banda district (southern U.P.), and 50 villages and 9 towns in Banda district. Two of these districts, Aligarh and Dehra Dun, have been resurveyed in 1979—80, with the results shown in Table 6.

Table 6. Rhesus surveys in villages and towns, northern India, 1959—60 and 1979—80

Region		Number of villages (V) or towns (T) surveyed	Number and percentage with resident rhesus groups	
			1959—60	1979—80
Aligarh (western Uttar Pradesh)	V	108	12 (11%)	0
Dehra Dun (northern Uttar Pradesh)	V	53	6 (11%)	0
Aligarh (western Uttar Pradesh)	T	16	11 (69%)	8 (50%)

In 1959—60, 11% of the villages in Aligarh and Dehra Dun had resident rhesus groups; in 1979—80, none of these villages had resident rhesus. We found that rhesus had been completely trapped or driven away from villages, usually at the insistence of the villagers who no longer wanted rhesus monkeys in or around their village. Social and religious tolerance no longer protected rhesus, which were considered to be too much of a nuisance and agricultural problem.

In towns, rhesus still occurred in 1979. In Aligarh district in 1959—60, 11 of 16 towns had resident rhesus, and in 1979—80, 8 of these towns still had rhesus. In towns where there is a less direct and immediate link between residents and the ownership of agricultural land, rhesus continue to exist much the same as stray dogs. Towns also often have temples, parks or public areas, where rhesus are sometimes permitted to exist. Town-dwelling rhesus are not particularly healthy or desirable animals, however, and they have high prevalences of tuberculosis, amoebic and bacillary dysentery, and viral and helminthic infections in common with people.

In addition to the extensive population surveys reported above, we have maintained more frequent census studies of a smaller population of rhesus monkeys in Aligarh district. Starting with a population of 17 groups of rhesus in October, 1959, we have censused these groups every 4 months for 20 years, obtaining accurate records of group changes, birth rates, infant mortality rates, and disappearance rates for juveniles and adults (Southwick and Siddiqi 1977, Southwick et al. 1980). This population, consisting of typical groups in a variety of habitats, is the one previously referred to which declined throughout the 1960s, but partially recovered and showed a 44.8% increase from 1970 to 1978 (Fig. 1).

We hoped that the export ban on rhesus monkeys, enacted in April 1978, would result in further improvement of the population, since it would theoretically reduce trapping. Rhesus trapping is still permitted in India to obtain monkeys for research within India, but the greatest percentage of rhesus were trapped for export.

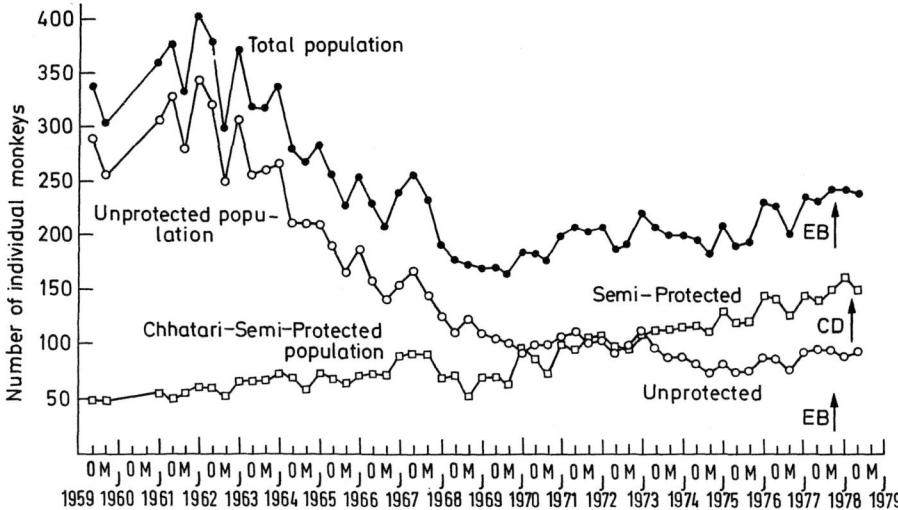

Fig. 1. Population changes of rhesus monkeys in Aligarh district, northern India. *O*, October; *M*, March; *J*, July; *EB*, Export ban

Continuing census data on the Aligarh district population have shown, however, that the rhesus population has declined slightly in the past 2 years. In 2 years prior to the ban, this population increased 22.3%; in 2 years since the ban, from April 1978 to April 1980, it declined 3% (Table 7).

Table 7. Status and changes in Aligarh district rhesus population before and after export ban of April 1978

	Census dates					
	Mar. 1970	Mar. 1976	Mar. 1977	Mar. 1978	Mar. 1979	Mar. 1980
Total population sizes	163	193	202	236	231	229
Percent population changes	⊢————————————→			+ 44.8%		
		⊢————————→		+ 22.3%		
			⊢——→	+ 16.8%		
				⊢——→	− 2.1%	
				⊢————————————→		− 3.0%

This has been a surprising result, and we do not have a fully adequate explanation for it. It may be a peculiarity of the Aligarh population, and not representative of rhesus throughout northern India. A period of 2 years may also be too short a time for beneficial effects to occur, and we might hope for improvements in future years. Other explanations might also exist and several of these will be considered in the following discussion.

Discussion

These results have shown a very serious decline of rhesus populations in agricultural habitats of northern India over the past 20 years. This decline has averaged 75% to 80% in roadside and canal bank habitats, even greater in some localities. In 161 villages surveyed in western and northern U.P., no resident rhesus were found — a 100% loss of rhesus groups in and around village habitats. In many areas, rhesus have disappeared as a common feature of the north Indian agricultural environment. The greatest losses have occurred in eastern and central U.P., some of the best and most modernized agricultural areas of northern India.

We hasten to add that these survey results deal only with agricultural environments and not forests. Hopefully, a better picture may emerge from forest studies now being conducted by the Zoological Survey of India. It must be recognized, however, that mixed deciduous forests in India, among the prime habitats of rhesus, have been subject to extensive deforestation. Some reforestation has occurred, but this has been primarily with eucalyptus which is unsuitable for rhesus habitat. Even in 1960, there were only 15,877 square miles of forest in Uttar Pradesh, and forest surveys at that time showed only 0.7 rhesus groups/square mile. Hence, in 1960, forest habitats were estimated to contain only 100,000 rhesus in U.P. If we rely totally on forest habitats for conserving the rhesus population, we are dealing with a relatively small area and one that is shrinking.

Several factors are undoubtedly responsible for the decline of the rhesus population in north India. Excessive trapping of animals for biomedical research and pharmaceutical production throughout the 1950s and 1960s was certainly a factor. It is unlikely, however, that this has been the main factor in the past 6 to 8 years. If this were true in recent years, we would have expected continued decline of the Aligarh rhesus population, and greater depletion of the rhesus in western U.P. It is more likely that the main factors causing loss of the rhesus population have been changing attitudes of the people of India, changing habitats, increasing population pressures, and intensification of agriculture.

Throughout our field studies, we have seen a change in social attitudes among villagers of India from tolerance for rhesus monkeys to intolerance. Villagers and farmers throughout U.P. tell us they no longer want rhesus around. They consider them too much of a pest, and they ask us to get rid of them. The pressure for food production is increasingly severe in India, where human population growth continues with a net increase of more than 1 million people per month.

In other words, the religious aura of the monkey in traditional Hinduism no longer provides sufficient protection for rhesus in the climate of economic and social pressures which India faces. Langurs, the true image of Hanuman, are more protected than rhesus, but even langurs are under agricultural pressure and have declined substantially in some areas (Sugiyama and Parthasarathy 1978).

The rhesus situation is somewhat analogous to that of the gibbon in China in the fourteenth and fifteenth centuries. The gibbon was traditionally revered in China, and extensively represented in literature, art, religious mythology, drama, and music, much as the rhesus formerly was in India. Nonetheless, the gibbon became extinct in most of China by the seventeenth century, as a result of loss of habitat, changes in social attitudes, and population pressures (van Gulik 1967). A comparable process can occur with the rhesus in India, with or without protection from trapping for export. Such a process has occurred, in fact, in much of West Bengal and Bangladesh, where rhesus are almost completely extirpated from the agricultural regions of the Gangetic Delta. Rhesus remain in the mangrove islands of the Sundarbans and in the forests or tea plantation areas of the hills, but between Calcutta and Dacca and the Himalayan foothills of Darjeeling District, rhesus monkeys are rare or nonexistent in agricultural areas (Green 1978, Oppenheimer et al. 1980). West Bengal and Bangladesh are characterized by very high human population densities.

Regarding the changes in the Aligarh District population, most of the previous population increase occurred in two groups at Chhatari-do-Raha, a rural agricultural setting 22 km north of Aligarh. These groups had increased from 50 monkeys in the early 1960s to 80 by 1970, to 149 by 1978. Local villagers around Chhatari-do-Raha were willing to tolerate 80 and even 100 monkeys, but when the monkey population reached 120, we heard frequent complaints and entreaties to have the monkeys taken away. The monkeys were, in fact, driven from the core area of their home range, a school yard and grove of trees, and forced into less desirable habitat along a sparsely vegetated roadside. This provided an example of social and attitudinal changes in local people; their previous tolerance and support of the rhesus shifted to intolerance when the local monkey population became too large.

Although people of India rarely kill monkeys, they frequently harass them in an effort to break up groups and drive them away. This is often tantamount to assigning the death warrant to the monkeys. Rhesus in agricultural areas show a remarkably strong home range attachment so long as they are permitted to exist in one area. For 21 years, we have been able to return to the same exact locations, Chhatari, for example, and find the same group of rhesus, often in the same trees. If such groups are deprived of their traditional home ranges, they split up and become itinerant. They represent a fugitive population. In such a state, they have difficulty in finding food and establishing stable social patterns necessary for survival and reproduction. They come into frequent conflict with people and dogs, and are kept "on the run". Usually they disappear and cannot be found again; their mortality rate is high and their reproduction virtually nil.

Since agricultural environments represent the greatest portion of the north Indian landscape, and since they formerly contained over 80% of the rhesus population of north India, we feel a positive conservation and management program is necessary. It is not the purpose of this paper to discuss this issue fully, but some of the principles that should be considered are the following: (1) an active program of relocation and reestablishment of rhesus should be undertaken in areas where favorable habitat exists and where local people can be persuaded to allow the monkeys to live, (2) the latter will require a positive incentive program, such as payments to villagers to subsidize crop losses and provide food for the monkeys, and (3) positive management programs in which a conservative number of juveniles are removed each year from expanding groups in order to prevent excessive population growth of monkeys, limit crop losses, and provide direct payments to villagers for animals. In other words, rhesus can be treated as a natural and scientific resource, of direct economic value to villagers, provided villagers do not have to bear the entire burden of agricultural losses. This idea was presented by Bermant and Chandrasekar (1971) almost 10 years ago, and it still represents a scientifically sound idea in rhesus conservation.

Recommendations for programs of strictly regulated harvesting are also based on the well-established facts that rhesus populations can increase at the rate of 15% to 16% per year if provided favorable habitat, but such rates of increase are incompatible with human needs in India for agricultural production. Furthermore, recent studies on rhesus in India and Nepal have shown that the removal of a percentage of juveniles each year has the effect of increasing reproductive rates of adults, improving the health and condition of all monkeys, reducing aggressive behavior within groups, and reducing both infant and adult mortality. The principle of scientifically regulated harvest is well established in wildlife conservation for a variety of species with relatively high reproductive rates, such as ring-necked pheasants and white-tailed deer. Although many primate species do not have the requisite behavioral and ecological characteristics to permit regulated harvest, rhesus monkeys clearly do. From a scientific standpoint, the best conservation program for rhesus in India would combine the following: (1) Providing some sanctuary areas in which monkeys are totally protected from all trapping and harassment, (2) designating other areas in which well-regulated cropping and management are practiced, and (3) reestablishing rhesus in suitable habitats from which they have been depleted. A balanced program involving several approaches affords the best prospects for primate conservation, and the best assurances that human and nonhuman primate populations can continue to coexist in India.

Summary

Although small populations of rhesus monkeys in northern India have shown some increase in a few localities over the past 5 to 10 years, in most of northern India, rhesus populations have declined seriously over the past 20 years. Declines in roadside, canal bank, and village habitats averaged 75% to 80% between 1959 and 1979, and in specific parts of central, northern, and eastern U.P., declines have been 90% to 100%. A survey of 161 villages in 1979 has shown that all resident rhesus in these villages have disappeared. Rhesus monkeys are rapidly disappearing as a common feature of the north Indian landscape.

The major reasons for this decline have been loss of the traditional protection of rhesus afforded by the people of India, alterations and destruction of habitats, increasing human population pressure, intensification of agriculture, and excessive trapping of rhesus for export in 1950s and 1960s.

Hopefully, the ban on export of rhesus may help forest populations, but there is little or no evidence as of April 1980 that it will provide the necessary help for rhesus in agricultural environments.

A conservation program for rhesus is urgently needed, and should consist of several aspects: (1) complete protection of rhesus in certain reserve areas, (2) reestablishment of rhesus in suitable habitats where people will permit them to live, (3) subsidies and reimbursements to villagers for food and crop losses, and (4) a scientifically managed harvest of juvenile rhesus in designated areas utilizing the best techniques of modern wildlife conservation. An essential need is the recognition of the rhesus as an important natural resource worthy of a positive conservation and wildlife management program.

Acknowledgments. These field studies were begun through the cooperation of Dr. M. Babar Mirza, former chairman of the Department of Zoology, Dr. M. Rafiq Siddiqi, and Dr. M.A. Beg, of Aligarh Muslim University, and Dr. Olive Reddick of the U.S. Educational Foundation in India, when the senior author held a Fulbright Postdoctoral Fellowship at Aligarh. We have enjoyed the continued support and encouragement of many colleagues in India, including Drs. S.M. Alam, current Chairman of Zoology at Aligarh University, Drs. K.K. Tiwari, R.P. Mukherjee, T.N. Ananthakrishnan, and M.L. Roonwal of the Zoological Survey of India, And Dr. R.K. Lahiri, Chief Wildlife Officer of West Bengal. Drs. D.G. Lindburg, M. Neville, P. Dolhinow, M. Bertrand, C. Louch, B.C. Pal, and M.Y. Farooqui have all accompanied us on various field trips and we have appreciated their assistance.

Drs. C. Frey, R. Audy, K. Meyer, F.B. Bang, C. Wallace and T. Simpson have all provided essential administrative support. Since 1961, the field work has been supported by U.S.P.H.S. Grants No. RO 7-AI-10048, MH-18440, and RR-00910 to Johns Hopkins University, and RR-01245 to the University of Colorado.

References

Bermant G, Chandrasekar S (1971) Rescue plan for Indian monkeys. Science 171:628–629

Green KM (1978) Primates of Bangladesh: A preliminary survey of population and habitat. Biol Conserv 13:141–160

Gulik RH van (1967) The gibbon in China. E.J. Brill, Leiden, Holland, 123 pp

Oppenheimer JR, Akonda AW, Husain KZ (1980) Rhesus monkeys: effect of habitat structure, human contact and religious beliefs on population size. Mimeo Rep, Dacca, 15 pp

Seth PK, Seth S (1980) Population dynamics of the free ranging rhesus monkeys in north and northwest India. Antropol Contemp 3 (2):268

Southwick CH, Siddiqi MR (1966) Population changes of rhesus monkeys in India, 1960–1965. Primates 7:303–314

Southwick CH, Siddiqi MR (1968) Population trends of rhesus monkeys in villages and towns of northern India, 1959–1965. J Anim Ecol 37:199–204

Southwick CH, Siddiqi MF (1976) Demographic characteristics of semi-protected rhesus groups in India. Yearb Phys Anthropol 20:242–252

Southwick CH, Siddiqi MF (1977) Population dynamics of rhesus monkeys in northern India. In: HSH Prince Rainier, Bourne G (eds) Primate conservation. Academic Press, London New York, pp 339–362

Southwick CH, Beg MA, Siddiqi MR (1961a) A population survey of rhesus monkeys in villages, towns and temples of northern India. Ecology 42:538–547

Southwick CH, Beg MA, Siddiqi MR (1961b) A population survey of rhesus monkeys in northern India: II. Transportation routes and forest areas. Ecology 42:698–710

Southwick CH, Beg MA, Siddiqi MR (1965) Rhesus monkeys in north India. In: DeVore I (ed) Primate behavior: Field studies of monkeys and apes. Holt Rinehart and Winston, New York, pp 111–159

Southwick CH, Siddiqi MR, Siddiqi MF (1970) Primate populations and biomedical research. Science 170:1051–1054

Southwick CH, Richie T, Taylor H, Teas HJ, Siddiqi MF (1980) Rhesus monkey populations in India and Nepal: Patterns of growth, decline, and natural regulation. In: Cohen MN, Malpass RS, Klein HG (eds) Biosocial mechanisms of population regulation. Yale Univ Press, New Haven London, pp 151–170

Sugiyama Y, Parthasarathy MD (1978) Population change of the Hanuman langur *(Presbytis entellus)*, in Dharwar area, India. J Bombay Nat Hist Soc 75:860–867

Taiwan Macaques: Ecology and Conservation Needs

F.E. POIRIER [1]

Introduction

Macaca cyclopis is relatively unknown and this is the first attempted population survey. In June—December 1978 I spent 31 days in various mountain regions surveying the monkey population. Prior to this attempt, it was widely considered on Taiwan that the monkeys were not observable and that most had probably been killed. The following were the certainties: (1) monkey once existed on Taiwan in great numbers, (2) if they still existed they would be difficult to find, (3) habitat destruction had a major impact on all animal populations in forested areas, and (4) there would be little official encouragement of this study.

Taiwan is rapidly industrializing and there is exploitation of much of the remaining stand of virgin timber. Deforestation has had a major impact upon the wildlife (see McCullough 1974). The situation on Taiwan meets the concerns expressed in the IPS's 1979 Report on Conservation which states that development plans often conflict with conservation priorities, and increasing human population growth poses immense problems for conservation. The IPS report also notes that habitat destruction through development schemes is most likely to affect first politically stable countries. This fits the Taiwan situation.

One brief report, published in Japanese in 1940 (Kuroda 1940), listed locales where monkeys were observed. Every locale in the 1940 Japanese study, then reported to have had monkey populations, now supports villages or cities. Japanese sightings helped demonstrate the impact of economic development and human population pressures.

Monkey Habitats

Monkey populations are found within the following forest types.

Bamboo. Most Bamboo forests are cultivated and spreading throughout the island at the expense of natural vegetation. Bamboo forests occur at low to medium altitudes and contain scattered monkey populations. Destruction of natural forests to be

1 Department of Anthropology, The Ohio State University, 208 Lord Hall, 124 West 17th Avenue, Columbus, OH 43210, USA

replanted in bamboo deprives the monkey population of its natural foods and drives
the animals higher into the mountains.

Conifer Forests. Five types of conifer forests are distinguishable at altitudes above
2000 m. Monkey are found throughout the conifer forests. The cypress type is the
most economically valuable and the monkey's future is precarious in these forests if
they are heavily logged. Monkeys are also found in the hemlock, fir-spruce and pine
types of conifer forests. However, these seem to be secondary or refuge habitats.

Conifer-Hardwood Forests. Monkeys are frequently encountered within the conifer-
hardwood forests which occupy about 2.8% of the total forest land and are found in
a belt along the contour at elevations of 2300–2500 m.

Temperate-Hardwood Forests. The highest concentrations of monkeys are found
within temperate hardwood forests which occur from sea level to 2000 plus m and can
be subdivided into tropical, subtropical and temperate zones. Monkeys are not commonly found in the tropical hardwood forest zone except in the underpopulated east
coast area.

The total forested land on Taiwan is approximately 2 million ha in area. In 1977
about 55% of the land was covered by forest, a reduction of 10% compared to the
1973 figure (Forestry in Taiwan 1973, 1977). In terms of economic exploitation,
a measure of the possible amount of ecological damage which can occur, 66.6% of the
forest land is considered operable, and 29.5% is considered to be economically inaccessible.

Human Predation

Aboriginal populations have gradually increased, changing their attention from a
decreasing faunal component to shifting cultivation of millet and potatoes. These
practices significantly alter the character of large forest tracts and affect resident
wildlife. Recently, rapid population increases and disruptions have been followed by
expansion of farmlands from lowlands upward to the mountain slopes. The forest has
repeatedly been ruined.

Most monkeys are hunted for their supposed medicinal value and as a source of
protein (Kuroda 1940). Hunting has reduced the size and distribution of the monkey
population. McCullough (1974) suggests that the macaque has the greatest potential
for management on a sustained kill basis and it is now accorded special species status
with hunting seasonally permitted (Eu 1969).

Monkeys are often hunted in the summer months when infants are carried by their
mothers. Hunting usually kills the mother, and often the infant, the object of the hunt.
Infants are sold to laboratories or local animal markets where they are resold as pets.
Hunting pressures have subsided because of the ban on hunting and closing of the
animal specimen shops in 1973. McCullough (1974) estimated that 200 macaques were
sold each year. In 1973 Peng et al. reported that the number of monkeys consumed

annually is approximately 960. Of this number, 300 are used for medical research, 400 for preparation of monkey skeletons to be exported to Japan, 200 for Chinese medicants, and 60 infants are exported live to Japan.

Trapping: Most animals are taken by purchased metal leg-hold traps and home-made wire snare traps. Both are destructive, the leg-hold traps often leave animals without a leg or foot. Animals caught in wire snare traps are often strangulated as they struggle to free themselves. Leg-hold traps are typically located along rather easily discernible travel routes and are cleverly disguised and buried in dead logs of holes in the ground. Home-made wire snares are usually nooses affixed to trees close to ground level. Although traps are rotated about every 2 weeks along travel routes, the monkeys regularly return to these routes. Previous trapping experiences do not seem to alter their behavior.

Animal shops: Many animal shops were visited; they are sad affairs with animals crowded into very small cages with little consideration being given to the animal's size or health condition. All forms of wildlife are sold, with monkeys among the favored stock. The market for live monkeys or monkey bones is excellent, with demand far exceeding supply. The price for live monkeys depends upon: (1) Whether they are indigenous or imported. Imported monkeys, such as pig-tailed and crab-eating macaques, are cheaper than local Taiwan macaques [1]. This indicates the latter's declining supply. (2) The animal's size and age: young monkeys are much more expensive than older animals. (3) Physical condition. Almost all adult monkeys in animal shops had a right foot or leg missing as a result of the leg-hold traps. Prices varied from U.S. $ 56.00–$ 840.00. In 1973 McCullough reported that Taiwan macaques sold for U.S. $ 16.00–$ 84.00. Obvious cost inflation has accompanied the declining supply.

Cage conditions in the animal shops are pathetic, as is the physical state of most animals. Crowded conditions and the diseased state of some of the animals can lead to the spread of communicable diseases, not only among the caged animals, but from the animals to humans and vice versa. Although the situation is a potential health hazard, there appears to be little recognition of this fact.

The use and sale of animals, especially monkeys and snakes, is an important feature of Chinese culture. Monkeys are favored as a food source. The whole body may be consumed, but the brain is favored. Monkey meat is considered to be of medicinal value. Monkey bones sell for U.S. $ 22.40 per kilo in local animal shops. The bones are considered to be a general tonic useful in combating weakness (see Poirier 1971 for a discussion of the use of Nilgiri langurs as a medicant). The medicant is prescribed for the elderly, pregnant women, and young children. The use of monkeys in the medicinal system has greatly declined as the number of animals available has declined. However, if the bones and meat were available, so would be the customers.

I estimate a figure of 1000–2000 animals as annually captured and/or destroyed (Poirier and Davidson 1979). Unfortunately, this figure is probably low. Do keep in mind that this figure represents the number of captured or destroyed animals of a species on a locally protected species list!

---

1 Baboons and orangutans were also seen in such animal shops

Preservation

In recent years Taiwan has made commendable efforts toward furthering wildlife preservation. However, population pressures, logging and agricultural interests, and cultural practices are strong. While trapping and hunting must be regulated, or stopped, there is a strong need for better enforcement of existing laws.

The most important variable in any conservation scheme is habitat protection and forest management. Efforts are being made to stop or control deforestation, and we were successful in having two small plots set aside as "monkey protection" areas. Although current logging policy favors retention of small belts of protected forest in logged areas, these are of little benefit to the monkeys. The area of the protected forests in such belts is too limited to be of value to anything but some smaller animals and avian forms.

Forest destruction has resulted in the dislocation of many local monkey populations, leading to colonization of what may be secondary or substandard habitats. This may lead to local overpopulation and ultimately lowered reproductive rates. Destruction of large hardwood forest tracts has restricted food sources and disrupted migratory patterns. One problem in establishing requirements for protected areas is the fact that this study generated minimal data concerning food and space requirements for sustained population growth.

Acknowledgments. Many people directly and indirectly aided this study. I would like to thank the following. Funding for this project came from the following sources: Pacific Cultural Foundation, Institute of Zoology, Academia Sinica (Tai Pei), Graduate School, College of Social and Behavioral Sciences and the Department of Anthropology, The Ohio State University.

I thank the following individuals at the Pacific Cultural Foundation: Dr. Li and Mr. Chao. I wish to thank the following individuals of the Institute of Zoology, Academia Sinica (Tai Pei): Dr. Chang, Dr. Chow and Mr. Ger, my research assistant. I also thank Mr. Koh of the Recreation department of the Taiwan Forestry Research Institute and Mr. Liu of the Taiwan Forestry Research Institute. Professor Kao and Mr. Ho of the Institute of History and Philology, Academia Sinica, and Professor Li of the Institute of Ethnology, Academia Sinica provided much assistance, office space, logistical support and friendship. The following provided much needed help: Dr. Huang, and Dr. Lin of the Department of Zoology, National Taiwan University and Professor Huang, Department of Geography, National Taiwan University. Mr. Eu of the Economic and Planning Council of the Executive Yuan was extremely helpful and provided many sorts of much needed logistical help and friendship. Dr. B. Gallin, Michigan State University and Mr. R. Davis, Princeton University provided information about Chinese culture and friendship for many long hours. I also wish to thank my colleague and friend Dr. Chen, Department of Anthropology, The Ohio State University.

References

Anonymous (1973) Forestry in Taiwan. For Inf Bull No 21. Taiwan Forestry Bureau, Tai Pei
Anonymous (1977) Forestry in Taiwan. Republic of China, Taiwan Forestry Bureau, Tai Pei
Eu H (1969) Forest recreation and wildlife management in Taiwan. China-UNDP-FAO Forest and
    forest industry development project. Tai Pei
Kuroda M (1940) A monograph of the Japanese mammals. The Sanseido Co Ltd, Tokyo (in Japanese)
Li H (1963) Woody flora of Taiwan. Livingston Publ Co, Narbeth Pa

Liu T (1970) Studies of the classification of the climax vegetation communities of Taiwan. Proc
    Natl Sci Counc 4 (Part II):1–36
Liu T (1972) The forest vegetation of Taiwan. Q J Chin For 5:57–85
Liu T, Koh C-C, Yang B-Y (1961) Ecological survey on Taiwan important forest types. Taiwan
    Forestry Research Institute, Tai Pei
McCullough D (1974) Status of larger mammals in Taiwan, Republic of China. Tourism Bureau,
    Tai Pei
Murie J (1982) Observations on the macaques. III. The Formosan or round-faced monkey. Proc
    Zool Soc London:771–774
Peirse H (1969) Draft final report UNDP-FAO Project 160. Forest and forest industry development.
    Taiwan, Republic of China, vol I. Tai Pei
Peng M, Lai Y, Yang C, Chiang H (1973) Formosan monkey *(Macaca cyclopis)*. Present situation in
    Taiwan and its reproductive biology. Exp Anim 22 Suppl:447–451
Poirier F (1971) The Nilgiri langur . . . A threatened species. Zoonooz 44:10–17
Poirier F (1977) The human influence on subspeciation and behavior differentiation among three
    nonhuman primate populations. Yearb Phys Anthropol 234–241
Poirier F, Davidson M (1979) A preliminary study of the Taiwan macaque *(Macaca cyclopis)*.
    Q J Taiwan Mus XXXII(3–4):123–191
Wang K (1962) Some environmental conditions and responses of vegetation on Taiwan. Biol Bull
    No 11. Tung Hai Univ, Tai Chung
Wang K (1968) The coniferous forests of Taiwan. Biol Bull No 34. Tung Hai Univ, Tai Chung

Prospects for a Self-sustaining Captive Chimpanzee Breeding Program

B.D. BLOOD [1]

The need for a self-sustaining captive breeding program for chimpanzees *(Pan troglodytes)* has been well established. The Interagency Primate Steering Committee — in the National Primate Plan — proposes support for the nationally coordinated chimpanzee production program in the United States (IPSC 1978). A special task force of the Steering Committee on the use of and the need for chimpanzees recommended that "a national, coordinated long-term program be designed and implemented to maximize the breeding potential of chimpanzees already available in the United States." The task force report signaled the need to assess the overall present chimpanzee breeding capability in the United States and to develop "plans to adjust and augment the present breeding capacity as indicated" (NIH 1978).

In order to develop an information base on chimpanzees in the United States, the Primate Steering Committee engaged the services of the International Species Inventory System (ISIS), a unique data system designed for collection, analysis and dissemination of information essential for the genetic and demographic management of wild species over many generations in captivity. The ISIS started this project in mid-1978.

In June 1979, and again in June 1980, the Steering Committee assembled special panels of experts who studied the preliminary census and inventory data presented by ISIS, identified and analyzed numerous variables which effect breeding and rearing of offspring, and made recommendations for a national chimpanzee breeding program.

The information that we now have about the captive chimpanzee in the United States is indicative of the potential — as well as the problems — for attaining a self-sustaining population.

As shown in Table 1, the number of institutions inscribed in the chimpanzee inventory — as of 31 December 1979— was 75, of which 68 are in the United States and 7 in other countries. Fifty-six are zoos or safari parks, and the remaining 19 are shown as "non-zoo" — i.e., institutions that hold animals for reasons other than exhibition (Seal 1979, Grahn, personal communication). They are mostly research centers, although some of the animals are in holding and breeding stations.

The total number of chimpanzees in the inventory is 1388 (Table 2). There are 289 in zoo and safari park collections — of which 261 are in the United States and

1 Former Executive Director, Interagency Primate Steering Committee, National Institute of Health, Bethesda, Maryland, USA. Present address: 1210 Potomac School Road, McLean, VA 22101, USA

Table 1. Institutions holding chimpanzees – 31 December 1979,
United States and selected other countries

	Zoos	Non-zoos	Total
United States	50	18	68
Other	6	1	7
Total	68	19	75

Source: International Species Inventory System (ISIS)

Table 2. Chimpanzees in captivity – 31 December 1979,
United States and selected other countries

	Zoos	Non-zoos	Total
United States	261	1012	1273
Other	28	87	115
Total	289	1099	1388

Source: International Species Inventory System (ISIS)

Table 3. Size of chimpanzee colonies (Reporting to ISIS
– 31 December 1979)

	Number of colonies	Number of animals in each colony
Zoo collections	2	25–30
	5	9–13
	14	5– 8
	17	1– 4
Non-zoo collections	2	150–200
	2	100–149
	3	50– 99
	5	25– 49
	7	less than 25

28 in other countries. The average number of the animals in these collections is slightly over four. The distribution of chimpanzees according to size of the collections is shown in Table 3.

There are 1012 chimpanzees in non-zoo collections in the 18 United States institutions; one major collection outside the United States brings the total of non-zoo animals in the inventory to 1099. We believe that almost all of the chimpanzees in the United States are inscribed in this inventory. There are perhaps as many as 100 additional animals in circuses and other trained animal facilities that are not included. This – added to the 1273 now registered – would give something like 1350 to 1375 for the grand total.

As for other countries, certainly there are a number of captive chimpanzees in zoos, research centers and other facilities that are not reported to ISIS. However, for an assessment of the potential for a self-sustaining breeding population, we believe it most practical to limit our consideration to the chimpanzees in United States institutions — and perhaps also those few other places now taking part.

It is not our purpose to go into further specific figures on the captive chimpanzee population, but you will be interested in some general determinations. About a third of United States chimpanzees are known to be captive-born; they make up much of the youngest segment of the population. Another third of the animals are known to have been wild-caught. And the other third are still of uncertain origin. Some of these will be traced — eventually — because the records on every animal are being checked and cross-checked. But in many cases, the origin will remain unknown.

In order to assure a permanent captive population of chimpanzees, special attention must be given to three fundamental points: reproduction, health, and genetics, not necessarily in that order of priority. It is not possible to say which is most important. Without special attention to all three, any program of captive breeding would be a failure.

According to ISIS data presented recently (Seal 1980) thre were 92 reported births in 1979. The mean number of births over the past 5 years has been 82 per year. Many of these died at an early age, and there is a question of just how many of the survivors will eventually become successful breeders. Chimpanzees do not reach breeding age until about 10 years old, and this long period between generations is another of the difficulties in designing and determining the effectiveness of a breeding program.

Of the 1273 United States chimpanzees in the inventory, the sexes are about equally divided. However, there are only 180 living females that have successfully bred, and just 83 of the males have sired offspring. It is a cause of considerable concern to find that only four captive-born males in the United States today have bred successfully. This illustrates one of the many reproductive problems for the long-term breeding program: that of male infertility. Plans are now being made for a workshop on male ape infertility to be held late this year.

The second fundamental requirement for successful captive breeding is to maintain the population in good health — and protected against accident and disasters. This involves all aspects of individual and group management, housing, nutrition, prevention and treatment of diseases and injuries, and general husbandry. The fact that the population is in numerous locations that are widely dispersed — even though it has serious disadvantages — does offer some protection against epidemics and other disasters (such as fire and storms) that may not always be prevented.

The third consideration for a long-term breeding program is genetics. Emphasis must be given to preserving genetic variability by minimizing the relationship between parents. It is important to remember that it is the *effective* number of parents rather than the *inventory* number of parents which is critical in relation to levels of inbreeding. The effective number is made up only of those individuals which actually contribute genes to the succeeding generation. Not only is it important to have adequate numbers of breeding females, there are important genetic reasons to have as many breeding males as possible.

The problem of inbreeding in exotic animal populations in captivity is receiving increased attention. Recent studies have documented the detrimental effects of even second generation inbreeding, in sibling crosses or backcrosses, with resultant decrease in fecundity and neonatal survival. Analyses recently reported by Ralls et al. (1979) of the National Zoo in Washington, show that juvenile mortality of inbred young was higher than that of noninbred young. This was true of 15 of the 16 species of ungulates that they studied. Their data show a clear tendency for a larger percentage of young to die when the female was mated to a related male than when she was mated to an unrelated male.

Cooperative breeding arrangements, including stock pooling, transfers, loans, and deposits, are the necessary and practical solution to the inbreeding problem. All of the resources that we have will not add up to species survival so long as each institution goes its own way, making its own selections and ad hoc transfers.

There is reason to believe, based upon analysis of available data, that the size and character of the chimpanzee population now in the United States is adequate for a self-sustaining captive-breeding program. Such a program will require a commitment by chimpanzee owners to take part in a coordinated overall breeding plan. The program will be possible only if and when enough institutions holding enough chimpanzees are prepared to make such a commitment and to give the breeding program highest priority when necessary — for their own enlightened good as well as for the good of all concerned.

References

Grahn L (1980) ISIS, Personal communication

Interagency Primate Steering Committee (1978) National Primate Plan. DHEW Publ No (NIH) 80-1250

NIH (1978) Report of the task force on the use of and need for chimpanzees. NIH, Bethesda, 25 pp

Ralls K, Brugger K, Ballou J (1979) Inbreeding and juvenile mortality in small populations of ungulates. Science 206:1101–1103

Seal US (1979) Census and inventory of chimpanzees in the United States: Development of a studbook format. ISIS Rep Nat Inst Health, Interagency Primate Steering Committee

Seal US (1980) ISIS report presented at meeting on chimpanzee breeding. Tanglewood NC

Part B
Symposium Reports

Miocene Hominoids and New Interpretations of Ape and Human Ancestry

R.L. CIOCHON [1] and R.S. CORRUCCINI [2]

It is widely believed that study of the hominoid primates of the Miocene will yield a better understanding of the nature of the last common ancestor of the apes and man, the time and place of the hominoid-pongid divergence, and the adaptive nature and reason for the initial differentiation of hominoids from pongids. The last 5 years have witnessed a reconceptualization of the affinities of the Miocene hominoids vis à vis the modern apes and man. It is for this reason that we organized a pre-Congress symposium at the VIII IPS Congress entitled "Miocene Hominoids and New Interpretations of Ape and Human Ancestry."

Our position is that this reconceptualization of Miocene Hominoidea has been brought about by at least five somewhat independent factors. These can be summarized as:

1. New fossil discoveries which have increased doubt about the hominid affinities of *Ramapithecus.*
2. The discovery of many additional new Miocene hominoid specimens, especially *Sivapithecus*, including several partial skulls from various geographical regions in the Old World particularly in China (see Xu Qinghua and Lu Qingwu 1980 and Wu Rukang 1981).
3. Recent reinterpretations regarding the postcranial data in Miocene hominoid evolution, particularly the more derived upper Miocene material from Potwar and Rudabanya as opposed to the more primitive remains known from lower Miocene deposits in Africa.
4. The increasing acceptance or influence of biomolecular data for understanding the timing and relationships of hominoid cladogenesis.
5. The incredible discovery of a large sample of 3—4 million-year old hominids from Eastern Africa that have unexpectedly chimpanzee-like characteristics (see Johanson and White 1979).

The participants in our pre-Congress symposium specifically addressed many of these points. It is our hope that when their discussions are combined with those of a group of additional contributors who could not make the journey to Florence, and

1 Department of Sociology and Anthropology, University of North Carolina, UNCC Station, Charlotte, NC 28223, USA
2 Department of Anthropology, Southern Illinois University, Carbondale, IL 62901, USA

published by Plenum Press of New York in Fall 1982 in a volume entitled *New Interpretations of Ape and Human Ancestry,* this reconceptualization of the Miocene hominoids will become better understood.

At the pre-Congress symposium we took full advantage of meeting in a small working group in a round-table format. Previous examples of such meetings, the Neanderthal centenary (von Koenigswald 1958) and hominid classification (Washburn 1963) symposia, demonstrated prominently the great advantages accruing to the field from timely meetings of leading authorities focused on specific, currently unresolved problems. In these particular examples, the proceedings of the symposia were speedily published and had considerable impact on our field. Specifically, we encouraged the leading workers in the field to assess the possibility of attaining some consensus on the crucial questions of hominoid taxonomy, hominid origins and the nature of the last common ancestor of pongids and hominids (Fig. 1). Their attention was directed to the following topics, in approximate order of priority:

1. Morphological description and taxonomic interpretations of individual fossils that have recently come to light.
2. Systematics of the Miocene hominoids, with special reference to the many generic and specific names applied to fossils which may be reduced through synonymy. Several participants brought casts representing important recent discoveries to be directly compared. The conference participants came to some agreement about the basic outlines of a much-needed revision of Simons and Pilbeam's (1965) influential but now outdated systematic revision of Dryopithecinae. The resurrection of

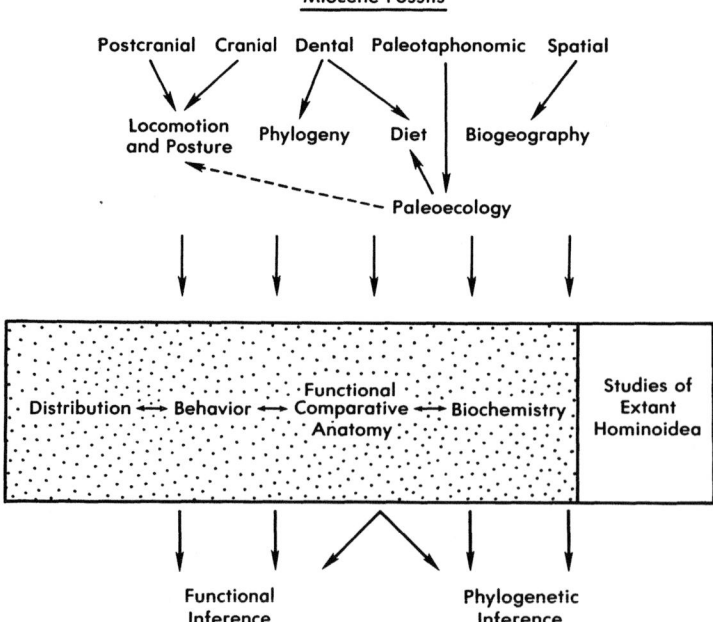

Fig. 1. Some aspects of Miocene to Recent hominoids considered by symposium participants

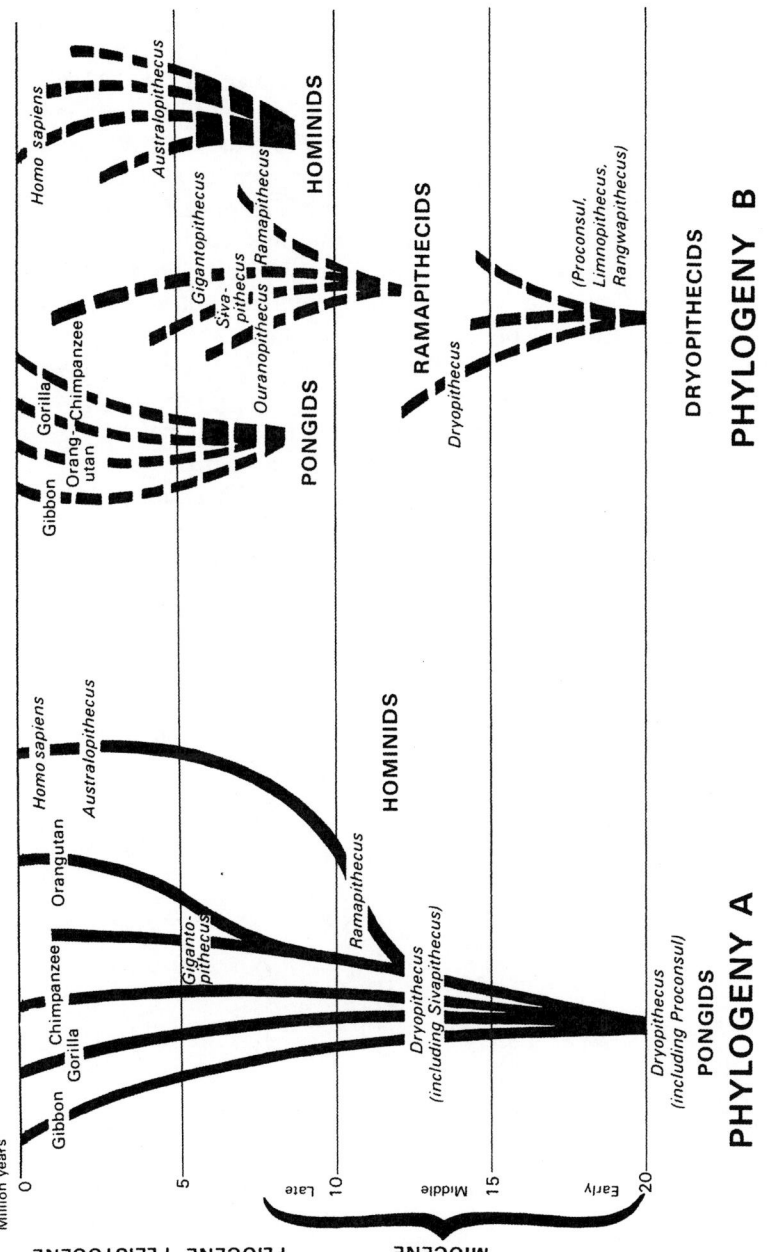

Fig. 2. Illustrates two alternative phylogenies of Miocene to Recent hominoid primates. *Phylogeny A* depicts the more "classical" view of hominoid evolution which was the traditionally accepted hypothesis until the mid 1970s. *Phylogeny B* represents a plausible alternative interpretation recently proposed by Pilbeam (1978, 1979). Note how the number of hominoid families increased from two in *Phylogeny A* to four in *Phylogeny B*. Part of this change is due to new fossil discoveries but a greater factor is a sweeping reconceptualization of the relationships of Miocene hominoids to extant apes and humans. (Modified after Pilbeam 1978)

Proconsul and *Sivapithecus* to full generic rank received general acceptance. The willingness to raise old generic categories to higher rank, such as Sivapithecidae or Ramapithecidae (see Fig. 2) to accomodate all the Eurasian thick-enameled forms (Pilbeam et al. 1977, Pilbeam 1979) contrasts with the harsh criticism initally experienced by L.S.B. Leakey when he proposed setting the African dryopithecines apart in a family Proconsulidae (Leakey 1968, Simons 1969). This latter family may be a more valid construction than the Dryopithecidae (Pilbeam 1978, 1979).

3. The putative ancestor-descendant relationship of specific Miocene and extant hominoid taxa has been a source of rapidly changing opinions. From the time of Pilgrim, Hopwood, and Lewis up through the 1960s, it was commonplace for new finds to be assigned a place in the direct ancestry of a living species (Fig. 2), and no one has carried this practice further than E.L. Simons (1972). The most recent tendency is to steer away from such specific evolutionary hypotheses in favor of the view that, statistically, most fossil species can be expected to be "evolutionary dead ends" originating in successive adaptive radiations (Eldredge and Gould 1972, Andrews 1974, Pilbeam 1979). The evidence for the following ancestor-descendant pairs was doubted: (a) *Limnopithecus* (or *Dendropithecus*) to *Hylobates;* (b) *Proconsul africanus* or *P. nyanzae* to *Pan;* (c) *Proconsul major* to *Gorilla.* The phyletic relationship between *Sivapithecus* and *Pongo* was considered more substantial.

4. The ancestor-descendant hypothesis of greatest interest, that linking *Ramapithecus* to *Australopithecus* and *Homo,* was also considered. Through the 1960s the opinion championed by Simons (1961, 1964, 1968, 1977) was widely accepted, but recent opinions have tended to be more cautious (Andrews 1977, Pilbeam 1978, 1979). Figure 3 illustrates the changing morphological configurations of *Ramapithecus* based on the discovery of new fossil remains. Andrews and Pilbeam discuss the

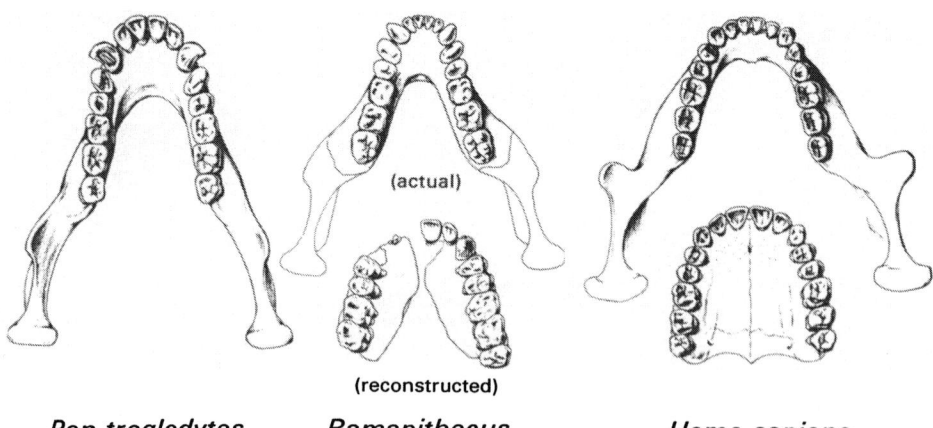

(actual)

(reconstructed)

Pan troglodytes   **Ramapithecus**   **Homo sapiens**

Fig. 3. Changing views of the morphology of *Ramapithecus.* Compares the parabolically shaped human-like reconstruction of the upper jaw of *Ramapithecus* originally proposed in the early 1960s with the actual fossil evidence of a V-shaped ape-like lower jaw of *Ramapithecus* discovered when D. Pilbeam recovered a nearly complete jaw from the Potwar Plateau in 1976. Chimpanzee and human jaws are included in this figure for comparison. (Adapted from Zihlman and Lowenstein 1979)

characters linking *Sivapithecus, Gigantopithecus,* and *Ramapithecus* and distinguishing these from other dryopithecines, but Andrews states "unfortunately, there is insufficient evidence to link this group with the hominids on the basis of this character complex, and it is quite possible that it forms an adaptive lineage parallel to a contemporary hominid lineage" (Andrews 1977, p. 54). This parallel hominid lineage could then have originated at any point in time up to the early Pleistocene, an opinion in keeping with those who see extant chimpanzees as essentially unchanged representatives of man's and *Pan*'s most proximate ancestor (Washburn 1963, Zihlman et al. 1978). Speculation surrounding *Ramapithecus* centers on its ancestor (*Rangwapithecus, Proconsul, Sivapithecus,* and *"S. africanus"* have been suggested) and its place of origin, Africa (Leakey 1962) and Asia (Andrews 1977) having been suggested.

5. The conference delegates varied in concurrence with recently increasing opinions that hominid divergence from pongids could be relatively late, and from an ancestor not resembling *Ramapithecus* (Robinson and Steudel 1973, Greenfield 1979, Zihlman et al. 1978, Zihlman and Lowenstein 1979). The extent to which any extant ape can adequately characterize the primitive morphotype, and whether that morphotype is more apelike or humanlike (Løvtrup 1978, Kortlandt 1974) are also valid questions.

6. Finally, more detailed consideration was given to relationships among the Miocene fossils themselves, and opinions were compared concerning the following relationships: Fayum anthropoids (particularly *Oligopithecus, Propliopithecus* and *Aegyptopithecus*) to *Proconsul, Proconsul* to *Rangwapithecus, Proconsul* species to *Dryopithecus, Sivapithecus* to *Proconsul, Ramapithecus* to *Sivapithecus, Gigantopithecus* to *Sivapithecus,* and the various new nomina to established groups (e.g., *Graecopithecus, Ouranopithecus, Hispanopithecus*).

The format of the conference included two full days for presentation of papers, with ample time left for discussion after each talk, followed by a round-table comparison of fossils. In what follows, we shall list the speakers and some of their more interesting and innovative facts and speculations.

P. Andrews spoke on "An African origin for *Sivapithecus* and *Ramapithecus.*" He stated that perhaps African and Asian *Ramapithecus* are significantly different, especially in dental arcade shape. He feels his and Walker's reconstruction of *R. wickeri* (Fig. 4) is accurate with its more parallel tooth row, contrasting with the V-shaped mandibles from the Potwar Plateau. This might resurrect the generic nomen *Kenyapithecus,* especially if the *"K. africanus"* specimen is reassigned to the African *Ramapithecus*-like group, but Andrews emphasized the tentative nature of these speculations. Andrews reacknowledged the need to move away from considering various *Proconsul* species direct ancestors of modern African apes. The Miocene genus is unique in at least some autapomorphic characters, such as the form of medial and distal maxillary cusp ridges. If Sivapithecidae is valid, then Proconsulidae may be an equally valid family-level taxonomic entity. This would constitute another of many instances of the late L.S.B. Leakey being right (in retrospect) for the wrong reasons.

L. de Bonis presented "Phyletic relationships of Miocene hominoids and higher primate classification." In his view, the Oligocene and early Miocene anthropoids were only forest-dwelling (A. Kortlandt, however, disputed this in his talk), while a new

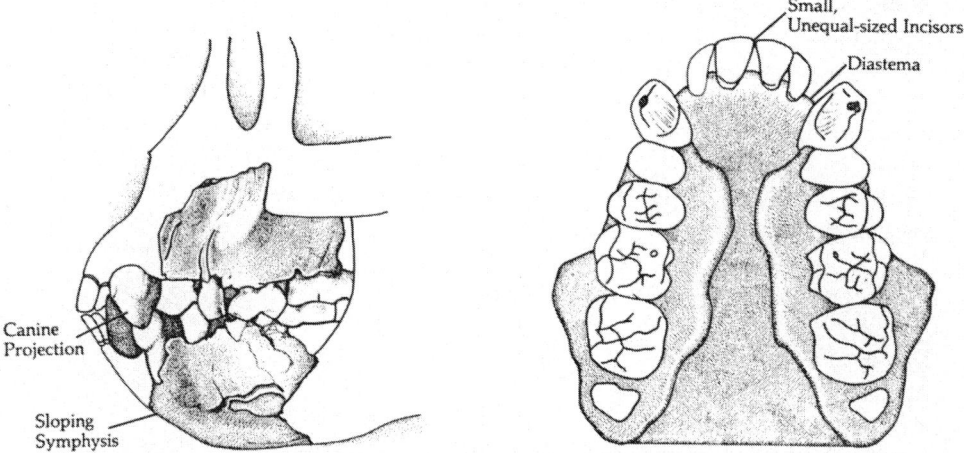

Fig. 4. Illustrates the upper and lower jaw fragments of the Fort Ternan *Ramapithecus wickeri* specimen *(left)* and a reconstruction of the upper jaw form of this specimen *(right)* developed by Walker and Andrews (1973). Note the parallel-shaped apelike dental arcade form which resulted from this reconstruction. (Reproduction of figure is courtesy of M.H. Wolpoff 1980 and the Alfred A. Knopf Publishing Company)

type of higher primate, the sivapithecines, arose in the middle Miocene open woodlands. Thickened enamel on jugal teeth is the basal synapomorphy of the Hominidae (sensu lato) among which true hominids and sivapithecines are sister groups. He expressed relationships in a cladogram based on both evidences of molecular biology and paleontology, in which the uncertain relationships of sivapithecines to dryopithecines and of pongines to panines are reflected in a three family division of living Hominoidea: Hominidae (including sivapithecines), Panidae, and Pongidae (*Pongo* only). The subsequent discussion centered on the problem of character reversal in enamel thickness if *Pan* and *Gorilla* indeed share common ancestry with other members (hominid and pongid) of the clade showing this trait. The general opinion was that such reversal was not excessively unlikely, for in most other cranial and postcranial characters the sivapithecine group is a better structural ancestor for both apes and man than other known groups. This is particularly well illustrated by a distal humerus (GSP 12271) recently described by Pilbeam et al. (1980). Figure 5 presents a comparison of this specimen with that of a female gorilla. The derived "extant-hominoid-like" morphological configuration of GSP 12271 is striking.

D. Gantt discussed "The enamel of Neogene hominoids: structural and phyletic implications." Although the ramapithecids (or sivapithecids) all are thick-enameled, there are three enamel prism patterns within the clade. The prism pattern of *Ramapithecus* is *not* that of humans, as had previously been stated, but a different type shared with *S. indicus* (but not *S. sivalensis*), and *Gigantopithecus*. *Pongo* is most similar to *S. indicus* and *Ouranopithecus* implying shared ancestry (at the roundtable comparison it was opined that *Ouranopithecus* is most significantly similar to *S. meteai*).

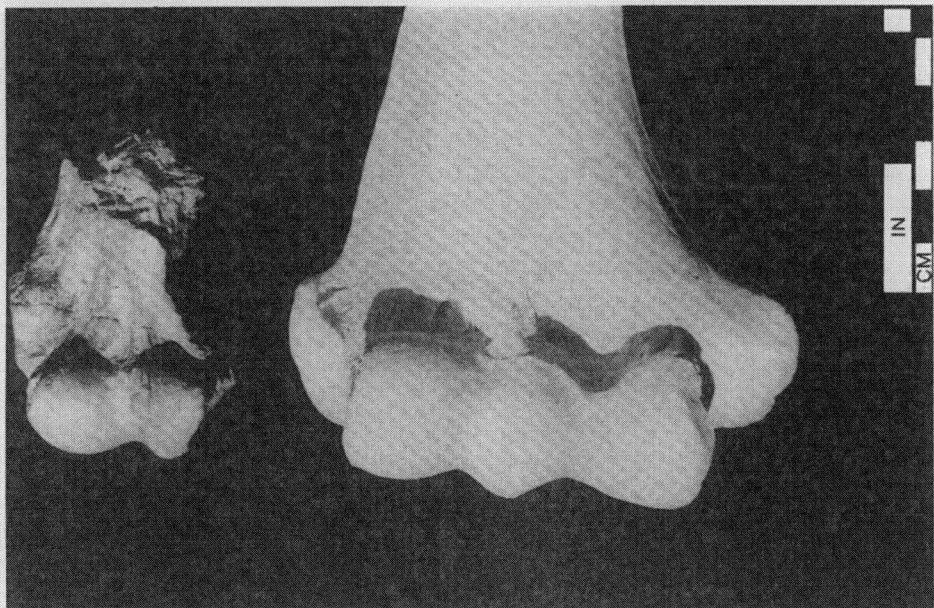

Fig. 5. Compares the partial distal humerus (GSP 12271) of an important new Miocene hominoid specimen from Potwar described by Pilbeam et al. (1980) with the distal humerus of a female gorilla. Note the remarkable similarity in the form of the globular capitulum and well-pronounced lateral trochlear ridge. The morphological configuration of GSP 12271 indicates that by Nagri times a major portion of the derived upper limb specializations which characterize all extant hominoids had evolved. This contrasts dramatically with lower Miocene African hominoid postcranial remains. (Photograph comparison presented here is based on casts and is courtesy of T.D. White)

D. Falk revised opinions concerning "Phyletic affinities of the endocasts of *Dryopithecus africanus.*" She compared sulcal patterns of the best-known Miocene endocast and extant primates, concluding the dryopithecine pattern, particularly the rectus sulcus, is in the primitive state and not similar to the (derived) living hominoids. It was commented subsequently that *Proconsul* is really unique in showing such uniformly primitive morphology in brain, face, postcranium, and dentition (cingulum), and that on the basis of shared derived characters its justification as hominoid can scarcely be maintained. There was also some agreement that *Pliopithecus* lacks hominoid synapomorphies.

General discussion at the end of day one indicated a sanguine attitude toward the hominid (i.e., human ancestral) status of *Ramapithecus;* all agreed that at least some specimens are female *Sivapithecus.*

G.H.R. von Koenigswald honored the conference organizers the second day by presenting "Significance of some undescribed material of fossil hominoids from the Siwalik Hills with remarks on stratigraphy and sedimentation." He started by documenting the change from forest to open dry savanna in the sedimentation of the Chinji to Dhok Pathan strata, opining that the latter development provoked bipedal locomotion and other facets of hominization. He described fossils recovered in the 1964–1965

Utrecht expedition to Pakistan, including teeth referable to probable new species of *Dryopithecus* and *Sivapithecus, S. lewisi,* and *Ramapithecus.* This last (a deep, thin mandible with weathered M_1 and P_4) he considers close to the original type specimen *"R. brevirostris"* and indicates possible relations to early Asiatic hominids. From the Nagri he recovered what was at that time the most complete maxilla of *Sivapithecus* with associated mandibular parts. He showed how the elongated M_3 with enameled obliteration of the *Dryopithecus*-pattern in the occlusal fovea in some *Sivapithecus* is a trait linking them with *Pongo.* This is also true of specimens from India.

A. Kortlandt spoke next on the topic of "Some fallacies of Oligocene and Miocene ape habitats." He presented an excellent critique of how our conventional views of ape habitats, concerning both living forms and those of the Miocene, have been distorted. According to Kortlandt too much emphasis has been placed on characterizing the living apes as inhabitants of lush closed-canopy rain forests. Instead, he offered much behavioral and ecological evidence indicating that an open woodland/savanna ecotype might be the preferred habitat for the African apes and also for the hominoids of the Miocene. Furthermore, he cautioned that interpretations of past habitats should be made more precise by incorporating all the data and analogies that modern climatology, ecology, zoology, and botany can provide.

M. Baba, one of the two speakers representing the opposing Berkeley and Wayne State camps of molecular anthropology, spoke on "The bearing of molecular data on the cladogenesis and times of divergences of hominoid lineages." Her points, plus those of the ensuing speaker, were very notable in representing something of a rapproachement between formerly irreconcilable viewpoints. Following a general review of molecular systematics based on immunology and protein sequencing, she showed evidence of evolutionary slowdown in hemoglobin serum proteins. However, she indicated that this finding may pertain only to those particular macromolecules, and may reflect functional responses in evolution (i.e., not all neutral mutations are neutral), leaving open the possibility that other data such as immunology of other molecules may be more validly fitted to a molecular clock model. Subsequent to her presentation in Florence, actual evidence indicative of a modest evolutionary slowdown in higher primate evolution has been published (Corruccini et al. 1980).

J. Cronin, in discussing "Apes, humans, and molecular clocks: a reappraisal," employed a synthetic approach to hominid cladogenesis using various immunological, sequencing, DNA hybridization, and chromosome morphology data. He derived an approximate date for the hominid-pongid divergence of 4.5 million years ago from all this, possibly as high as 5.5 million years ago; more than suggested by previous "clock" estimates (Sarich 1971). He further noted that the Afar skeleton "Lucy" is phenetically as ape-like as *Ramapithecus,* but 10 million years later in time. This would seem to reduce the probability of *Ramapithecus* being a completely evolutionarily stagnant hominid for so long and would somewhat reconcile the biological vs molecular differences separating *Pan* and *Homo* (Fig. 6). Cronin concluded that the molecular picture of a close correspondence between man and chimp definitely received added impetus with the announcement and description of the geologically oldest species of *Australopithecus, A. afarensis* (Johanson and White 1979). The significance of *A. afarensis* for considerations regarding Miocene hominoid relationships lies in the following three points: (1) *A. afarensis* is demonstrably distinct from *A. africanus* and is likely the

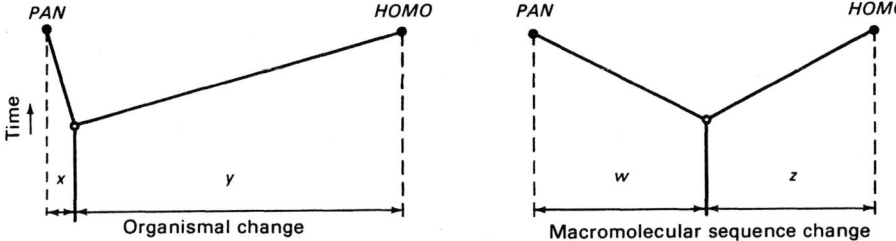

Fig. 6. Illustrates how the biological and molecular views concerning the evolutionary divergence of *Pan* and *Homo* differ. The amount of organismal change (structural-functional transformation) that has taken place along the human lineage *(y)* appears to far exceed that which is thought to have occurred along the chimpanzee lineage *(x)*. Evidence derived from macromolecular studies indicates that approximately the same amount of change has occurred in the genes along the chimpanzee lineage *(w)* as in the genes along the human lineage *(z)*. New fossil evidence supporting a later hominid-pongid divergence date *and* the unexpectedly chimpanzee-like features apparent in the hominid *A. afarensis* might force a redrawing of the organismal change diagram. (After King and Wilson 1975)

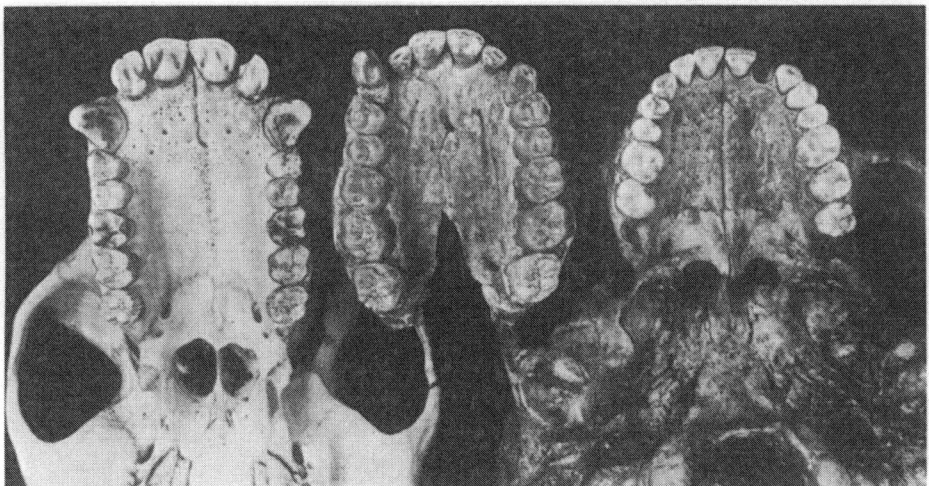

Fig. 7. Compares the most complete maxilla (AL 200) of the new hominid species *Australopithecus afarensis (center)* with maxillae of *Pan troglodytes (left)* and *Homo sapiens (right)*. Note how AL 200 either resembles *Pan* or occupies an intermediate position between *Pan* and *Homo* in the following features: (1) form of the dental arcade, (2) large projecting canine (however, note that right canine of AL 200 is actually out of socket), (3) canine-incisor diastema, and (4) substantial alveolar prognathism (dramatically evident in a superior view of AL 200). The phyletically primitive apelike features exhibited by this and other *A. afarensis* hominid specimens do add support to a later divergence date of hominids and pongids and bring into question the hominid affinities of *Ramapithecus*. (Photograph courtesy of D.C. Johanson, T.D. White, W. Kimbel, and the Cleveland Museum of Natural History)

common ancestor of both that species (plus its later descendants, the robust australopithecines) and of the lineage leading to genus *Homo;* (2) *A. afarensis* is the oldest demonstrable hominid; (3) the dental, facial, and temporal (but not postcranial) morphology of the species is perhaps more chimplike (see Fig. 7) than *Dryopithecus*-like, leading Johanson and White to question its descent from *Ramapithecus.*

S.R.K. Chopra next presented "Significance of new hominoid discoveries from the Siwalik Hills of India." Included in his descriptions of the latest material was a single tooth of a hylobatid named *Pliopithecus krishnaii* recovered from Nagri strata. Discussants suggested that the molar appeared to be that of a true gibbon (Hylobatinae) despite its inclusion in the genus *Pliopithecus.* If this judgement proves correct, then the line leading to the extant gibbons had become separate by at least 10 million years ago in India. He also presented a molar tooth described as *Sivasimia chinjiensis* with unusually strong orang-like features in the enamel, crown pattern, and crenulations.

Finally, B. Senut closed out the formal part of the conference with her paper, "Outlines of the distal humerus in hominoid primates: applications to some Plio-Pleistocene hominids." She again emphasized the intermediate status of the Afar hominids to *Pan* and *Gorilla* on the one hand and *Homo* on the other.

The pre-Congress symposium "Miocene Hominoids and New Interpretations of Ape and Human Ancestry" was successful in making a necessary first step toward defining a reconceptualization of the hominoids of the Miocene vis à vis the modern apes and man. It is hoped that this reconceptualization will become a primary basis for the study of hominoid evolution in years to come. Certainly, the publication of the papers from this symposium together with an additional set of papers in the volume *New Interpretations of Ape and Human Ancestry* (Plenum, New York, 1982) will do much to establish this new paradigm.

Acknowledgments. We gratefully acknowledge the L.S.B. Leakey Foundation for funding the costs of our pre-Congress symposium in Florence, Italy. We also thank the International Studies Program of the University of North Carolina at Charlotte and the National Science Foundation for funding some travel costs of symposium participants.

References

Andrews P (1974) New species of *Dryopithecus* from Kenya. Nature (London) 249:188–190
Andrews P (1977) Taxonomy and relationships of fossil apes. In: Chivers DJ, Joysey KA (eds) Recent advances in primatology. Academic Press, London New York, pp 43–56
Corruccini RS, Baba M, Goodman M, Ciochon RL, Cronin JE (1980) Non-linear macromolecular evolution and the molecular clock. Evolution 34:1216–1219
Eldredge N, Gould SJ (1972) Punctuated equilibria: an alternative to phyletic gradualism. In: Schopf TJM (ed) Models of paleobiology. Freeman Cooper and Co, pp 82–115
Greenfield LO (1979) On the adaptive pattern of *"Ramapithecus"*. Am J Phys Anthropol 50: 527–548
Johanson DC, White TD (1979) A systematic assessment of early African hominids. Science 203: 321–330
King M-C, Wilson AC (1975). Evolution at two levels in humans and chimpanzees. Science 188: 107–116
Koenigswald GHR von (ed) (1958) Hundert Jahre Neanderthaler. Kemink en Zoon, Utrecht

Kortlandt A (1974) New perspectives on ape and human evolution. Curr Anthropol 15:427–448

Leakey, LSB (1962) A new lower Pliocene fossil primate from Kenya. Ann Mag Nat Hist 4:686–696

Leakey LSB (1968) Lower dentition of *Kenyapithecus africanus*. Nature (London) 217:827–830

Løvtrup S (1978) Review of ontogeny and phylogeny (by SJ Gould). Syst Zool 27:125–130

Pilbeam D (1978) Rearranging our family tree. Hum Nat 1 (6):38–45

Pilbeam D (1979) Recent finds and interpretations of Miocene hominoids. Annu Rev Anthropol 8:333–352

Pilbeam D, Meyer GE, Badgley C, Rose MD, Pickford MHL, Behrensmeyer AK, Ibrahim Shah SM (1977) New hominoid primates from the Siwaliks of Pakistan and their bearing on hominoid evolution. Nature (London) 270:689–695

Pilbeam DR, Rose MD, Badgley C, Lipschutz B (1980) Miocene hominoids from Pakistan. Postilla (Yale) 181:1–94

Robinson JT, Steudel K (1973) Multivariate discriminant analysis of dental data bearing on early hominid affinities. J Hum Evol 2:509–528

Sarich VM (1971) A molecular approach to the question of human origins. In: Sarich VM, Dolhinow P (eds) Background for man. Little Brown and Co, Boston, pp 60–81

Simons EL (1961) The phyletic position of *Ramapithecus*. Postilla (Yale) 57:1–20

Simons EL (1964) On the mandible of *Ramapithecus*. Proc Natl Acad Sci USA 51:528–535

Simons EL (1968) A source for dental comparison of *Ramapithecus* with *Australopithecus* and *Homo*. S Afr J Sci 64:92–112

Simons EL (1969) Late Miocene hominid from Fort Ternan, Kenya. Nature (London) 221:448

Simons EL (1972) Primate evolution. Macmillan, New York

Simons EL (1977) *Ramapithecus*. Sci Am 236:28–35

Simons EL, Pilbeam DR (1965) Preliminary revision of the Dryopithecinae (Pongidae, Anthropoidea). Folia Primatol 3:81–152

Walker A, Andrews P (1973) Reconstruction of the dental arcades of *Ramapithecus wickeri*. Nature (London) 244:313–314

Washburn SL (ed) (1963) Classification and human evolution. Aldine, Chicago

Wolpoff MH (1980) Paleoanthropology. A Knopf, New York

Wu Rukang (1981) First skull of *Ramapithecus* found. China Reconstructs 30 (4):68–69

Xu Qinghua, Lu Qingwu (1980) The Lufeng ape skull and its significance. China Reconstructs 29 (1):56–57

Zihlman AL, Lowenstein J (1979) False start in the human parade. Nat Hist 88:86–91

Zihlman AL, Cronin JE, Cramer DL, Sarich VM (1978) Pygmy chimpanzee as a possible prototype for the common ancestor of humans, chimpanzees and gorillas. Nature (London) 275:744–746

Infanticide in Langur Monkeys (Genus Presbytis): Recent Research and a Review of Hypotheses [1]

G. HAUSFATER [2] and C. VOGEL [3]

Introduction

The purpose of this symposium was to evaluate current hypotheses about the reproductive strategies of male and female langur monkeys (Genus *Presbytis*) in the light of new data from long-term field studies of this genus. To this end, scientists representing several different long-term studies of langurs, as well as theoreticians and researchers familiar with the behavior of other Colobines, met on 4–6 July 1980 at the Istituto di Antropologia e Paleontologia Umana in Pisa. A list of symposium participants is given in the last section of this report; Fig. 1 shows the study sites represented by these individuals. The meeting resulted in a productive and congenial exchange of ideas on langur social behavior, although all participants regretted that financial and personal reasons prevented the attendance of those colleagues from India, Japan, and the United States who are engaged in research on langur behavior at sites other than those represented at the symposium.

Through the efforts of the scientists and research groups mentioned above, langur monkeys have been studied at over a dozen different localities in India, Sri Lanka, and Nepal (Hrdy 1977a, Oppenheimer 1977, Vogel 1977). The various study sites shown in Fig. 1 differ markedly in their ecological characteristics and physical geography; the sites also show important differences in seasonal patterns of rainfall and vegetation abundance and in population structure, especially the adult sex ratio within troops. Thus, the first aim of the symposium was to review available data on langurs and to formulate new research designs such that information from this diversity of habitat types and population structures could be used to test specific hypotheses concerning the social strategies of langur monkeys.

The second aim of the symposium was to identify particular areas of langur biology where data were lacking and to agree upon data collection procedures and definitions such that future comparisons of behavior and ecology between study sites would be

---

1 Report of the pre-Congress Satellite Symposium "Theoretical Models on Male and Female Reproductive Strategies with Respect to the Actual Behavior of Langurs", 8th Congress of the International Primatological Society, Florence, Italy
2 Section of Neurobiology and Behavior, Cornell University, Ithaca, NY 14850, USA
3 Lehrstuhl für Anthropologie, Universität Göttingen, 3400 Göttingen, FRG

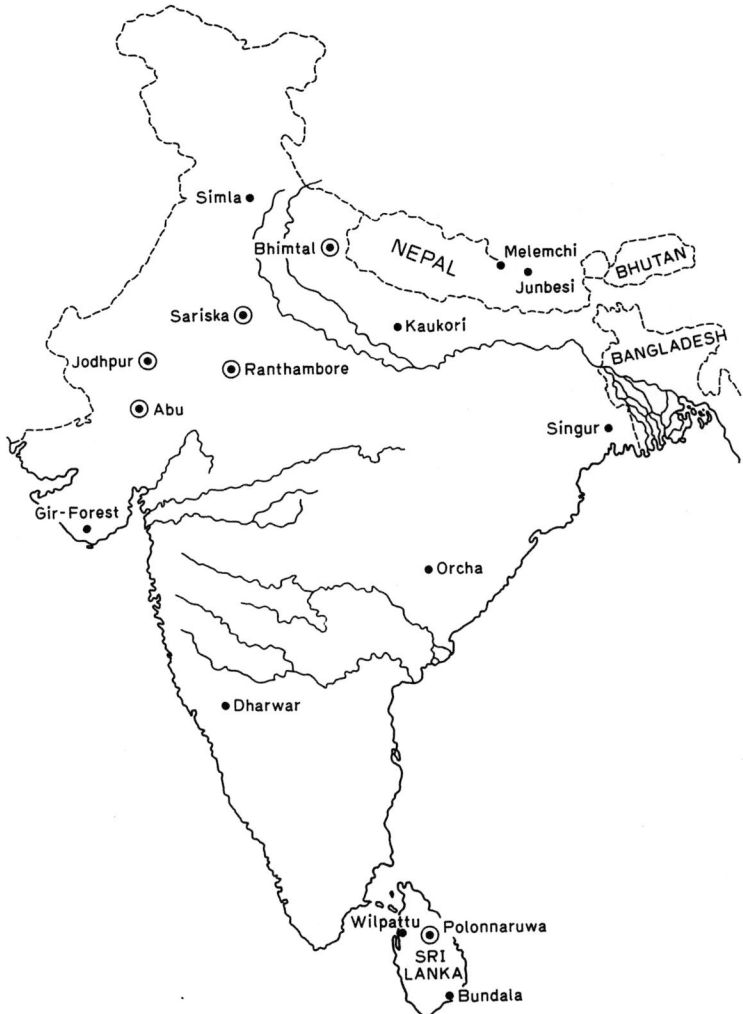

Fig. 1. Map of the Indian subcontinent showing sites at which langur monkeys (Genus *Presbytis*) have been studied. ⊙ indicates study sites represented at this symposium

facilitated. In this regard, discussions at the conference focused most closely upon the kinds of data required to test hypotheses concerning infanticide in langurs and thus the current report also deals most extensively with this phenomenon. However, many other aspects of langur social behavior and ecology were also discussed at the symposium, for example female counterstrategies to infanticide, allomothering and the abusive handling of infants, and age effects on social rank, reproduction and survival in both sexes.

Throughout this report, the term strategy has been used as a convenient label for certain inferences about the long-term patterns (and goals) of behavior by male and female langurs. The short-term decisions and/or actions which contribute to such long-term patterning, we refer to as "tactics". All symposium participants were in agreement, however, that neither term should be construed as an explanation of behavior nor as a substitute for exact hypotheses amenable to testing and falsification with empirical data.

Langur Social Organization and Infanticide

Langur populations typically contain two distinct types of social groups (Sugiyama et al. 1965, Vogel 1975). The first group type, usually referred to as "bisexual troops", is composed of one or more adult males, several adult females and their young. The second type of group contains exclusively adult and immature males and is therefore usually referred to as an "all-male band". In several well-studied populations, most bisexual troops contain only one adult male, called the "resident male", although recent evidence from long-term studies indicates that almost all bisexual troops periodically enter a multimale configuration as a result of immigration from a neighboring "all-male band" (Vogel 1979). The events surrounding these changes in male composition of bisexual troops show considerable variability from case to case (Hrdy 1977a), but frequently culminate in the migrant males attacking and expelling the resident males from the bisexual troop (Sugiyama 1965). Some time after the expulsion of the resident male from the troop, one of the migrant males expels all of the others from the troop and thereby restores the troop to one-male status with himself as the new resident male (Hrdy 1977a, Vogel 1979).

The process of replacement of the resident male in bisexual troops is often further accompanied by aggression directed by the new male toward females and their dependent infants. In particular, new resident males at several study sites have been reported to stalk females carrying unweaned infants and eventually to inflict fatal bite wounds upon those infants (Hrdy 1977a). Females appear mostly to be unsuccessful in defending their unweaned offspring from an attacking male and usually resume sexual cycles (or at least show behavioral signs of estrus) shortly after the death of an unweaned infant (Sugiyama 1965, Mohnot 1971, Hrdy 1974). Gradually, the male's aggression toward both females and their offspring diminishes so that infants born approximately a month or more after the male replacement are not harmed by the new resident male (Hrdy 1977a). A similar pattern of infanticide following changes in the male composition of captive groups of crab-eating monkeys *(Macaca fascicularis)* has been reported by Angst and Thommen (1977) who also reviewed suspected cases of infanticide in other primate species; yet more extensive reviews of primate infanticide are given by Hrdy (1977b, 1979).

Substantial controversy and debate surrounds the assertions that langur males intentionally or habitually kill infants of their own species and that this behavior is somehow a natural or normal part of their behavioral repertoire constituting an adaptive reproductive strategy (Curtin 1977, Hrdy 1977b, Curtin and Dolhinow 1978,

Boggess 1979). Furthermore, the evolutionary origins, immediate causation and expected frequency of occurrence of this behavior are also the subject of strong disagreement among the scienticists studying langurs (Hausfater 1977, Ripley 1980). Below, some aspects of the controversy and evidence concerning infanticide in langurs are discussed in more detail and then hypotheses concerning the frequency and function of infanticide are evaluated in the light of recent findings from the long-term studies of langurs represented at this symposium.

The Infanticide Controversy

Controversies in primate behavior most commonly result from differences in opinion as to the function and causation of a particular behavior. The controversy surrounding langur infanticide, however, may well be one of the few cases in which at one level, the disagreement reduces simply to whether or not the behavior occurs at all under natural conditions. At one extreme in this debate are individuals who suggest that most reports of infant killing by adult male langurs are at best the product of incomplete observations subject to imaginative interpretation. At the other extreme are researchers for whom any harsh gesture by an adult male toward an infant is taken as evidence of the actual or potential infanticidal proclivities of that male. Yet there can be little doubt that infanticide does indeed occur in some natural populations of langur monkeys, for there are now several published eyewitness reports of adult males biting infants to death around the time of an adult male replacement (Sugiyama 1965, Mohnot 1971, Hrdy 1977a, see also Wolf and Fleagle 1977 for *Presbytis cristata*). On the other hand, the claim based on such cases that infanticide is "latent" in all male primates, not just langurs, is somewhat akin to the statement that the ability to, say, climb trees, is latent in all carnivores; the proposition may indeed be true, but otherwise tells us little about the behavior of any particular species, for example, aardwolves or otters.

There are, however, more fundamental issues underlying the controversy over langur infanticide. The first of these issues concerns the completeness and quality of the data set offered to document the occurrence of infanticide, and the second issue concerns the expectations against which this data set is then evaluated. With respect to the first issue, it is unfortunate that infanticide as a label for a particular constellation of behaviors has been used in a relatively nonspecific and uncritical fashion. Thus, reports of "infanticide" in the literature include cases in which an adult male was actually observed to deliver a fatal bite to an infant, cases in which an infant disappeared shortly after a new male entered the troop, cases in which a wound was received by an infant and it died only months later, and nearly every other possible permutation of an adult male replacement, accompanied by aggression toward mothers and infants by the new male and the eventual disappearance or death of one or more infants (Boggess 1979, Vogel 1979). In all but a few cases there is some question as to whether the wounding of the infant was by the new resident male and whether such wounds were a direct cause of the infant's subsequent disappearance from the group, although in most such cases both suppositions do seem likely to be true.

The point to be gained from the above considerations, however is not that descriptions of infanticide have been lacking or of inferior quality, but rather that all such reports should include detailed information on the events actually observed as well as information on the precise timing of the new resident male's entry into the troop, the time of onset and duration of aggression toward infants, and the age of infants subjected to this aggression. In the case of the death or disappearance of an infant, precise information should be given on the interval between wounding and death or disappearance of the infant, and on the reproductive condition of the mother at the time of the infant's disappearance and the interval from this disappearance to the mother's next behavioral estrus, conception and birth. As the summaries prepared by Boggess (1979) and Vogel (1979) demonstrate, there is considerable variability between studies in the completeness and accuracy of this record.

On the other hand, those individuals critical of the idea that infanticide may be a normal part of the behavioral repertoire of langurs have sometimes imposed standards of observation completeness and quality that far exceed those applied to other classes of comparably rare events, e.g., predation. Thus, if a leopard were observed to jump into the middle of a monkey group and then to flee with adult males in pursuit, holding a female in his mouth, few observers would seriously question that an act of predation had occurred, particularly so when later the adult males, but not the female, were observed returning to the group. Yet in analogous descriptions of infanticide, it has been seriously suggested that the mother may have killed her own infant in the confusion surrounding her struggle with an adult male attacker. In sum, although it is obviously important to obtain precise descriptions of presumed cases of infanticide, and to clearly differentiate between the actual observations and speculations, it is equally important to avoid imposing standards of evidence on this particular class of behavior which are so strict as to rule out the possibility of ever obtaining a "convincing" or satisfactory description of the behavior itself.

The second issue in the infanticide controversy concerns expectations of the frequency with which this behavior should be observed in natural populations. In several review articles, conclusions about the causation and frequency of infanticide have been based upon a simple tabular listing of study sites from which the behavior has, or has not, been reported (Curtin 1977, Boggess 1979, Hrdy 1979). Since infanticide has thus far been observed at only a relatively small number of sites, the conclusion has been drawn that infanticide is not a very common or important phenomenon among langurs and that the behavior probably only occurs in specific aberrant populations (Curtin and Dolhinow 1978). In fact, arguments or conclusions based simply on this type of "balance sheet" approach to infanticide are bound to be incorrect and are also directly contrary to the insights into social strategies provided by recent advances in evolutionary theory (Wilson 1975). In the past, the selection of study sites for langurs was probably more strongly influenced by the food, water and shelter requirements of the observer than by any aspect of the habitat of importance to the monkeys themselves. Thus, there is no reason to presume that the study sites entered in the balance sheet analyses provide a representative sample of langur populations, habitats, or social behavior and, in fact, they most probably do not.

Additionally, given an average tenure of 2–5 years for the resident males in bisexual troops (Sugiyama 1965, Hardy 1977a), as well as the small number of troops monitored in most field studies, one would not really expect a very high frequency of observations of infanticide in any but a very long-term study. Thus, in some sense, infanticide must be treated as an expectedly rare behavior, and in the case of any such behavior, the actual number of field observations of its occurrence really provides very little information as to whether the behavior is in some sense a natural or adaptive part of the species's biology. Again by analogy, very few attacks upon monkeys by terrestrial carnivores have ever been observed, yet there is still good reason, and much circumstantial evidence, to believe that such attacks constitute an important aspect of the ecology of most primate species under natural conditions.

There is another, much deeper, vein to this issue, however. Briefly, most surveys, reviews and criticisms of langur infanticide have implicitly assumed that infanticide must be an "all or none" phenomenon. Thus, if one male in a population commits infanticide, then the presumption is made that all males in that population should be expected to do so. By extension, if infanticide occurred in one such population, then it is further presumed that it should be expected to occur in all populations living in the same type of habitat; and vice versa, if in several cases of observed male replacements no infanticide has been committed, then it has been concluded that infanticide does not occur in the respective population or habitat. In fact, in the literature on langur social behavior, surprisingly little consideration has been given to the possibility that infanticide might exist in langur populations as a behavioral polymorphism or as a "mixed" evolutionary stable strategy (ESS) (Maynard Smith 1974). Such a polymorphism need not be based only on genetic differences among males; behavioral polymorphisms based on developmental differences among males, or differences in the immediate characteristics of the troop into which males transfer or from which they came are also realistic possibilities for langur monkeys.

Were infanticide to exist in langur populations as a behavioral polymorphism, then the actual proportion of infanticidal males in that population (usually called the equilibrium frequency of infanticidal males), could logically take on any value from 1% to 99% of all males. In fact, males need not be always infanticidal or always not so; they might as likely pursue a "mixed" or variable strategy in which they engaged in infanticide following some replacements, but not others, again depending on the immediate social or demographic characteristics of the group or population, e.g., number of females and/or their reproductive condition, sex ratio, etc. In sum, variance among males and among populations in the frequency of infanticide does not make this behavior in any way suspect as unnatural. Rather, such variance provides some of the most important information for elucidating the causation and function of infanticide, and also provides the means of making a direct test of the various theoretical formulations of infanticide as a reproductive strategy (see p. 171).

Hypotheses About Infanticide

Table 1 presents a list and brief statement of hypotheses concerning the nature and function of infanticide in langurs grouped according to whether the underlying causal mechanism involves resource availability, population density, psychosocial factors or reproductive characteristics of langurs. Within each of these categories are combined hypotheses which provide an explanation for infanticide at the level of proximate and ultimate causation, for any hypothesis at one level typically implies a hypothesis at the other level as well. Below, each hypothesis is explained in more detail and evaluated in light of new data from long-term studies of langurs.

Table 1. Summary of hypotheses concerning the function and causation of infanticide in langurs

Hypotheses

1. Resource availability

    1.1 Elimination of competition hypothesis: Infanticide eliminates (unrelated) infants and juveniles who would be potential competitors of the new male's own offspring.

2. Population density

    2.1 Population regulation hypothesis: Infanticide is considered to be a mechanism for decreasing recruitment to an expanding population.

    2.2 Social stress hypothesis: Individuals living in langur populations at high densities or otherwise overcrowded conditions are hypothesized to suffer from severe social stress which results in infanticide and other forms of maladaptive or pathological behavior.

3. Psychosocial factors

    3.1 Social bonding hypothesis: Infanticide produces a rapid return to estrus by the mother and her resultant consortship with the new male helps to establish social bonds which solidify his position in the troop.

    3.2 Mixed emotions hypothesis: A new resident male is considered to experience simultaneous sexual frustration and rage which releases itself as infanticide.

    3.3 General disturbance hypothesis: Replacement of the adult male in a bisexual troop is considered to produce a general breakdown of social relations which leads to increased aggression by all individuals in the troop. Infants, as a vulnerable age-class, suffer a disproportionately high frequency of wounds, some of which lead to death.

4. Reproductive characteristics

    4.1 Reproductive advantage: The death of an unweaned infant results in the abrupt termination of its mother's amenorrhea. In this way, the infanticidal male reduces his waiting time to insemination of the female and thereby gains a reproductive advantage compared to his noninfanticidal counterpart.

    4.2 Elimination of competitor's offspring hypothesis: Even in the absence of the effects described in hypotheses 1.1 and 4.1 above, a male can proportionately increase his own reproductive success by eliminating the offspring of his competitors.

Resource Availability

Hypotheses involving resource availability as a causal factor in infanticide are based on the assumption that there is significant competition for food or other resources in langur troops. Such hypotheses further state that it will thus always be to the advantage of a new resident male to eliminate from this troop the infants and juveniles sired by the previous male, for these immature individuals will most likely be in competition with his own progeny for food resources, maternal care and eventually, mates.

The primary difficulty in evaluating the above hypothesis is the critical lack of information on the ecology of langur monkeys. Despite several studies of different aspects of langur diet and feeding behavior (Ripley 1970, Hladik and Hladik 1972, Winkler in prep.), the extent of competition for food (and other resources) between and within age-sex classes has yet to be analyzed for any one langur population. Fortunately, an excellent model for this type of analysis is provided by the work of Dittus (1977) on the toque monkey *(Macaca sinica)* in Sri Lanka. Hopefully, future studies on langurs will provide the sort of data required to critically evaluate the resource competition hypothesis.

Population Density

The density and demographic characteristics of langur populations have also been implicated as factors underlying infanticide in langurs. One of the first hypotheses linking population density and infanticide was the suggestion by Eisenberg et al. (1972, see also Rudran 1973) that infanticide functioned as a density-dependent mechanism to control population growth. Simply stated, at high population densities, infanticide was believed to lower recruitment to the population through birth and thereby to maintain the population in balance with its resources. Although initial statements of this hypothesis emphasized infanticide as a means to achieve small seasonal adjustments in group size (Rudran 1973), subsequent presentations have emphasized the potential of major limitations of numbers through infanticide and have sought a direct correlation between infanticide and population density (see e.g., Hrdy 1979, Ripley 1980). Unfortunately, current evidence on the relationship between population density and the occurrence of infanticide is inconclusive, particularly so since very different estimates of population density have been published for the same populations (e.g., Dharwar and Jodhpur). In particular, it is necessary to distinguish between density estimates based on the area of a particular geopolitical province, e.g, Rajasthan, the town of Jodhpur, etc., and estimates based upon the number of square units of habitat within that province that are actually available to, and suitable for use by, langurs. In certain very dry regions and in certain urban zones, estimates based on geopolitical boundaries are likely to differ considerably from the more ecologically meaningful estimates based on usable habitat only (see Muckenhirn 1975 and Vogel 1977 for discussions of this point).

It is also important to emphasize that strict density dependence implies a very exact relationship between population density and rate of infanticide: Specifically, that the rate of infanticide is a monotonically increasing function of population density. Currently available data are inconclusive or negative with respect to this specific relationship and hypothesis, but more precise estimates of density as well as other population parameters are badly needed and should be a major focus of future langur research.

It is still possible that population density may be a critical factor in the occurrence of infanticide even if not operating in the strict density-dependent fashion outlined above. By analogy with research on rodent societies, it has been argued that individuals living in primate populations at high densities suffer severe social stress. One result of this presumed social stress is hypothesized to be a breakdown of the normal pattern of social organization and the emergence of various social pathologies, especially infanticide (Curtin 1977). In contrast to the above density-dependent hypothesis, infanticide in the present hypothesis is viewed as an aberrant and maladaptive behavior indicative of a highly disturbed population and otherwise unrelated to control of population size. In recent reviews of langur infanticide (e.g., Curtin and Dolhinow 1978), provisioning of human food to langurs — which enjoy protected status as sacred creatures in India — as well as deforestation and reduction of natural predators, have been suggested as the basis of a rapid expansion of langur populations, especially in the area of human settlements, and thereby as the cause of overpopulation and associated infanticide.

In fact, however, langurs have only rarely been studied at sites where interactions with humans were entirely absent or negligible. Most of the studied populations engage in some crop-raiding or otherwise receive food from humans. More importantly, langur troops maintained in enclosures in captivity, such as the facility at the University of California at Berkeley, have not been reported to engage in aberrant or maladaptive behavior, although captivity is certainly an unusual situation in which effective population density is high and food resources both provisioned and abundant. In sum, the high density/human disturbance hypothesis seems unlikely to provide much insight into the origins and function of infanticide in langurs.

Psychosocial Hypotheses

One of the first hypotheses concerning infanticide in langurs focused on the fact that death of an infant sped the mother's return to breeding condition (Sugiyama 1964). This event was considered important not primarily from a reproductive standpoint (see below), but because the female's estrus provided an opportunity for the new resident male to consort with the female and thereby establish the necessary social bonds to solidify his leadership position in the troop (Sugiyama 1965, 1966). More recently, this same hypothesis has been used by Fossey (manuscript) to explain several cases of infanticide in the mountain gorilla *(Gorilla gorilla beringei),* but in most recent reviews of langur infanticide the social bonding hypothesis has not received the serious consideration which it deserves.

Another hypothesis in this group focuses on the psychological characteristics of migrant males, including the new resident male of a bisexual troop. These males, as previous residents of an all-male band, have had little access to females. Additionally, the take-over of a bisexual troop and replacement of the resident male involves a large amount of aggression. Thus, Mohnot (1971) hypothesized that the new resident male in a bisexual troop could be characterized as experiencing simultaneous sexual frustration and rage which then vents itself in infant killing. Such a hypothesis, although as difficult to test as any psychological hypothesis, would account for the great intermale variability in infanticide and also for the fact that males sometimes continue to aggress against females even after their infants are dead. As in the case of the social bonding hypothesis above, the present hypothesis deserves more careful consideration than it has received.

Finally, in an extension of the above hypothesis, it has been suggested that the upheaval associated with the replacement of a resident male has a generally disruptive effect on troop organization and leads to increased aggression by all individuals, not just the new male. It is further argued that infants, as the most vulnerable targets of aggression from all individuals, thus suffer proportionately high mortality around the time of an adult male replacement. Given the comparatively abusive treatment of langur infants by some classes of allomothers, this hypothesis cannot be immediately dismissed, but available evidence does not support the idea that reported fatal wounds to infants were delivered by individuals other than adult males.

Reproductive Characteristics

The hypothesis that infanticide confers a reproductive advantage upon the new resident male in a bisexual troop of langurs has received the most widespread and detailed consideration (Hrdy 1974, 1977a). This hypothesis emphasizes, as in the case of the social bonding hypothesis, that infanticide reduces the waiting time to a mother's return to estrus condition. However, in the present case, the importance of the female's return to estrus is that the waiting time to the new resident male's first insemination of that female is likewise reduced and the new male thereby gains a reproductive advantage compared to his noninfanticidal competitors. In fact, even in the absence of the reproductive advantage due to shortening of waiting time to first conception, a male would still produce a relative increase in his own reproductive success through the death of the offspring of his competitors.

As part of a mathematical examination of the reproductive advantage hypothesis, Chapman and Hausfater (1979) analyzed the interaction of various demographic and reproductive factors in determining the reproductive advantage obtained by infanticidal males. The Chapman-Hausfater model demonstrated that under most conditions infanticidal males do obtain a substantial reproductive advantage compared to their noninfanticidal counterparts. However, when an infanticidal male is himself replaced infanticidally after a length of residency, or tenure, of 8–19 months, 32–40 months, 54–61 months, or 77–80 months inclusive, then the reproductive advantage obtained by the first male through infanticide is completely eliminated by the actions of his successor; the first male will thus produce fewer offspring than would have a noninfanticidal male replaced under the same conditions.

However, tenure length in the bisexual troop was not the only important determinant of the reproductive advantage accruing to infanticidal males. The advantage obtained by such males was also strongly influenced by the characteristics and length of the interval between successive conceptions by females under natural conditions. In particular, the longer the duration of lactational amenorrhea, and the greater the proportion of the interconception interval spent in this state, the greater the reproductive advantage obtained by infanticidal males. Among langurs, recent data from the long-term study of Vogel and co-workers at Jodhpur indicate that females in this population show a surprisingly short lactational amenorrhea of only 2–4 months, measured as the time from birth of a surviving infant to the first signs of behavioral estrus by the mother (Vogel and Loch, in prep.). Moreover, the total time required to proceed from a live birth to the subsequent conception was calculated as only 8.8 months and the total interbirth interval averaged only 18–19 months, when estimated from

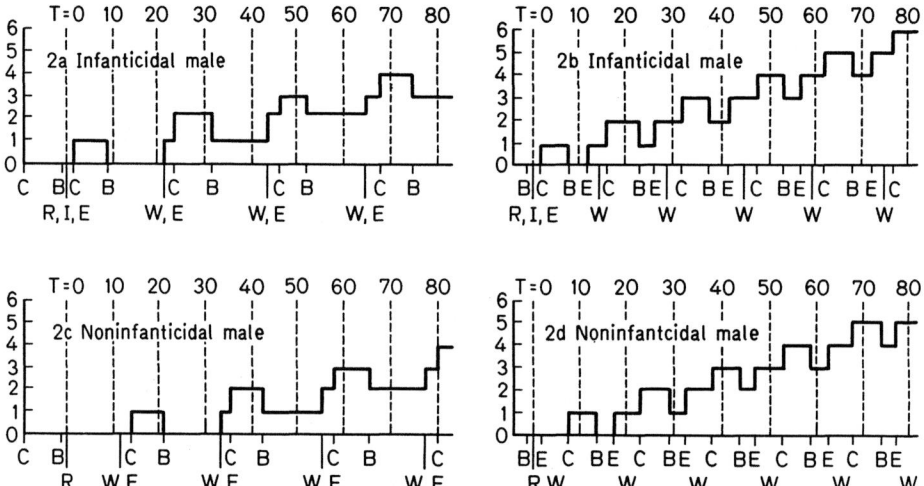

Fig. 2 a–d. Cumulative number of offspring *(vertical axis)* that are sired by an infanticidal (a, b) and noninfanticidal (c, d) male with a given female and that survive subsequent replacement of their sire by an infanticidal male. a and c are drawn using estimates of interconception interval length and the timing of reproductive events reported in cross-sectional studes of langurs. b and d are drawn similarly, but using revised estimates of interconception interval length and the timing of reproductive events obtained from the long-term study at Jodhpur. The letters *C, B, R, I, E,* and *W* along the *horizontal axis* represent conception, birth, adult male replacement, infanticide, resumption of sexual cycles and estrus, and weaning, respectively. Various time intervals after the adult male replacement, each a potential value of *T*, are shown across the *top* of the figure and are further demarcated by *broken vertical lines.* See text for further explanation

a population-wide census, and 15.3 months, when estimated from a sample (n = 24) of females of known reproductive history.

Figure 2 compares the offspring production schedule of an infanticidal male replaced infanticidally, as estimated in the Chapman-Hausfater formulation, and similar calculations made by Winkler using revised estimates of langur reproductive parameters obtained from the Jodhpur study. The shortened duration of amenorrhea and of overall interconception interval for the Jodhpur population compared to previous estimates of these parameters from cross-sectional studies of langurs results in a much smaller reproductive advantage, on average, accruing to infanticidal males and also much shorter temporal "windows" during which infanticide would be expected to occur following adult male replacement in bisexual group. However, it has been argued that these reproductive data were obtained during a period when the main Jodhpur study troops approximately doubled in size and thus may not be the most representative parameter values to use in testing models of infanticide in langurs. Nevertheless, the Jodhpur study has certainly provided the most exact estimates of langur reproductive parameters yet obtained from any field study of this genus.

In sum, it is premature to rule out nearly any of the above hypotheses concerning infanticide in langurs. Nevertheless, as a result of several recent review papers on infanticide, mathematical analyses of the sort discussed below, and the present symposium, a much clearer statement of the alternative hypotheses has emerged as well as a much clearer understanding of the kinds of data both necessary and available to test these hypotheses. Continuation of the ongoing long-term studies of langurs seems a prime requisite for obtaining the kinds of data required to differentiate among the above alternatives and, of course, to generate new and more precise hypotheses. Thus, it is particularly unfortunate that shortly after this symposium, the Government of India decided not to issue extensions for the long-term studies of Indo-German and American research teams. Hopefully, the appropriate Indian authorities and agencies will reconsider this action in the near future and allow these important and productive research programs to continue.

Equilibrium Models of Infanticide

Regardless of further revisions in the estimates of langur reproductive parameters, it seems likely that one basic conclusion of the Chapman-Hausfater analysis will remain valid: that once infanticide is introduced into a population, it will spread among the males of that population even if some individuals are replaced at tenures otherwise disadvantageous for the trait. In a recently completed model of infanticide as a polymorphic strategy, Hausfater et al. (in press) have demonstrated the necessary and sufficient conditions for infanticide to reach fixation in a population and also the conditions under which the trait will persist in the population at an equilibrium frequency below 100% of all males. Briefly, this model showed that when the expected reproductive success $(\widetilde{R}_{II})$, averaged over all tenures, of infanticidal males who are themselves replaced infanticidally equals or exceeds the expected reproductive success $(\widetilde{R}_{NI})$, averaged over all tenures, of noninfanticidal males replaced under the same conditions, then the male population will rapidly move to fixation for infanticide. This end point was further shown to be mathematically stable. In contrast, when the expected reproductive success of infanticidal males replaced infanticidally was less than that of their noninfanticidal counterparts, then the trait will remain present in the population at an equilibrium frequency between zero and one and this end point is also mathematically stable. The specific equilibrium frequency (P_o) of infanticidal males reached under this second set of conditions can be calculated by the formula:

$$P_o = \frac{\widetilde{R}_{IN} - \widetilde{R}_{NN}}{\widetilde{R}_{IN} - \widetilde{R}_{NN} + \widetilde{R}_{NI} - \widetilde{R}_{II}}$$

where $\widetilde{R}_{IN}$ and $\widetilde{R}_{NN}$ denote the expected reproductive success, averaged over all tenures in the noninfanticidal replacement condition, of infanticidal and noninfanticidal males, respectively, and $\widetilde{R}_{II}$ and $\widetilde{R}_{NI}$ are defined as above.

Using several different, but realistic estimates of the expected reproductive success of infanticidal and noninfanticidal males under each replacement condition, this new model yielded equilibrium frequencies for infanticidal males as low as 3%, e.g., when males were replaced at an average tenure of 19 months, to over 95%, e.g, when males were replaced at an average tenure of 56 months. In general, only 100 to 600 generations were required for infanticide to reach the predicted equilibrium frequency when the dynamics of langur populations were simulated in a related computer model. In sum, this new model of infanticide as a polymorphic trait not only again emphasizes the need for basic data on reproductive and demographic parameters of langur populations, but also provides a potential explanation for the differing frequency of infanticide among such populations.

Female Counterstrategies

Given the lack of sufficient information to test most hypotheses concerning infanticide as a male reproductive strategy in langurs, it is obviously even more difficult to evaluate hypotheses concerning potential female counterstrategies: To evaluate the effectiveness of a counterstrategy, one must first have a good understanding of the basic tactics that the countermoves are intended to discourage. Nevertheless, it may be helpful to review briefly the potential female counterstrategies which were discussed during the symposium.

The most direct and effective counterstrategy to infanticide, regardless of its origins, would be for females in a bisexual troop to band together and wage an active and successful defense of their dependent infants from the attacks of the new resident male. Alternatively, the females might act in concert to defend their current resident male and thereby to forestall his replacement by a migrant individual. In fact, female cooperation in defense of their infants has been observed at two study sites (Hrdy 1977a, Loch, in prep.). A similar tactic would be for females to structure their behavior, both aggressive and nonaggressive, in such a way as to precisely control the timing of an adult male replacement, and in fact, female control of the timing of male replacement, particularly when coupled with reproductive synchrony by the females, would produce one of the most effective counterstrategies to infanticide. Thus, if adult male replacements occurred exclusively when all females in the group were in estrus or were pregnant, there would be no infants killed at all, regardless of the infanticidal tendencies of the new resident male (see Chapman and Hausfater 1979 for details of this argument).

In relation to the social bonding and reproductive advantage hypotheses, a effective tactic would be for females to enter a "pseudoestrus" coincident with an adult male replacement. Then, by consorting with the new male, the female might successfully forestall the aggression to themselves and their offspring that might otherwise be expected had they not rapidly exhibited estrous behaviors. Although "pseudoestrous" behavior has been reported in a few cases (Hrdy 1977a), there exists a question as to how "pseudo" such behavior actually was; the revised estimates of reproductive parameters from Jodhpur (see p. 169) make it seem likely that the reported cases of pseudoestrus may, in fact, have been true estruses.

In a small number of cases, a female with an unweaned infant has been observed to leave the bisexual troop during the time of an adult male's replacement (Hrdy 1977a). These observations have led to the suggestion that if infanticide represents a transient psychological or social condition on the part of the new resident male (or the entire troop), females and their unweaned infants can escape infanticide by a tactic of temporarily abandoning the bisexual troop during the expected time of increased male aggression or until the infant is beyond the age of greatest vulnerability to attack.

Obviously, the fertile mind can generate many more potential female counterstrategies to infanticide by male langurs. Nevertheless, there is as yet no evidence that female langurs in any study population regularly or consistently engage in specific behaviors that could reasonably be interpreted as a counterstrategy to infanticide. This statement should not be taken to mean that counterstrategies have been disproved or that their existence is unlikely. Rather, merely that it is still too early to draw any conclusions whatever concerning the existence or need for female counterstrategies to infanticide.

Other Aspects of Female Social Behavior

An important characteristic of langur social organization, as in many other Colobines, is a high frequency of infant transfer within the troop and of allomaternal care of infants by both related and unrelated individuals. Allomothering and infant transfer in Colobines has been analyzed by Horwich and Manski (1975), Hrdy (1976, 1978) and Vogel (1979) and there is close agreement between these studies concerning the nature and frequency of allomothering within bisexual troops. There is also good agreement between researchers on the potential costs and benefits of allomothering for the mother, the infant and the allomother, but quantitative assessment of the hypothesized costs and benefits of this behavior remain to be completed.

Briefly, the most direct benefit to the mother of infant transfer is supposed to be that the burden of infant care is reduced through the participation of allomothers. The allomother is believed to benefit by "learning to mother", while presumably the infant benefits through increased protection and enriched social experience (for discussion see Hrdy 1976, Vogel 1979). However, allomothering also has some relatively high costs, especially so for the infants, in that some classes of allomothers are extremely abusive in their treatment of infants (Hrdy 1978, Vogel 1979). Generally, the most solicitous allomothers are younger nulliparous females. The least solicitous, often abusive allomothers are females whose own infants are near to weaning or newly independent. Pregnant females and females with newborn infants are generally careful allomothers (Hrdy 1978).

Hopefully, through long-term studies of langurs and other Colobines, quantitative data will be obtained on the survival and social development of infants who have received differing amounts of allomaternal care. Similarly, long-term studies of langurs should eventually provide information for comparison of the reproductive performance and maternal success of adult females who themselves experienced and/or gave different amounts of allomaternal care prior to their own maturation.

Another important characteristic of langur social organization is relatively stable female dominance relationships (Hrdy and Hrdy 1976). In contrast to the so-called nepotistic dominance system of female macaques and baboons (Hausfater 1975), dominance rank among female langurs is believed to be age-dependent with younger adult females generally dominant to older adult females (Hrdy and Hrdy 1976).

This latter finding has been disputed by Dolhinow et al. (1979) who did not find linear and age-dependent female rank orders in two colony troops. Further complicating this matter is the finding that different langur troops at Jodhpur apparently show different types of female rank orders (Loch and Vogel, personal communication). Unfortunately, data are as yet insufficient to evaluate rank (or age) effects on female reproductive success, offspring survival, etc. in the wild. Also, further data and analyses are required to more fully understand the complex interplay of social and demographic processes in producing age-graded patterns of rank relations among langurs (Hausfater et al. 1981) as well as in producing other special characteristics of langur social organization.

Outlook

The study of langur social behavior has reached a critical point in its development. Recent advances in ethological and sociobiological theory have produced an array of explicit hypotheses and quantitative models concerning the social strategies of male and female langurs. At the same time, long-term and longitudinal field studies of langurs in their natural environment have begun to provide the appropriate data for testing such hypotheses and models. Thus, studies of social behavior in the genus *Presbytis* seem likely to provide extremely important insights into the evolutionary significance of variability in the social behavior of individuals, troops, and populations. Hopefully, as a result of this symposium, cooperation between theoreticians and field workers, and between members of various research teams, will be increased and research efforts more closely directed toward filling the gaps in our knowledge of the social, demographic, and ecological influences on the social behavior of male and female langurs.

Symposium Participants

The following is a list of the participants in the I.P.S. pre-Congress satellite symposium entitled "Theoretical models on male and female reproductive strategies with respect to the actual behavior of langurs":

Christian Vogel (Symposium Organizer), University of Göttingen (Germany)
Silvana Borgognini-Tarli (Local Organizer), University of Pisa (Italy)
Glenn Hausfater (Rapporteur), Cornell University (United States)
Ann Baker-Dittus, Smithsonian Institution Primate Project (Sri Lanka)
Wolfgang Dittus, Smithsonian Institution Primate Project (Sri Lanka)
Axel Goldau, University of Göttingen (Germany)
Sarah Blaffer Hrdy, Harvard University (United States)
Jutta Küster, University of Göttingen (Germany)
Hartmut Loch, University of Göttingen (Germany)
James Moore, Harvard University (United States)
John Oates, Hunter College (United States)
Paul Winkler, University of Göttingen (Germany)

Acknowledgments. The assistance of Paul Winkler in the preparation of this report, especially Figs. 1 and 2, is gratefully acknowledged. We are indebted to the Organizing Committee of the 8th Congress of the IPS for their assistance in organizing this symposium as well as to the various institutions funding our field projects. We are extremely grateful to all participants in this integrative symposium and particularly to Dr. Silvana Borgognini-Tarli for her cordial sponsorship.

References

Angst W, Thommen D (1977) New data and a discussion of infant killing in old world monkeys and apes. Folia Primatol 27:198–229

Boggess J (1979) Troop male membership changes and infant killing in langurs *(Presbytis entellus)*. Folia Primatol 32:65–107

Chapman M, Hausfater G (1979) The reproductive consequences of infanticide in langurs: a mathematical model. Behav Ecol Sociobiol 5:227–240

Curtin R (1977) Langur social behavior and infant mortality. In: Boaz NT, Cronin JE (eds) The Kroeber Anthropological Society Papers, No 50. Berkeley papers in physical anthropology. Kroeber Anthropological Society, Berkeley, California, pp 27–36

Curtin R, Dolhinow PC (1978) Primate social behavior in a changing world. Am Sci 66:468–475

Dittus WPJ (1977) The social regulation of population density and age-sex distribution in the toque monkey. Behavior 63:281–322

Dolhinow P, McKenna JJ, Vonder Haar Laws J (1979) Rank and reproduction among female langur monkeys: aging and improvement (they're not just getting older, they're getting better). Aggr Behavior 5:19–30

Eisenberg JF, Muckenhirn NA, Rudran RL (1972) The relation between ecology and social structure in primates. Science 176:863–874

Hausfater G (1975) Dominance and reproduction in baboons *(Papio cynocephalus):* a quantitative analysis. S Karger, Basel

Hausfater G (1977) To the editors. Am Sci 65

Hausfater G, Aref S, Cairns SJ (in press) Infanticide as an alternative male reproductive strategy in langurs: A mathematical model. J Theor Biol

Hausfater G, Saunders CD, Chapman M (1981) Computer models of primate life-histories. In: Alexander RD, Tinkle D (eds) Natural selection and social behavior. Chiron Press, New York

Hladik CM, Hladik A (1972) Disponibilités alimentaires et domaines vitaux des primates à Ceylon. Terre Vie 2:149–215

Horwich RH, Manski D (1975) Maternal care and infant transfer in two species of Colobus monkeys. Primates 16:49–73

Hrdy SB (1974) Male-male competition and infanticide among the langurs *(Presbytis entellus)* of Abu, Rajasthan. Folia Primatol 22:19–58

Hrdy SB (1976) Care and exploitation of nonhuman primate infants by conspecifics other than the mother. In: Rosenblatt JS, Hinde RA, Shaw E, Beer C (eds) Advances in the study of behavior, vol VI. Academic Press, London New York, pp 101–158

Hrdy SB (1977a) The langurs of Abu. Harvard Univ Press, Cambridge

Hrdy SB (1977b) Infanticide as a primate reproductive strategy. Am Sci 65:40–49

Hrdy SB (1978) Allomaternal care and abuse of infants among Hanuman langurs. In: Chivers DJ, Herbert J (eds) Recent advances in primatology, vol I: Behavior. Academic Press, London New York, pp 169–172

Hrdy SB (1979) Infanticide among animals: a review, classification, and examination of the implications for the reproductive strategies of females. Ethol Sociobiol 1:13–40

Hrdy SB, Hrdy DB (1976) Hierarchical relations among male Hanuman langurs (Primates: Colobinae, *Presbytis entellus*). Science 193:913–915

Maynard-Smith J (1974) The theory of games and the evolution of animal conflicts. J Theor Biol 47:209–221

Mohnot SM (1971) Some aspects of social changes and infant-killing in the hanuman langur *Presbytis entellus* (Primates: Cercopithecidae), in Western India. Mammalia 35:175–198

Muckenhirn NA (1975) Non-human primates. Usage and availability for biomedical programs. National Academy of Sciences, Washington

Oppenheimer JR (1977) *Presbytis entellus,* the Hanuman langurs. In: HSH Prince Rainier III of Monaco, Bourne GH (eds) Primate conservation. Academic Press, London New York, pp 469–512

Ripley SC (1970) Leaves and leaf-monkeys. In: Napier JH, Napier PH (eds) Old world monkeys: evolution, systematics and behavior. Academic Press, London New York, pp 481–509

Ripley S (1980) Infanticide in langurs and man: adaptive advantage or social pathology. In: Cohen MN, Malpass RS, Klein HG (eds) Biosocial mechanisms of population regulation. Yale Univ Press, New Haven London, pp 349–390

Rudran R (1973) Adult male replacement in one-male troops of purple-faced langurs *(Presbytis senex senex)* and its effect on population structure. Folia Primatol 19:166–192

Sugiyama Y (1964) Group composition, population density and some sociological observations of Hanuman langurs *(Presbytis entellus)*. Primates 5:7–37

Sugiyama Y (1965) On the social change of Hanuman langurs *(Presbytis entellus)* in their natural condition. Primates 6:381–418

Sugiyama Y (1966) An artificial social change in a Hanuman langur troop *(Presbytis entellus)*. Primates 7:41–72

Sugiyama Y, Yoshiba K, Pathasarathy MD (1965) Home range, mating season, male group and inter-troop relations in Hanuman langurs *(Presbytis entellus)*. Primates 6:73–106

Vogel C (1975) Soziale Organisationsformen bei catarrhinen Primaten. In: Kurth G, Eibl-Eibesfeldt I (eds) Hominisation und Verhalten. G Fischer, Stuttgart, pp 159–200

Vogel C (1977) Ecology and sociology of *Presbytis entellus.* In: Prasad MRD, Anand Kumar TC (eds) Use of non-human primates in biomedical research. Indian National Science Academy, New Delhi, pp 24–45

Vogel C (1979) Der Hanuman Langur *(Presbytis entellus),* ein Parade-Exempel für die theoretischen Konzepte der Soziobiologie? In: Rathmayer W (ed) Verhandlungen der Deutschen Zoologischen Gesellschaft 1979 in Regensburg. G Fischer, Stuttgart, pp 73–89

Wilson EO (1975) Sociobiology, The New Synthesis. The Belknap Press of Havard Univ Press, Cambridge Massachusetts London

Wolf KE, Fleagle JG (1977) Adult male replacement in a group of silvered leaf-monkeys *(Presbytis cristata)* at Kuala Selangor, Malaysia. Primates 18:949–955

Recent Advances in the Study of Tool-Use by Nonhuman Primates

W.C. McGREW[1]

Introduction

This paper reports on a symposium on "Tool-Use by Nonhuman Primates", organized by the author, and held in Florence on 5 July 1980. It was one of the satellite symposia commissioned by B. Chiarelli, as part of the VIIIth Congress of the International Primatological Society. It was hosted by the Istituto Italiano di Preistoria e Protostoria. Seven papers were presented, and these will be published together in 1982, as a special issue of the *Journal of Human Evolution.*

This report will be done in four parts: (1) general background knowledge on the phenomenon; (2) recent developments in research; (3) summary and synthesis of the papers presented; (4) suggestions for direction of future research. In none of these parts can the author be comprehensive; interested readers are referred to the forthcoming publication of the symposium for specific details and to Beck's (1980) excellent volume for a general overview.

Background

Before the early 1970s, studies of tool-use by nonhuman primates were mainly of two kinds: First, in the wild, these were largely limited to descriptive accounts. Many of these were anecdotal by-products of more general field studies (e.g., Beatty 1951, Jones and Sabater Pi 1969). Such reports continue to provide useful information (e.g., Plooij 1978, Nishida and Uehara 1980), but by their brief nature, they often raise more questions than they answer. More useful are the more extensive reports, which represent natural history in the best sense of the term. Best known of these are the publications of Jane Goodall (1964, 1968) on the tool-use of wild chimpanzees in the Gombe National Park, Tanzania. These revealed a whole new side of daily life, the existence of which had previously been completely unsuspected, e.g., using probes to fish for termites for food. These also indicated underlying cognitive capacities of some complexity, e.g,, the ability to *make* tools, in advance of their use. These findings rightly received much attention, and the textbooks on human evolution were altered accordingly.

1 Department of Psychology, University of Stirling, Scotland

Second, in zoological gardens and laboratories, studies of tool-use proceeded in an entirely different vein. Primarily, these were embedded in the comparative psychology of problem-solving. Many of the problems were replications of those devised many years before by such founding fathers as Köhler and Yerkes. Examples of these are such classic standard procedures as using a hoe-like implement to rake in food items which are horizontally out of reach, or stacking boxes to obtain access to food items suspended overhead (e.g., Schiller 1952). Such studies used tool use as a measure of intelligence and examined the ontogeny of behaviour in which nature and nurture obviously interact. The design of such experiments tended to be operationally rigorous, in contrast to their unsystematic counterparts in the field. However, no attempt was made to relate such abilities to the natural selection which had shaped them as adaptations in nature. (This is, of course, understandable, since at that time such adaptations had yet to be recognised by primatologists.)

Recent Developments

From the early 1970s, new avenues of research on tool-use began to be explored; these reflected the new sorts of questions being asked. At least three lines of inquiry emerged, and the papers presented in the symposium will be used to exemplify these.

Both in captivity and nature, there has been the realisation that the occurrence of tool use is not phylogenetically graduated on some simple, progressive scale. Even invertebrate forms may use tools impressively in a variety of modes (Beck 1980). Similarly, the mental capacities underlying tool-use are recognised as not being discrete from those involved in many other behaviours. The frustrating search for a logically and biologically satisfying definition of tool-use shows this aptly. Chimpanzees may be the predominant tool-users, but there are some apes which never use tools and some nonprimates which do so regularly. This has led to efforts to look at species of primates other than *Pan troglodytes* and at nonmammalian species which exhibit other types of object manipulation. Such studies still concentrate on feeding, but shelter construction has also been examined (e.g., Lethmate 1977). The papers presented at the symposium by Beck and Lethmate illustrate these recent tendencies.

In studies of captive primates, emphasis has shifted to the social and cognitive operations underlying the actions observed. More taxing tasks have been set in more flexible conditions (Döhl 1973). Attention has focused on the social transmission of skills in tool-use, even to the extent of the communication and cooperation involved (e.g., Beck 1973). This has been related to increased interest in the ontogeny of tool-using behaviours. Most importantly, there has been a growing recognition that studies in captivity should complement those in the wild, that tool-use in the laboratory might elucidate processes, which because of lack of controlled conditions in nature, would always be confounded there (e.g., Wright 1972). In this sense, tool-use is an obvious type of behaviour upon which to undertake an ethology of cognition (Savage-Rumbaugh et al. 1978). In this symposium, the papers by Kitahara-Frisch and Norikoshi and by Jordan typify this trend.

In field studies, growing emphasis has been placed on quantified, systematic study of the factors which shape the behaviour and on the adaptiveness of tool-use in both proximate and ultimate senses. Notice has been taken of environmental constraints, whether these be in the diversity of prey, availability of raw materials, or nutritional aspects of diet. For those tool-using behaviours which are predatory, primarily on social insects, there has been increased interest in the biology of the prey. It is now clear that one cannot understand termite-fishing, e.g., without taking the termites' point of view as well as the chimpanzee's. Finally, there have begun to be comparative studies of different populations of the same species, especially in chimpanzees (McGrew et al. 1979). If these studies were being conducted of the material cultures of various human societies, we would call them cross-cultural. The papers in the symposium by Nishida and Hiraiwa and by McBeath and McGrew are examples of this.

Wider Perspectives

In the opening paper, Beck (1982) spoke of *chimpocentrism*, i.e. the biassed focusing of attention on tool-use by *Pan troglodytes*, to the exclusion of others. To show how this has diverted attention from other taxa, he compared tool-use by chimpanzees in obtaining social insects with another form of object manipulation: shell-dropping by herring gulls *(Larus argentatus)*.

He described how the bird picks up a living mollusc in its beak, flies some distance, drops it from a height, and repeats this until the shell breaks. It then eats the soft, inner parts of the mollusc. He compared the chimpanzees and gulls on eight grounds, and both show: concentration and perseveration, selectivity in habitat-choice, transport of object to site-of-use, appropriate modification of performance, flexibility in tactics, ontogenetic improvement in technique, probably observational learning, and inter-populational differences.

This is not to say that chimpanzees' termite-fishing and gulls'-shell dropping are equivalent on all points: For example, in the former tools are *made*, as well as *used*. But it shows that, on the basis of the performance, we have virtually equal grounds for inferring the mental capacities underlying the behaviour. This neatly recalls Beck's original point about chimpocentrism. On the basis of this and other arguments, Beck concluded that such subsistence is less important in the evolution of higher intelligence than are the selective pressures of social life with all its attendant complexities (e.g., Humphrey 1976).

Lethmate (1982) provided a synthetic paper (read by McGrew) on the tool-using skills of two species of great apes. This was based on a wide review of tool-use in captive and wild chimpanzees compared with captive, wild, and rehabilitated orangutans *(Pongo pygmaeus)*. It emerged from this that captive and rehabilitated orangutans perform virtually all types of tool-use known for chimpanzees, plus some new ones, e.g. the spontaneous making of tools by combining different sorts of elements. However, this proven ability makes all the more puzzling the almost complete absence of the behaviours in wild orangutans, especially when wild chimpanzees show such a wide repertoire. Why should this ability be so seldom used by orangutans in nature? Lethmate speculates that orangutans may be desocialised apes, and that their ancestors had a richer social life like other great apes which selected for high intelligence (e.g., Humphrey 1976). In more recent times, they were forced to adapt to arboreal foraging in patchy environments of primary forest, which for a large-bodied form requires near-solitariness. The cognitive capacities are still expressed in food finding and processing, but tool-use is large irrelevant in arboreal fruit-eating.

Studies in Captivity

More has been printed on the tool-use of common chimpanzees than on all other non-human species combined. It is therefore ironic and regretable that so little is known of this activity in its sib species, the pygmy chimpanzee or bonobo, *Pan paniscus*. Jordan (1982) has studied the captive pygmy chimpanzees in Frankfurt, Stuttgart, and Antwerp for 6 years. Her paper (read by Tutin) presented the first observations of spontaneous tool-use in this species. These were presented for direct comparison with the common chimpanzee, using the classification of modes given by Beck (1975). It transpired that Jordan has recorded 14 of the 20 modes. These range from the simpler modes of *brandish, club,* and *throw* to the more sophisticated ones of *probe, lever,* and *absorb.* One concludes from this that no obvious difference in tool-using ability exists between the two species, and that further studies seem likely to fill in the remaining gaps.

A more specific study focussing on one of Beck's modes was that of spontaneous use by captive chimpanzees of tools as absorbent devices for liquids. This was reported by Kitahara-Frisch in a paper co-authored with Norikoshi (1982), based on studies in the Tama Zoo, Tokyo. Two young chimpanzees were presented with the analogue of leaf-sponging, as reported in Tanzania by Goodall (1968). In the wild, chimpanzees pluck, crush, dunk, and suck on leaves which they insert into tree-holes to obtain drinking water.

In Kitahara-Frisch and Norikoshi's study, fruit-juice was made available in a narrow-mouthed container attached to the cage. Green branches were provided as potential raw materials. Over a series of 21 testing sessions, one chimpanzee showed progression from: insertion of a finger → insertion to leaf → insertion of short twig → insertion of longer twigs → splitting of ends of inserted twigs → chewing on ends of twigs to produce a frayed, brush-like object. The final stage, which occurred in the last six sessions, was demonstrably more efficient than the unmodified twigs as an absorbent device. The behaviour developed individually, without shaping, in a captive-born subject. This serves as a reminder that social transmission of techniques, which is so often presumed in discussion of nonhuman culture (e.g., Bonner 1980, McGrew and Tutin 1978), is not always necessary in the acquisition of a skilled behaviour.

Studies in Nature

Nishida presented a paper (coauthored with Hiraiwa 1982) on tool use by wild chimpanzees at Kasoje in Tanzania to obtain wood-boring ants (*Camponotus* spp.). The procedure involves probing with tools made from vegetation into the arboreal nests of the ants, i.e., ant-fishing (Nishida 1973). The new findings covered a variety of topics, such as: Chimpanzees show traditional exploitation of particular, individual trees which contain the ants. They use six distinct types of tools to do so, and these evidence selectivity in raw materials. There is competition between chimpanzees for tools and fishing sites. The ontogeny of tool-use in ant-fishing is delayed by comparison with termite-fishing apparently as a function of the fiercer defensive tactics of the prey. The weight of ants consumed per bout is very small, i.e., only about 2 g. This contrasts with other types of insect-eating shown by the same chimpanzees and suggests that *Camponotus* have nonnutritional significance in the diet. Given their well-developed formic acid glands, they may represent a spicy "junk food". The latter finding is of particular interest to those who question the existence of arbitrary social traditions in nonhuman forms.

Finally, a paper by McBeath and the author (1982) attacked a more specific problem, as part of ecological studies of wild chimpanzees at Mt. Assirik, Senegal. The study centred on the influence of habitat on the performance of termite-fishing. We collected 279 tools from five types of habitat and found that 67% of the tools came from a type of habitat occupying only 5% of the study area. Why should the tools be so concentrated in the ecotone between short-grass plateau and woodland?

Certain potential explanations were easily eliminated: Neither the abundance nor quality of termite-mounds was greater in ecotone. Chimpanzees were *not* spending time in the ecotone for other reasons, that is, the effect was not simply an artifact. The abundance and array of types of raw materials, e.g., twig, vine, leaf-stem, etc. were *not* greater in the ecotone, nor was the abundance of individual plants of tool-producing species. Instead the concentration was apparently due to the differential distribution of *one* species of plant, *Grewia lasiodiscus,* which provided 80% of the tools, and which was most commonly found at mounds in the ecotone.

Directions for Future Research

Primatologists studying tool-use should be more attentive to studies of the phenomenon in other taxa. Here, as in other areas of primatology, there has been a tendency to ignore other forms, especially mammalian ones, whilst concentrating on human primates for comparison (Eisenberg 1973). For example, it would be useful to see a detailed study of the ontogeny of hammer-stone usage by sea otters *(Enhydra lutris),* with longitudinal studies conducted in the field as well as making use of existing captive individuals.

Laboratory workers should be encouraged to link their studies to naturally occurring behaviours, e.g., to set adaptively meaningful tasks to their subjects. For instance, it would be helpful to have a study of fishing-tool-use in conditions where exposure to models could be controlled. This might elucidate the mechanisms involved in the acquisition and possible transmission of the behaviours. Several zoological gardens have artificial termite-mounds where such research could be done (Hancocks 1980).

Field-workers should strive to relate the tool-use seen to environmental constraints, such as the biology of the prey, phenology of raw materials, etc. One looks to further results from Boesch's (1978) study of stone tool-use by chimpanzees in the Tai Forest, Ivory Coast. It appears that the apes there choose stones of appropriate size and kind, according to the species of hard-shelled fruit to be smashed open. When possible, field experimentation should be added to observation, e.g., in the way that Norton-Griffiths (1967) so ingeniously did in testing forces applied in techniques used by oyster-catchers *(Haematopus ostralegus)* in opening molluscs.

Cross-cultural comparisons amongst different populations of chimpanzees should be expanded. In addition to contrasting widely separated populations in different parts of Africa, cross-subcultural comparisons should be made of neighbouring communities within the same regional population. Exchanges of visits between field-sites and co-operative field-studies should be encouraged, so that raw data, and not just results, can be juxtaposed.

Finally, at some point, a meeting should be convened which brings together archaeologists, anthropologists studying the material cultures of preliterate peoples, and primatologists studying tool-use. Obvious grounds for such a meeting already exist, e.g., in Wynn's (1979) speculative reconstruction of the cognitive abilities needed for Acheulean lithic technology, in the comprehensive taxonomy of subsistence tools used by living foraging peoples, devised by Oswalt (1973). Such a multidisciplinary group could tackle questions of common concern such as definitions of tools, typology, constraints of raw materials, dissemination of skills, etc.

That such an overlap exists was brought home most pointedly in a short paper by Prasad (1982), delivered by Baldwin. Prasad described a single, quartzitic pebble-tool recovered from late Mio-Pliocene (9–10 million years B.P.) sediments in India. The same formation has yielded remains of *Ramapithecus,* so that this tool, if genuine, may represent the earliest known implement used by a primate. One can only hope that further studies produce more examples of this provocative find.

References

Beatty H (1951) A note on the behavior of the chimpanzee. J Mammal 32:118

Beck BB (1973) Cooperative tool use by captive hamadryas baboons. Science 182:594–597

Beck BB (1975) Primate tool behavior. In: Tuttle RH (ed) Socioecology and psychology of primates. Mouton, The Hague, pp 413–447

Beck BB (1980) Animal tool behavior. Garland STPM Press, New York

Beck BB (in press) (1982) Chimpocentrism: Bias in cognitive ethology. J Hum Evol

Boesch C (1978) Nouvelles observations sur les chimpanzes de la forêt de Tai (Côte d'Ivoire). Terre Vie 32:195–201

Bonner JT (1980) The evolution of culture in animals. Univ Press, Princeton, New Jersey

Döhl J (1973) Gedächtnisprüfung eines Schimpansen für erlernte komplizierte Handlungsweisen. Z Tierpsychol 33:204–208

Eisenberg JF (1973) Mammalian social systems: Are primate systems unique? In: Menzel EW (ed) Precultural primate behavior. S Karger, Basel, pp 232–249

Goodall J (1974) Tool-using and aimed throwing in a community of free-living chimpanzees. Nature (London) 201:1264–1266

Goodall J v L (1968) The behaviour of free-ranging chimpanzees in the Gombe Stream Reserve. Anim Behav Monogr 1:161–311

Hancocks D (1980) Bringing nature into the zoo: Inexpensive solutions for zoo environments. Int J Study Anim Probl 1:170–177

Humphrey NK (1976) The social function of intellect. In: Bateson PPG, Hinde RA (eds) Growing points in ethology. Univ Press, Cambridge, pp 303–317

Jones C, Sabater Pi J (1969) Sticks used by chimpanzees in Rio Muni, West Africa. Nature (London) 223:100–101

Jordan C (in press) (1982) Object manipulation and tool-use in captive pygmy chimpanzees *(Pan paniscus)*. J Hum Evol

Kitahara-Frisch J, Norikoshi K (in press) (1982) Spontaneous sponge-making in captive chimpanzees. J Hum Evol

Lethmate J (1977) Nestbauverhalten eines isoliert aufgezogenen, jungen Orang-Utans. Primates 18: 545–554

Lethmate J (in press) (1982) Tool-using skills of orang-utans. J Hum Evol

McBeath NM, McGrew WC (in press) (1982) Tools used by wild chimpanzees to obtain termites at Mt. Assirik, Senegal: The influence of habitat. J Hum Evol

McGrew WC, Tutin CEG (1978) Evidence for a social custom in wild chimpanzees? Man 13:234–251

McGrew WC, Tutin CEG, Baldwin PJ (1979) Chimpanzees, tools, and termites: Cross-cultural comparisons of Senegal, Tanzania, and Rio Muni. Man 14:185–214

Nishida T (1973) The ant-gathering behaviour by the use of tools among wild chimpanzees of the Mahali Mountains. J Hum Evol 2:357–370

Nishida T, Hiraiwa M (in press) (1982) A natural history of a tool behaviour by wild chimpanzees in feeding upon wood-boring ants. J Hum Evol

Nishida T, Uehara S (1980) Chimpanzees, tools, and termites: Another example from Tanzania. Curr Anthropol 21:671–672

Norton-Griffiths MN (1967) Some ecological aspects of the feeding behaviour of the oystercatcher *Haematopus ostralegus* on the edible mussel *Mytilus edulis*. Ibis 109:412–424

Oswalt WH (1973) Habitat and technology: The evolution of hunting. Holt, Rinehart and Winston, New York

Plooij FX (1978) Tool-use during chimpanzees' bushpig hunt. Carnivore 1:103–106

Prasad KN (in press) (1982) Was *Ramapithecus* a tool-user? J Hum Evol

Savage-Rumbaugh ES, Rumbaugh DM, Boysen S (1978) Linguistically-mediated tool use and exchange by chimpanzees *(Pan troglodytes)*. Behav Brain Sci 1:539–554

Schiller PH (1952) Innate constituents of complex responses in primates. Psychol Rev 59:177–191

Wright RVA (1972) Imitative learning of a flaked stone technology – the case of an orangutan. Mankind 8:296–306

Wynn T (1979) The intelligence of later Acheulean hominids. Man 14:371–391

Primate Communication in the 1980s: Summary of the Satellite Symposium on Primate Communication

C.T. SNOWDON[1], C.H. BROWN[2], and M.R. PETERSEN[3]

Introduction

More than 15 years have passed since the last major symposium on primate communication was held (Altmann 1967). At that first symposium primate communication was in its infancy; the techniques for the analysis of communication were just beginning to be developed and had not been widely applied to primate species. In fact of the 17 papers in the Altmann volume only 5 dealt with the actual structure of communication; the remaining 12 dealt with reproductive behavior, agonistic behavior, and social dynamics.

The ensuing 15 years have seen the rapid development of techniques designed to analyze communication signals. We can now feel confident in providing physical descriptions of most auditory and visual signals, many chemical signals, and a few tactile signals. With the physical description of signals now relatively simplified, students of primate communication can now attend to other major problems: What information are the various signals providing about the communicator? How is the information from a signal interpreted and acted upon by the recipient of the signal? What is the relationship of signal structure to the environment in which it is used? What rules govern the sequencing of signals within an individual and between individuals? How do animals acquire the proper use of signals? Do animals perceive the physical aspects of signals in the same way as we human beings do?

We had two major reasons for organizing this symposium. The first was to determine to what extent the preceding questions had been answered and to develop some directions and goals for research on primate communication in the next decade. The second reason for the symposium was to showcase some of the excellent work on primate communication which has been overshadowed by the recent work on language analogs with great apes. It is our opinion that the research on great apes has demonstrated the capacity of these animals for complex cognitive abilities similar to those involved in language. However, these studies do not speak to the evolution of language. Evolutionary processes can best be studied in the spontaneous signals emitted by

1 Department of Psychology, University of Wisconsin, 1202 West Johnson Street, Madison, WI 53706, USA
2 Department of Psychology, University of Missouri, 210 McAlester Hall, Columbia, MO 65211, USA
3 Department of Psychology, Indiana University, Bloomington, IN 47405, USA

animals in normal social circumstances. All of the participants in this symposium, whether they study primates in the field or in captivity, focused their work on the natural, spontaneous communication of monkeys and apes.

Participants in the symposium were investigators who had already made major contributions to the understanding of primate communication in the last ten years or they were young investigators with promising ideas or techniques that seemed valuable for future work. As the symposium progressed, it became clear that there was a striking convergence of ideas, of problems, and of technical solutions to hitherto unsolvable problems. This convergence was all the more remarkable for its appearance among investigators who had had no prior formal contact with one another.

In this summary we identify the major themes of the symposium and provide a brief summary of the points made by individual participants. We are including in this summary the results of some who were unable to attend the symposium, but who are contributing to the published version. The complete text of the symposium will be published as a separate volume (Snowdon et al. 1982).

There were three major themes of the symposium: (1) the definition of the signal of communication, (2) the pragmatics of communication, and (3) the ontogeny of communication.

The Definition of the Signal

Although technical advances in the last 15 years now allow us to define with precision the physical structure of a communicative signal, there are still several major problems of signal definition. The most important of these is that the animal perceiving the signal may not be making the same physical analysis of the signal that our instruments are making nor is it necessarily making the same analysis of the signal as our human perceptual systems do. The most important theme to emerge from the symposium was the definition of the signal from the animal's point of view, and several promising approaches were suggested. The question of signal definition will be divided into seven subtopics: (1) Analysis of signal structure. (2) Psychophysical aspects of signals. (3) Perceptual classification of signals. (4) Syntactic aspects of signals. (5) Hedonic aspects of signals. (6) Semantic aspects of signals. (7) Single versus multidimensional signal systems.

Analysis of Signal Structure

Steven Green introduced the symposium with a discussion of various approaches to the study of primate communication. He distinguished between the functional approach which analyzes communication in terms of the effect the signals have on the responses of the recipients, the informational approach where the focus is on the communicator and the information about the communicator which is available in the signals, the structural approach where the physical properties of the signal are examined, and the neurological approach which examines the brain structures involved in the production and reception of signals. He commented upon two new, exciting approaches to the study of animal vocalizations. The first, which could be labeled the acoustic properties approach is concerned with the relationship between signal morphology, the environment in which it is given, and its function. Several recent studies have presented data on what types of signals are best transmitted in different habitats, which are best for long distance, which best for short distance communication. A related series of studies concerns the psychophysics of signal sensation and perception by animals.

The second new approach involves the application of experimental models and paradigms from linguistics and psycholinguistics to the study of primate communication. This linguistic approach has led to studies on whether monkeys perceive their signals categorically, to studies on the grammatical sequences of calls and to studies of semantics in primate communication. Work relating to all of these points was presented at the symposium and will be reviewed below.

In order to analyze communication with each of these five approaches Green suggested that investigators make increased use of experimental techniques such as playbacks of signals to animals in a social context, more careful determination of the context in which calls are given, including social, experiential, and physiological contexts, and more attention to sequences of calls and the timing of silent intervals between calls. Brief silences have recently been shown to be of great importance in certain aspects of human speech.

Several other papers touched on aspects of signal analysis. Smith, Newman, and Symmes demonstrated the use of discriminant analysis to determine which individual was calling within a group. With a sufficiently large number of individually identified calls for each animal in a group, a set of discriminant functions can be written which will assign subsequent calls to a particular individual with great accuracy.

Gautier and Gautier reported on a different technique for determining the identity of an individual vocalizer within a group using telemetry devices attached to each individual. The transmitted calls are directly recorded by separate pens of a recorder giving an exact temporal record of the calls of each individual. Snowdon described the development of quantitative techniques for discriminating between call types in a repertoire. Sound spectrograms are placed under a calibrated graticule, and several parameters of each call are measured. Then multiple analyses of variance or discriminant analyses are used to determine whether similar calls are drawn from the same distribution or from different distributions. The sources of acoustic variance for a particular call can be partitioned into populational, individual, and contextual components.

Epple described the use of gas chromatography to determine the structure of various chemicals used for communication by monkeys. She could determine individual, gender, and populational differences in scent structure.

Psychophysics and Primate Communication

C. Brown has been studying the psychophysics of sound localization in macaques. Animals are trained to orient toward a sound from one speaker on the midline and then must indicate whether a second sound comes from the same midline speaker or through one of several speakers arrayed in an arc along the side of the monkey.

A variety of different types of stimuli have been used including natural calls of macaques. Brown showed that frequency modulation was an important cue for sound localization. Thresholds of angular separation could be reduced by half with as little as 400 Hz of frequency modulation. Frequency modulated "clear" calls of macaques were as localizable as broadband "harsh" calls. Thus, in macaques one would expect to find "clear" calls used more frequently for communication over distance when localizability is important. And one would expect to find greater frequency modulation in clear calls at greater distances.

Brown also studied localization in three dimensions. Here again broad bandwidth calls were much more localizable than narrow bandwidth calls, but a much greater frequency modulation was necessary for a significant reduction in the localization threshold. Calls with more than 2 kHz frequency modulation were easily localized. This suggests that forest-living monkeys might have more frequency modulation in their calls than animals living only on the ground.

Snowdon and Hodun presented data on pygmy marmosets that supported Brown's findings with macaques. Pygmy marmosets have three forms of trills that differ in the amount of frequency modulation. One trill has less than 1000 Hz modulation; another has 3–4 kHz of frequency modulation, and the last has greater than 5 kHz of modulation. Snowdon and Hodun mapped the home range of one group of pygmy marmosets in the Amazon and then recorded the instances of each type of trill along with the position of caller and recipient. The trill with the least frequency modulation was used only when animals were within 5 m of one another. The trill with the most frequency modulation, and thus the most locatable, was used at the furthest distances. Thus, the psychophysical relationships between signal structure and sound localization described by Brown are found in the design of calls for short- and long-distance communication in monkeys.

Perception of Monkey Vocalizations by Monkeys

M. Petersen presented several studies on how Japanese macaques perceive their own vocalizations. Several years ago Green showed that there were seven different variations of "coo" calls given by Japanese macaques. Petersen selected two of these variants, the "smooth early high" and the "smooth late high", for study. A group of Japanese macaques and a control group of other species of monkeys were trained to discriminate between several tokens of each type of call. The Japanese macaques were quickly able to discriminate between the two call types, regardless of which individual had produced the call. On the other hand the control animals either required several times the number of trials of the Japanese macaques or failed to learn the discrimination at all. In a second study Petersen sorted the calls in terms of pitch. The highest pitched smooth-early-high and smooth-late-high calls were combined into one group the lowest pitched calls of each type were combined into the second group. When this discrimination task was presented the results were completely the opposite. The contra species of monkeys readily learned a discrimination based on pitch while the Japanese macaques learned slowly or not at all. Petersen suggested that these studies provided parallels to studies on human speech perception. The discrimination of coo types is a "phonetic" distinction of importance to Japanese macaques, but presumably not important to the control animals. Thus, Japanese macaques have an advantage in discriminating the phonetic contrasts of their species-specific calls. The pitch discrimination task pitted an irrelevant variable against a relevant variable, the early or late high. For control animals the location of the high was not a relevant cue, hence they could show better performance on the pitch discrimination task.

One final line of support for the species-specific perception of calls came from analysis of the data based on which cerebral hemisphere primarily processed the sounds. When the sounds were presented to the right ear and hence were processed primarily by the left hemisphere, the Japanese macaques showed a better performance level than when sounds were presented to the left ear (or right hemisphere). This right ear advantage was present on the phonetic discrimination task but not on the pitch discrimination task. On the other hand, the control species showed no right-ear (left-hemisphere) advantage with either task. The results showing right-ear advantage on phonetically relevant tasks are exactly parallel to studies on laterality effects in human speech perception. Petersen's results suggest several points of similarity in the phonetic processing of Japanese macaques and human beings.

Snowdon also reported on a study of perception of species-specific calls by pygmy marmosets. These animals give several variations of trill-like calls, and two of the variants appear to be used in very different social situations. The two variants differed from each other only in duration. Snowdon developed a marmoset vocal synthesizer which could produce both the natural parameters of the two types of calls as well as intermediates between them. When these synthesized calls were played back to monkeys, their responses indicated a categorization of calls very similar to the categorization of speech sounds by human beings. That is, the monkeys treated as equivalent several synthesized variations of the short trill (ranging from 176 to 247 ms). However as soon as the call duration was increased beyond 250 ms the animals no longer responded as to the short call, but in a way appropriate to the longer call. The duration distributions of these two trills as produced by the monkeys fell on either side of a 250 ms boundary. Thus the animals had a perceptual boundary which exactly corresponded to their production boundary.

In another study, human beings were presented with both labeling and discrimination tasks with the same synthesized trills. The human subjects failed to categorize the pygmy marmoset calls; instead they showed continuous discrimination of the calls. These results support those of Petersen that there are species-specific adaptations for the perception of species-specific calls but that the mechanisms of these perceptual adaptations are quite similar across species.

Syntactic Aspects of Primate Communication

Several presentations dealt with various forms of rules that appeared to affect what type of signal was emitted, how sequences of signals were generated within an animal's utterances, and how animals sequenced the order of their calls within a group.

U. Jürgens described the technique of computing transitional probabilities between calls as a means of determining which calls were functionally similar to one another.

He reasoned that two calls appearing in close temporal sequence to one another are probably related to similar motivational or informational states. He has used this technique to divide an apparently graded vocal continuum. For example, one can construct a continuum of three calls of squirrel monkeys based on structural similarity going from an "alarm peep" to a "yap" to a "caw". These calls could be treated as three functionally different calls or as three functionally similar calls. A transitional probability analysis indicated that alarm peeps and yaps frequently occurred close together while caws rarely occurred with the others. Thus the main functional difference on the acoustic continuum can be easily specified.

Smith, Newman, and Symmes presented data on the calls of squirrel monkeys and found differences in the repertoire of each sex. Thus, males had some calls which were never given by females and vice versa. Furthermore, the distribution of the frequency with which different calls were given varied between males and females. Thus, gender differences in call structure are quite important.

Gautier and Gautier also looked at the distribution of calls given by different members of a monkey group. They measured the rate of calling for each animal in a group, they measured the diversity of calling (that is how many different call types each animal gave) and the time of day in which calling occurred. They found two peaks of vocal activity — in morning and evening. However, only infant animals contributed to both peaks. Adults called mainly in the morning and juveniles and subadults primarily in the evening. The adult male had the least frequent rate of calling as well as the least call diversity. The infant had the greatest rate of calling but was second only to the adult male in low call diversity. From the infant through the subadults through the adult female there was a continuing increase in call diversity with the adult female having the most diverse set of calls but not a high frequency of calling. Thus not only gender but age have an effect on the frequency and diversity of calls produced. In addition, time of day is an important variable with different individuals calling at different times.

Sequencing of several calls into phrases was discussed by Robinson and by Snowdon. Robinson has studied the titi monkey and finds several predictable sequences of calls. He has completed studies where sequences are played back in normal or in abnormal sequences. Monkeys react with disturbance to the playback of abnormal sequences but not to normal sequences, indicating that they perceive a proper grammatical structure to call sequences.

Snowdon reported on the repertoire of the cotton-topped tamarin, which, like the titi monkey, is characterized by a small set of elements which are combined to form a variety of different sequences. Most of the combinations formed are what Marler has called phonetic syntax. That is, the combination of elements produces a message that is different from what the sum of the indivudal units would communicate. However, the cotton-top tamarin does have a few examples of lexical syntax, where the sequences seem to communicate the sum of the individual units. This sort of syntax is also evident in some call sequences of the titi monkey as Robinson reported.

Snowdon also presented data on conversational syntax, rules governing which animals should call and in what sequence. A group of three pygmy marmosets was observed and the sequences in which trill calls were given was recorded. Sequences in which all three animals called once before one animal called a second time occurred much more frequently than chance. And one ordering of the sequence of three animals was much more frequent than other orderings. Thus, pygmy marmosets take turns when trill calling, and the sequence of turn taking is highly predictable.

Another sort of conversational calling is found in the calls of gibbons. B. Deputte presented data on the vocal exchanges of male and female gibbons. Early in the morning it is generally the female which calls first, giving a very simple call. The male responds to the female's call with a slightly more complex call. The female then responds with a still more complex call. Calls are exchanged becoming ever more complex until finally the male reaches a paroxysm of calling which may continue several minutes without pause. It appears that the calling of the female is important in inciting the male to his final stage of calling. There is a clear syntax that governs which animal calls when, and which governs the relative complexity of the calls given.

Hedonic Measures of Vocal Activity

Jürgens presented some very interesting data on the hedonics of vocal calling. The neuroanatomy of the vocal system of the squirrel monkey is quite well known and it is possible to place electrodes that will elicit a particular call type. The electrodes are not placed in the motor cortex, but in subcortical areas, and they thus probably tap the motivational systems associated with each call. Jürgens implanted electrodes which would elicit several vocalizations. He then placed animals into a two-chambered container. When the animal was in one chamber electrical stimulation was applied through a vocal eliciting electrode. When the animal moved to the other chamber, electrical stimulation ceased. From the response of animals in this situation Jürgens was able to rank calls on a hedonic continuum from pleasurable to neutral to aversive. From the use of transitional probabilities described above and the hedonic tests Jürgens could classify a large number of squirrel monkey calls into five functional categories which in turn had several call types at different loci on an hedonic continuum. This combination of techniques provides a novel method for the classification of vocalizations. Furthermore, from the hedonic measurements it was clear that certain physical aspects of calls are correlated with hedonic values. Thus, high pitched calls and calls with much frequency modulation tended to be aversive while nonmodulated, low-pitched calls tended to be positive in the stimulation test. Unfortunately, the technique for measuring hedonics can probably only be used with squirrel monkeys since the neuroanatomy of the vocal system is unknown for other primate species.

Semantic Aspects of Communication

D. Cheney and R. Seyfarth have been studying vervet monkeys in East Africa. Following up on earlier work of Struhsaker, they have focused on the alarm calls of vervets. Three quite different forms of alarm calls are given, and they appear to be specific to different predators. One is given upon seeing a leopard, another on seeing a martial eagle, and the third on seeing a constricting snake. In order to test whether these different calls symbolized and actually indicated a specific external referent, Cheney and Seyfarth conducted a series of playback experiments. From a hidden speaker they would broadcast one type of alarm call to a group of animals and observe their responses. Upon hearing an eagle alarm the monkeys would look up in the air and would run into brush where presumably they would not be seen from above. Upon hearing a snake alarm call the monkeys would look down at the ground and run into a tree. And upon hearing the leopard alarm call they would look out over the plain and run to a tree. Thus the mere hearing of the alarm call elicited an appropriate response. It was not necessary for a predator actually to have been present. This result provides the first clear evidence in primates that they can communicate about external referents and not simply about their internal motivational states. Secondly, the data indicate that monkeys can use vocalizations symbolically. That is, they reacted to the eagle alarm call as if an eagle were present, to the leopard alarm as if a leopard were present, and so forth. While symbolic abilities have been shown in great apes learning artificial communication systems, there has not been clear evidence before now of symbolic communication with the natural communication system of a monkey.

Single Versus Multidimensional Communication

Three papers dealt with the issue of the complexity of the communicative channels involved in communication of sexual state in female monkeys. C. Bielert presented the results of several studies on the cues for mating in chacma baboons. He found a greater incidence of masturbation in male baboons after they had participated in sex testing with females when the females were in the ovulatory phase of their cycle. He then set about to determine which cues of the female were most important in arousing the males. He placed a cycling female in a cage in the same room as several males and found increased masturbation as the female reached ovulation. In the chacma baboon sexual swelling also accompanies ovulation. Bielert next administered estrogen and progestin to ovariectomized females and found that increased doses of hormones led to increased sexual swellings and also increased rates of masturbation by the males. Finally, he placed a visual partition between the males and a hormonally treated female. Although olfactory and visual cues would still be available as potential arousing stimuli, male masturbation decreased markedly. The results indicated that the visual cues provided by the anogenital swellings are important as arousing stimuli to male chacma baboons. Although olfactory stimuli may be important for consummation of mating, they would appear to be effective only at a very close range. Thus, a single cue — the sexual swelling of females — communicates sexual readiness to male chacma baboons.

E.B. Keverne described similar studies on sexual arousal in rhesus macaques and in talapoin monkeys. In the rhesus monkey olfactory cues seem to be of importance in sexual arousal. Attempts to mimic the color changes of sex skin that correlate with the ovulatory cycle had no effect on changing male responsiveness to females. However, cyclic changes were found in the content of female vaginal secretions. Applying these secretions to the vaginas of ovariectomized females made them more attractive to males. When males were made temporarily anosmic, they no longer responded to receptive females as strongly. In a related study a group of five talapoin males were made temporarily anosmic. The changes in sexual behavior produced were quite individually variable with the first-ranked male showing no impairment, the second-ranked male showing some reduction of sexual activity, and the fourth-ranked male showing a cessation of sexual performance. The third-ranked male showed an increased sexual performance. The females, however, showed an increase in sexual solicitation. With both rhesus macaques and talapoins olfactory cues seem to be important communicative channels for sexual behavior. However, in both monkey groups the data were highly variable and individual-specific with some males responding to some females with or without olfactory cues and other males never responding to a given female no matter what her olfactory cues were. Thus the olfactory arousal of mating in monkeys cannot be viewed as a reflexive or instictive act. Prior experience, personal choice, and other subtle cues are probably affecting how males and females respond to one another.

D. Goldfoot also presented data on sexual communication in rhesus macaques. He argued for the development of a multiple channel model for communication based on several experiments he has performed. First he noted that there was great variability in the sexual responses of monkeys to one another. Some monkeys would copulate regardless of whether the female was ovariectomized or not or whether her hormonal status was ovulatory or not, while other pairs never mated, even when the female was ovulating. He found that vaginal odors were not terribly potent sexual attractness to males, but that a female who still carried ejaculate from a previous mating was often more attractive than one which had not experienced a recent ejaculation. Thus suggested that novel odors rather than a particular type of secretion might be effective in arousing males. Goldfoot mixed a synthetic substance with a smell of green peppers which upon application to a female was fully as arousing of male sexual activity as natural vaginal secretions from an ovulatory female. Thus it appears that vaginal olfactants are not the major nor the only channel for communicating a female's sexual readiness.

Further observations by Goldfoot indicated that subtle behavioral cues such as looking over the shoulder and reaching back were probably more important than olfactory cues in the initiation and maintenance of sexual activity. If initiation gestures from the female did not occur, then any mounting attempt by the male had a high probability of being rejected. When females were the first to initiate sexual activity, then there was a high probability of copulation through ejaculation. Thus, it is the behavior and communicative signals of the female rhesus monkey that appear to be most important in sexual activity.

Goldfoot argued for a multidimensional model of communication in sexual activity that took into account variables such as olfactory, visual, and auditory channels, which took into account subtle initiation on the part of the female, and which took into account the roles of prior experience with particular sex partners and particular cues. The simple stimulus-response model of classical ethology was said to be inadequate to account for the complexities of sexual communication in monkeys.

Pragmatics of Communication

A major concern in the study of communication has been the question of pragmatics. That is, what is the signaler communicating? Although the focus of much of the symposium was on problems relating to the definition of the signal in terms of how recipients perceive it and respond to it, many of the papers also dealt with questions of pragmatics. Three major areas were discussed: (1) Affiliative communication. (2) Sexual communication, and (3) Spacing of animals.

Affiliative Communication

Smith et al. noted that relatively little attention has been given to the signals involved in affiliative behavior, in part because they are often less intense and therefore more difficult for an observer to notice. As a matter of fact, though, much of the recent experimental work with monkey vocalizations has dealt with affiliative types of calls. The trills of pygmy marmosets, the coos of Japanese macaques, the sexual signals in monkeys are all funtionally affiliative type calls (though the sexual signals are given a separate category below).

In their squirrel monkeys Smith et al. described a "chuck" call which was given only by females. Using the discriminate analysis technique described above to identify individual callers, Smith et al. found that chuck calls were given primarily between females who had an established affiliative relationship. Thus, females which spent the most time with each other gave chuck calls to each other, but they did not give these calls with other females even on those occasions when they were sitting next to those females. Thus, the call does not specify affiliation in a general sense between any animals who are physically close to one another, but it serves to express a relatively long-lasting social relationship between two animals.

Gautier and Gautier showed from their data on vocalizations from individual animals within a group, that the adult female played the most important role in social cohesion and affiliation. When other animals called, it was generally the adult female which responded. One possible function of the female's greater diversity of calling might be to allow her to respond appropriately to each of the other members of the group. Deputte pointed to another form of affiliative communication in the antiphonal calling or duetting in gibbons. There is a close coordination between males and females in the alternation of calling and in the increasing complexity of the structure of the calls. It seems probable that the function of the duetting is not only to keep other pairs of gibbons at a distance, but also to strengthen the bond between the calling pair.

Sexual Communication

Bielert, Keverne, and Goldfoot each described the communication signals used in promoting sexual activities between pairs of monkeys. There appear to be many species differences in the cue systems that are important in sexual communication. Bielert showed the importance of the sexual swelling in chacma baboons, where at least at more than a couple of meters distance olfactory cues by themselves do not seem sufficient arousal cues. Keverne argued for olfactory cues resulting from the vaginal secretions of the female rhesus macaque as being effective at arousing sexual activity in the male, but he reported less clear olfactory effects in the talapoin where females have a sexual swelling similar to that of the baboon. Finally, Goldfoot showed that behavioral and visual invitational signals from female rhesus macaques are necessary before the male can copulate successfully. All of these species differences have been observed in the laboratory under somewhat differing test conditions. One possible resolution of the differences in results may be that visual cues like sexual swelling and invitational gestures are effective at the long ranges typically found in the field, while more subtle cues and olfactory stimuli produced from the female's vagina may be the most effective proximal cues for sexual activity. The most successful mating pair would be that in which both distal and proximal cues were consonant in the information communicated.

Spacing of Animals

Several papers dealt with the signals used by animals to maintain distance between unfamiliar groups of animals while keeping animals within a group close together. Three of the authors dealt with acoustic signals while the fourth dealt with olfactory signals.

P. Waser presented a paper on the evolution of loud calls in baboons and mangabeys. These two groups of monkeys represent a radiation of several species of Cercopithecines into a vast array of habitats in Africa. After the terminology of Gautier and Gautier (1977) he describes the type 1 loud calls of these animals noted for being quite loud and stereotyped. In the mangabeys and baboons Waser finds the type 1 loud calls to be strikingly homologous in structure and to appear in similar circumstances. Thus, the calls are frequently given following a group disturbance, following chorusing of calls by other conspecifics, and after losing contact with conspecifics. These calls are all given by adult males and not by other group members.

Although there is a striking similarity in the nature of the call structure, the type of animal calling and the situations in which the call is given across species, Waser pointed out that the responses to the calls were quite different across species. Thus, adult male mangabeys respond by rapidly approaching the source of a loud call, while baboon males seem not to notice the call. In *Cercocebus albigena* these loud calls seem to function in maintaining intergroup spacing, but similar functions have not been observed in other species. Several trends in signal form were found across species: within species the call structure is extremely stable in form; between species the calls differ most in their temporal structure; fundamental frequencies were constant over species; forest species give high frequency barks after calls; forest arboreal species show calls with the greatest sterotypy and discreteness; and individual identity is more readily coded in the temporal parameters of the calls of forest species than in the open-living species. The striking similarities in structure and context in which it is given across species which radiate so widely are evidence of an evolutionary conservatism in the structure and message of these calls. It is only in the response to these calls that species with different social structures and different habitats demonstrate differences. The effect of habitat differences and social structure on signal form may be secondary, while the responses to these calls may be more labile and reflect the different selection pressures of the different species.

J. Robinson presented work on the vocal systems regulating intergroup spacing in the wedge-capped capuchin monkey. There are three closely related calls which appear to be used in different aspects of spacing. There are two forms of the "arrawh": one given by animals separated from the rest of the group and softer more tonal version given by animals who are dropping behind in a group progression. The "huh" call appears to be a general contact call which has a general tonic effect on other members of the group. The third call the "heh" is a component of a threat display. Each of these calls showed intergrading forms with one another, and they can be given in short temporal relationship to one another. Thus, the arrawh of an isolated animal is answered by huhs from the rest of the group. Robinson suggested several hypotheses for the intermediate call, the huh, which he then tested through field observations. Among the predictions one could make about the function of a contact call are: that calling decreases spacing by bringing group members to food sources or by bringing animals back together after they have been foraging during the day; that calling increases spacing either within the group or by deterring the approach of neighbors; and that calls maintain interindividual distances. Robinson studied variations in calling with activity, with foraging success, with the type of food taken, with the gender of the vocalizer, with the presence of kin, with the presence of dominant animals, and with individual spacing. Animals called more when close together, especially when the neighbor was dominant and there was a possibility of competition for food. They also called more during rapid movement. Thus, huh call seemed really to have two basic functions: (1) to space animals farther apart when they were close together and competing for resources, and (2) to bring animals within a group closer together during travel or when one animal was separated. Thus, the contact call huh appears to have two different functions.

Snowdon described a similar system in the long calls of the cotton-top tamarin where the same call appeared to be used in intergroup encounters to maintain separation, to promote within-group cohesion, and by lost or separated animals. However, an analysis of the fine structure of the calls used in each of these situations indicated that, while similar, there were significant differences in structure between the calls given in each situation. Thus, each function was coded by a particular variant of the long call. Playback studies have supported the idea that the acoustic differences in the calls are sufficient to produce responses appropriate to the spacing, the cohesion, and the lost animal contexts.

Epple presented the results of several studies on the gonadal control of the signals involved in aggressive behavior in saddle-backed tamarins. She noted that in many rodent and primate studies a direct relationship between levels of gonadal hormones and aggressive behavior has been demonstrated. In the saddle-backed tamarin the basic social structure is monogamy, which appears to be maintained by aggressive activities of each sex toward same-sexed intruding animals. A variety of laboratory studies done by Epple and others have shown that a variety of threat and attack behaviors such as scent-marking, threat faces, arch posture, ruffling, and direct attacks will be given to an intruder. Scent marking from the suprapubic gland appears to be very important in distal threat displays.

Epple studied groups of animals which had either been gonadectomized prepubertally, gonadectomized in adulthood, or sham gonadectomized. They were then studied in a series of intruder conditions to determine their aggressive responses. In another study animals were given different forms of social experience to determine how this would affect their responses to intruders. Gonadectomy performed in adult animals had little effect on scent gland size nor on the response to intruders. Scent marking and all other forms of threat and aggressive behavior were unaffected by the gonadectomy. With prepubertal gonadectomy however, the results were different. When the animals were tested at 1 year of age, they showed significantly fewer signals of threat (threat face, arching, ruffling, and scent marking) although they did not differ from controls in direct aggression. When the same animals were tested again at 2 years of age the differences between gonadectomized and control animals disappeared with the exception of scent marking. Thus, prepubertal gonadectomy had a temporary effect on threat signaling to intruders, but social experience seemed to overcome this deficit.

When males which were castrated as adults were kept with their mates, they showed the same rate of scent marking as they had prior to castration. When they were removed and formed into a trio with one other male and a different female, their scent-marking rate decreased to low levels, possibly because they were subordinate within the trio. However, when they were returned to their original mate their scent-marking scores returned to normal. Thus the nature of the social relationship of an animal has more effect on scent marking to intruders than gonadal condition does. A further confirmation of this point came from studies where young animals were paired either with another young animal, with an adult animal of the opposite sex, or with a trio consisting of one young male, one young female, and one adult male. Only those animals living in pairs with adult animals of the opposite sex showed significant aggressive displays to intruders (threat facing, scent marking, and total aggression). Despite the evidence from several other species of a relationship between gonadal hormonal state and aggressive and threat behavior, the saddle-back tamarin shows only a minor effect of gonadal state on its aggressive behavior and that only if castration is prepubertal. Experience can overcome this deficit and, in general, the nature of the animal's social experience, including the experience of its mate, is of greater importance than its hormonal condition in determining aggressive signals to intruders.

Ontogeny of Communication

The question of ontogeny of communication in primates is one of the major problems to be solved with future research. At present we have a much more complete understanding of ontogeny in birds than we do with primates. Because relatively few investigations of ontogeny have been completed, the symposium did not have many presentations dealing with ontogeny. Three levels of approach to the ontogeny issue were discernible however: descriptive, analytical, and cognitive.

Descriptive Studies of Ontogeny

Gautier and Gautier as described above noted that young animals differed from older ones in the time of day of calling, in the rate of calling, and in the diversity of calling. Infants called at both calling peaks of the day, while older animals called at only one peak; infants called more frequently than other monkeys, but they had the least call diversity after the adult male. Call diversity showed a progressive increase over development. Snowdon reported that marmoset and tamarin infants are highly vocal, calling more frequently than older animals. Much of their calling contained imperfect elements of adult structures given in inappropriate context. Through development there is an increasing match to adult forms and an increasing use of calls in appropriate contexts.

Analytical Studies of Ontogeny

The work of Epple, showing that the role of experience was more important than hormonal state in the development of scent marking and other aggressive signals in the saddle-back tamarin, has been discussed above. Seyfarth and Cheney presented some ontogenetic data on the use and response to vervet alarm calls. Infants gave alarm calls to a much broader range of stimuli than the adults did, however the types of stimuli to which the animals gave alarm calls were of the same general class as the true predators. Thus, infants might call eagle alarm to any large bird whether a raptor or not, or might even give an alarm to a falling leaf, but the stimuli were always aerial. They might give a snake alarm to a sinuous branch on the ground, but they did give this alarm to only long thin objects on the ground. The definition of the stimuli to give alarm calls became more restricted as the animals got older. Nonetheless, there appeared to be a basic taxonomic categorization in the youngest infants. Seyfarth and Cheney suggested that much of the sharpening of predator category definition came through imitation of adult calling and through selective reinforcement by adults of only those infant alarm calls that were truly associated with a predator.

Newman and Symmes presented data on the ontogeny of the isolation peep in squirrel monkeys. They have found that there are consistent individual differences in the structure of alarm peeps which appear a few days after birth and which continue throughout development. In addition, the acoustic characteristics which separate the Gothic arch form of squirrel monkey from the Roman arch form also appear at a very early age and remain constant through development. Raising animals of one type with a group of the other type does not produce an alteration in the typical structure of the call. Thus, with the squirrel monkey at least, it seems that there is little role for experience in the modification of both the individual-specific and population-specific form of the isolation peep.

Cognitive Studies of Ontogeny

S. Chevalier-Skolnikoff has been applying the theories of Piaget for cognitive development in human infants to the study of cognitive and communicative abilities in monkeys and great apes. She has studied orangutans, gorillas, macaques, and capuchin monkeys both cross-sectionally and longitudinally. She has been able to develop a set of criteria for each of the stages in the sensorimotor period of Piaget which allow her to classify the stage of development of each primate. All of the primates which she has studied showed a remarkable consistency in the chronological age at which they passed from one stage to another. Thus, within her orangutan sample all animals passed from Stage 3 to Stage 4 within 3 weeks of one another. Communication appears to develop according to the same stages as cognitive development, and, like the data from human children, there is a remarkable consistency from animal to animal in the order of communicative development and in the rate of development. While great apes could progress through each of the six stages of the sensorimotor phase, the macaques and cebids only progressed through the fifth stage. The use of the Piagetian framework offers considerable promise for describing differences in communicative ability between highly different species, as well as relating the cognitive abilities of animals to their communication ability.

Summary

The symposium participants presented several lines of research that hold promise for the next decade of study of primate communication. First, the field has progressed from simple descriptive studies cataloging differences in primate signals to the use of experimental paradigms both in captive and wild populations. There has been development of more complex models and paradigms for the analysis of communication which reflect the acceptance that primate communication is a highly complex activity. There is an active interest in determining what characteristics of the signals are important to the primates rather than simply interpreting signals from an anthropocentric view. There is a greater sophistication in determining the contextual situations in which calls are given and in developing predictive models that lead to the determination of what and for what purpose a signal is used. Finally, the study of how adult communication is acquired is still in its infancy. While some promising approaches have been presented, the problem of ontogeny is clearly the question that demands solution in the 1980s. Are communicative signals genetically determined and stereotyped in form? Can we develop a primate model analogous to Marler's work with white-crowned sparrows? What is the relationship between the cognitive ability of an animal and the sophistication of its communication system? These are among the exciting questions for primate communication research in the next decade.

Acknowledgments. This paper was prepared by Charles T. Snowdon on behalf of the symposium organizers and participants. Preparation of this summary was supported by a USPHS National Institute of Mental Health Research Scientist Development Award. Address reprint requests to Charles T. Snowdon, Department of Psychology, University of Wisconsin, Madison, Wisconsin, 53706, USA.

References

Altmann SA (1967) Social communication among primates. Univ of Chicago Press, Chicago
Gautier J-P, Gautier A (1977) Communication in old world monkeys. In: Sebeok T (ed) How animals communicate. Indiana Univ. Press, Bloomington
Snowdon CT, Brown CH, Petersen MR (1982) Primate communication. Cambridge Univ Press, New York

Primate Locomotor Systems: Summary of Results of the Pre-Congress Symposium in Pisa

H. ISHIDA [1], R.H. TUTTLE [2], and S. BORGOGNINI-TARLI [3]

We organized the symposium on primate locomotor systems in order to share contemporary knowledge and to identify important problems for future research on primate positional behavior and functional morphology. It was held at the Institutes of Zoology and Anthropology at Pisa University, which contributed greatly to the success of the conference. On behalf of the participants, we would like to thank them for their efforts. Herebelow we present commentaries on each paper that was presented in the symposium.

The symposium was divided into four sessions. The first session, chaired by Dr. Ishida, was chiefly concerned with kinematographic, morphological, and behavioral studies on the locomotion of prosimian primates. Dr. Jouffroy opened the session with a paper entitled "A biomechanical study of the quadrupedal branch walking of *Perodicticus potto*". She pointed out that the potto can walk as fast as humans can (i.e., 0.75 m/s). This result is quite different from the commonly held view that pottos move very slowly. Dr. Jouffroy has employed sophisticated cineradiographic techniques in her studies. They revealed new details about the gait and forelimb movements of pottos during branch walking. She concluded that the gait of the potto can be included in Hildebrand's symmetrical-diagonal types of locomotion. Further it is a "forward-cross" type of walk, which is characteristic of primate quadrupedalism, and which contrasts with the "backward-cross" type, which is common in nonprimate mammals.

Dr. Jouffroy observed that the potto extends its forelimb markedly at the end of the swing phase of locomotion. As the hand nears contact with the substrate, the ulna is parallel with it and the forelimb is maintained just above the support. This allows the potto to make smooth movements of its body along a branch and to execute long strides.

In a paper entitled "Locomotor pattern and functional morphology of the biceps femoris muscle in the slow loris *(Nycticebus coucang)*" Dr. Kawabata reported that the slow loris always keeps its head forward as it moves versatilely in many directions. Slow lorises execute two types of bridging behaviors between discontinuous substrates. They are termed rapid and slow types. The rapid type of bridging occurs during horizontal movements from one substrate to another. The slow type of bridging was used when the loris moved in other (nonhorizontal) directions between discontinous substrates.

1 Osaka University, Faculty of Human Sciences, Suita, Osaka, 565 Japan
2 Department of Anthropology, University of Chicago, Chicago, IL 60637, USA
3 Instituto di Antropologia, Via S. Maria, 53, Pisa, Italy

Dr. Kawabata and his co-workers also studied the biceps femoris muscles of slow lorises electromyographically. Stronger EMG activities were observed as the subjects descended vertically than when they ascended vertically and moved horizontally. They concluded that the well-developed mass of the biceps femoris muscle in *Nycticebus coucang* can be attributed to the fact that they always move headfirst and are most challenged to control their movements against gravitational forces when they descend vertically.

Dr. Niemitz lectured on "Locomotor adaptations in the genus *Tarsius*". He began by discussing the functional meanings of the construction of the hand in tarsiers. He stressed that the species from Borneo and Sumatra have markedly elongate fingers. The morphology of the hand in *Tarsius bancanus* is probably related to the fact that infants must cling to their mothers from the first day of their lives since the species does not sequester the young in nests.

Dr. Niemitz also presented an analysis of quadrupedal walking by *Tarsius bancanus*. This type of locomotion is very rare in the most specialized tarsioid species, *T. bancanus*. The frequency of quadrupedal walking is less than 1% of its overall activity round. Its leaping behavior was analyzed from slow motion cine films. The acceleration phase of the tarsioid leap lasts only 90 ms or less. Hindlimb extension begins at the hip joint and is followed very rapidly by extension of the ankle and knee joints. The tarsioid foot is accelerated by seven times gravitation (or more), while the body is exposed to about four times gravitation (or slightly more). The acceleration of the entire body does not take place at a constant rate. It has two phases that are distinguished by contrasts in the rates of increase. The first phase appears in the first 20 ms of the acceleration phase and this is followed by a remarkable second phase during the last 30 or 40 ms. The latter is attributable to actions of the midtarsal joint of the tarsier's elongated foot.

Dr. Niemitz ended his paper with a discussion of the importance of the tail in the locomotion of tarsiers. The tail helps to stabilize the body axis during mid-flight, keeping it upright in order to prevent toppling over upon landing. The second function of the tail is to correct the spin which the body inaugurates during take off. If uncorrected, this spin would prevent precise pedal landing on the target substrate.

The general discussion of papers in the first session focussed on lorisine and tarsioid locomotion. It was debated whether the forelimb or the hindlimb is of greater importance in the propulsive movements of lorisine primates. In order to give a definite answer to this question, we must develop a new type of apparatus that can measure the forces produced by each limb more precisely during locomotion. It was agreed that force plate techniques should be used in future studies.

Five papers on primate bipedality and the evolution of human bipedalism were presented in the second session, which was chaired by Dr. Jungers. The first paper, entitled "Muscle recruitment and the evolution of bipedality", was presented by Dr. Vangor. She has attempted to elucidate problems on the evolution of human bipedality through the use of telemetered electromyography. She has analyzed EMG activities of hip and thigh muscles in *Ateles, Lagothrix,* and *Erythrocebus*. She concluded that (1) facultative bipedalism is not preadaptive for human bipedality because during facultative bipedalism, no particular primate species shows hindlimb muscular activities that are similar to human ones; (2) muscle activity of the hip and thigh in quadrupedal monkeys is not more humanlike than that observed in nonprimate mammals such as cats and dogs; and (3) vertical climbing appears to be preadaptive for bipedality. For many muscles, the recruitment pattern during vertical climbing is similar to, or the same as, that seen during bipedal walking by the same subjects.

Dr. Kimura lectured on the "Speed of bipedal gaits in humans and nonhuman primates". He and his co-workers analyzed biomechanical data from bipedal primates. They found that the foot force components acting on the substrate depend upon the speed of walking, though their characteristics are different among species. They also observed that among anthropoid primates there are two ways to increase speed during bipedal walking. Humans and Japanese monkeys gain speed by changing cadence, whereas chimpanzees and gibbons increase their stride length on some occasions and their cadence on other occasions.

Dr. Okada presented a paper on "Biomechanical characteristics of hylobatid walking on a flat surface". He and his colleagues conducted detailed experiments on the bipedal walking of gibbons and presented kinematic, kinetic, and electromyographic results from their studies. Unlike spider monkeys and chimpanzees, in gibbons appreciable knee extension occurs in the mid-stance phase of bipedal locomotor cycles without a delayed break of tarso-metatarsal contact with the substrate. As in humans, a change of the sagittal force from deceleration to acceleration often occurs in the middle of the hylobatid stance phase. These observations, along with EMG data, elucidate certain morphological and behavioral adaptations of *Hylobates*.

A paper entitled "Biomechanical evaluation of evolutionary models for prehabitual bipedalism" was presented by Dr. Yamazaki. He and his co-workers analyzed limb-joint motions by means of cinematography and then simulated joint moments and the forces produced by key muscles of the hindlimb, using a two-dimensional simplified mathematical model of the bipedal walking of 11 anthropoid subjects (a chimpanzee, two gibbons, three Japanese monkeys, two spider monkeys, and three men). Nine indices were calculated from the data on each species. A cluster analysis of the species, using the nine indices, showed that the bipedalism of gibbons is the most similar to that of humans, and that the bipedal walking of Japanese monkeys is far from that of humans. They concluded that a relatively small-bodied ape model for the ancestry of the Hominidae is reasonable. Although the number of subjects used was small, the results of this study are suggestive as to the evolution of prehabitual bipedalism. They plan to evaluate other models of locomotor evolution with this method.

The final paper of the second session was entitled "Climbing, bipedalism and anthropoid hindlimb muscles". Dr. Ishida and his colleagues analyzed the positional behavior, especially vertical climbing, and the hindlimb muscles of chimpanzee, gibbon, spider monkey, Japanese monkey, and Hamadryas baboon subjects kinesiologically and biomechanically. They observed that during vertical climbing, the cercopithecine monkeys employ two phases of knee flexion and extension and they maintain flexure of the elbow joints. The chimpanzee, gibbon, and spider monkey evince a single phase of knee extension and concurrently hoist themselves from their extended forelimbs. The difference in climbing correlates well with the morphological difference of the ischicruralis (cf. biceps femoris) muscles between the two groups of primates. The differentiation of nonhuman primate bipedalism could be related to transformations in the muscular system owing to locomotor behavior, especially vertical climbing. Because the human muscular system more closely resembles those of primates with the "ape type" of bipedal locomotion, they inferred that vertical climbing was probably an important component in the adaptive complex of the proximate ancestors of the Hominidae.

The third session was chaired by Dr. Niemitz. Dr. Matsunami presented a paper entitled "Ipsilateral control of primate limb movements". He and his colleagues recorded activites from 197 neurons in the motor cortices of two rhesus monkeys. They consisted of 13 ipsilateral, 50 bilateral, and 134 contralateral neurons. They showed that ipsilateral units control proximal muscles of the primate forelimb and that bilateral units are associated more with arm movements than with finger or wrist movements.

Dr. Negayama reported results from a "Developmental study of locomotor behavior in Japanese monkeys". He and his co-workers observed and classified the positional behavior of three neonatal subjects during the first 3 months after birth. The frequency of locomotion increased consistently with growth. Crawling and toddling were locomotor patterns at the neonatal stage, but they were replaced by other locomotor behaviors within the first month. New locomotor behaviors appeared almost abruptly. Comparing locomotor and postural development, elementary postural activities appeared somewhat earlier than the locomotor behaviors. It was concluded that the first month is a very critical period for Japanese monkeys in terms of their locomotor development.

Dr. Etter lectured on "Problems of functional adaptation in the shoulder joints of catarrhine primates". He compared the scapulae of a number of catarrhine primates and demonstrated a constant correlation between body size and the area of the scapula. He reached the following conclusions: The catarrhine monkey scapula is mainly under axial strain during quadrupedal posture and locomotion. The bony structures, which are constructed to resist this stress, are uniformly built and only vary according to body size. In great apes, the body weight is not transmitted through the scapulae during quadrupedal posturing and locomotion. Instead the upper body is supported by a sling, consisting of the pectoralis major muscles, which are attached to the proximal ends of the humeri. Their scapulae are built much more variably to accommodate a wide range of extended excursions by the arm. The scapulae of gibbons are monkey-like in some morphometric characteristics. In those features that are connected with mobility of the arm, they are very close to those of great apes. Finally, Dr. Etter emphasized that the morphologies of catarrhine scapulae are closely linked to their respective functions. This must be taken into account when critically appraising phylogenetic models.

Dr. Jungers presented a paper entitled "Implications of quadriceps femoris function in nonhuman primates: a comparative electromyographic analyis". His initial functional anatomical and electromyographic study of the quadriceps femoris muscle in *Lemur fulvus* revealed a division of labor among its components and demonstrated a close correspondence between habitual positional behaviors and observed muscular morphology. Dr. Jungers and his co-workers extended their EMG analysis of the quadriceps femoris muscle to include *Nycticebus, Pithecia, Lagothrix, Ateles, Erythrocebus,* and *Pan* as they engaged in a variety of positional behaviors. Anatomically,

Lemur, Pithecia, and *Erythrocebus* are basically similar to one another and *Nycticebus Lagothrix,* and *Pan* form a second group. However, their electromyographic patterns differed from those that might be expected on the basis of their anatomical similarities as follows: (1) Fusion of muscle bellies in *Nycticebus* and *Pan* has not led to synchrony in their activities. (2) Despite close similarities of muscle proportions and the presence of a superior patella in *Lemur* and *Pithecia,* muscle recruitment is strikingly different during their quadrupedalism, posturing, and leaping. (3) Despite great similarities in the morphology of the rectus femoris muscle in all species, enormous variation in EMG patterns was noted among species doing the same activity and within each species during different activities. On the basis of these results, they emphasized that attempts to deduce muscle function only from considerations of gross muscle architecture and the locations of origins and insertions of muscles are tenuous. They also suggested that knowledge of microscopic structure, kinematics, and kinetics should help to refine speculations about muscle function. Such studies underscore the potential of the experimental approach for testing hypotheses and for discovering unsuspected roles for individual muscles and muscle complexes.

The fourth session of the meeting, chaired by Dr. Jouffroy and Dr. Kimura, was devoted to general discussions of primate locomotor systems. Topics raised included (1) the similarities between the locomotor systems of *Perodicticus* and *Nycticebus,* (2) relationships between vertical climbing and hindlimb muscular morphology, (3) possible improvements of methods for biomechanical analyses, and (4) the importance of cinematographic analyses in primate locomotor studies.

It was generally agreed that there are still many unsolved problems regarding the nature of primate musculoskeletal systems, their mechanics and functions, and their relations to the nervous system, and their roles in natural environments. Many of these problems must be elucidated if we are to advance with models on the origin of hominid bipedalism.

Full Titles of Papers and Listing of Authors and Their Affiliations

1. A Biomechanical Study of the Quadrupedal Branch Walking of *Perodicticus potto*
   F.K. Jouffroy, S. Renous and J.P. Gasc
   Museum National d'Histoire Naturelle, Paris, France
2. Locomotor Pattern and Functional Morphology of the Biceps Femoris Muscle in the Slow Loris
   H. Ishida, N. Kawabata and S. Matano
   Osaka University, Osaka, Japan
3. Functional Morphology of the Limbs in *Tarsius*
   C. Niemitz
   Freie Universität Berlin, Berlin, West Germany
4. Muscle Recruitment and the Evolution of Bipedality
   A. Vangor
   University of Pennsylvania, Philadelphia, USA
5. Speed of Bipedal Gaits of Man and Non-Human Primates
   T. Kimura [1], M. Okada [2], N. Yamazaki [3], and H. Ishida [4]
   [1] Teikyo University, Tokyo, Japan; [2] Tskukuba University, Tsukuba, Japan; [3] Keio University, Tokyo, Japan; [4] Osaka University, Osaka, Japan

6. Biomechanical Characteristics of Hylobatid Bipedal Walking on a Flat Surface
   M. Okada[1], N. Yamazaki[2], H. Ishida[3], T. Kimura[4], R. Tuttle[5] and S. Kondo[6]
   [1] Tsukuba University, Tsukuba; [2] Keio University, [3] Osaka University, Osaka, [4] Teikyo University, Tokyo, [5] University of Chicago, Chicago, USA; [6] Kyoto University, Inuyama, Japan
7. Biomechanical Evaluation of Evolutionary Models for Prehabitual Bipedalism
   N. Yamazaki[1], H. Ishida[2], M. Okada[3], T. Kimura[4], and S. Kondo[5]
   [1] Keio University, Tokyo, [2] Osaka University, Osaka, [3] Tsukuba University, Tsukuba, [4] Teikyo University, Tokyo, [5] Kyoto University, Inuyama, Japan
8. Climbing, Bipedalism and Anthropoid Hindlimb Muscles
   H. Ishida[1], N. Yamazaki[2], R. Tuttle[3], T. Watanabe[4], M. Okada[5], S. Kondo[4], and S. Matano[1]
   [1] Osaka University, Osaka, [2] Keio University, Tokyo, [3] Japan University of Chicago, Chicago, USA, [4] Kyoto University, Inuyama, [5] Tsukuba University, Tsukuba, Japan
9. Developmental Study of Locomotor Behavior in Japanese Monkeys
   N. Negayama, K. Kondo and N. Itoigawa
   Osaka University, Osaka, Japan
10. Isilateral Control of Primate Limb Movements
    K. Matsunami and K. Hamada
    Kyoto University, Inuyama, Japan
11. Problems of the Functional Adaptation in the Catarrhine Shoulder Joints
    H. Etter
    Bäretswil, Switzerland
12. Implications of Quadriceps Femoris Function in Non-Human Primates
    W. Jungers[1], F. Jouffroy[2], J. Stern[1], and R. Susman[1]
    [1] New York State University at Stony Brook, New York, USA, [2] Museum National d'Histoire Naturelle, Paris, France

Results of the Pre-Congress Symposium on "Methods and Concepts in Primate Brain Evolution"

D. FALK [1] and E. ARMSTRONG [2]

The pre-Congress symposium on Methods and Concepts in Primate Brain Evolution organized by E. Armstrong and D. Falk, took place in Turin, Italy on 4–5 July 1980 at the Institute of Comparative Anatomy. Nine speakers participated in the scientific program: they were J. Allman from California Institute of Technology, USA; E. Armstrong, from Louisiana State University, USA. R. Bauchot from Université de Paris, France; B. Campbell from Walter Reed Institute, USA; D. Falk, University of Puerto Rico; H. Jerison, University of California, USA; R. Martin, University of London, England; and K. Zilles, University of Kiel, FRG.

Much attention is given to the evolution of primate brain because of interest in how nervous systems may evolve, how substrates for behaviors change, what fossil endocasts might tell us about past organisms' capabilities and behaviors, and, in an applied vein, how the neuroanatomy, physiology, and chemistry of humans and primates resemble and differ from other common laboratory mammals used in the development of medicine. The development of rigorous models is necessary to begin the analysis of the above factors since many primate species are endangered and thus unsuitable for experimental procedures. Both comparative and paleoneurological data must be analyzed for this new level of understanding.

Several common themes are seen through the papers. The size of a brain is an important consideration in describing and evaluating similarities and differences. Allometry is a powerful tool used to control for size by many of the participants (Armstrong, Bauchot, Jerison, Martin), but until we understand the *causes* of the scaling (Martin) this tool is less powerful than it could be. Analyses of constraints on brain size (Leutenegger) and descriptions of different sized endocasts (Falk) are necessary for the advancement of evolutionary hypotheses.

A second major theme involved the highly visual nature of primates. Several scientists discussed similarities and differences among primates in visual structures (Allman, Armstrong, Bauchot, Campbell, Zilles). Several other sensory and nonsensory areas were also analyzed for evolutionary hypotheses.

Each report is reviewed briefly below in alphabetical order and the main findings are summarized.

1 Boston University, Sargent College of Allied Health Profession, 36 Cummington Street, Boston, MA 02215, USA
2 Department of Anatomy, Louisiana State University Medical Center, 1100 Florida Avenue, N.O. LA 70119, USA

"Evolution of the Cortical Visual Areas in Primates" (J. Allman). According to Allman a large set of changes in the organization of visual structures occurred when primates emerged from the ancestral stock of placental mammals. Retinotectal visual projection systems are common to all primates but differ from other mammals. The primate skull and endocast of the 55 million year old *Tetonius homunculus* show that the development of large, frontally directed eyes and expanded visual cortex occurred by the early Eocene. A comparative study of visuotopically organized cortical areas in various groups of extant primates suggests that certain areas are common to all primates and were therefore probably present in the common primate ancestor. Those areas include the primary visual cortex (V-I), the adjacent second visual area (V-II), and the middle temporal (MT) and dorsolateral (DL) visual areas of the third visual tier. The MT neurons modulate the analysis of visual movement and DL neurons are involved in form perception. Certain visual areas are present in some, but not all species. Each area probably performs a distinct set of functions in visual perception and visuomotor coordination. Thus, an area possessed by one species or larger taxon, but not by another, will endow the possessor with special behavioral capabilities. Dr. Allman reported that a major task for the future will be to determine the distinctive functions of cortical areas and how they relate to the behavioral, social, and ecological specializations of the various primates.

"Inferences About Human Brain Evolution from the Thalamus" (E. Armstrong). Thalamic tissue volumes and numbers of neurons were compared among two gibbons, one chimpanzee, one gorilla, and three humans. The results were compared with Shariff's cortical data collected from five primates, including man, that is in the literature. Both thalamic and cortical visual areas have low levels of additions of neurons per gain in brain weight. The human thalamic visual relay nucleus (LGB) contains the same number of cells as do the great apes. This and other volumetric and physiological data suggested that the LGB in the common ancestor of man and great apes was made up of the same number of neurons found in these extant hominoids. Both thalamic and cortical motor areas show a much greater (about 2 X) growth gradient among nonhuman primates than do the visual areas, and in both regions nonhuman primate data predict more human neurons than observed. Association regions increase their numbers of neurons per gain in brain weight at a rate only slightly more than motor areas. Nonhuman primate data predict the human cortex to have more neurons than found, but in the thalamus the human pulvinar has a number of neurons that is allometrical expected for our brain size. It is not clear from these data whether the difference is the result of different taxa or regions being studied. Human thalamic limbic nuclei appear derived in that they contain more neurons than ape data predict. No comparable cortical data exist. Numbers of neurons give a more detailed picture of similarities and differences than do volumes. Conservative, allometrically expected and derived nuclei are all found within the thalamus. The juxtaposition of populations of neurons that appear differentially changed was interpreted to be best described by the concept of mosaic evolution. While maintaining an overall integrity, selection pressures increased thalamic nuclei separately.

"Brain Organizations and Taxonomic Relationship in Insectivora and Primates" (R. Bauchot). Volumes of 25 different brain structures were determined by measuring them in serially sectioned brains from 63 species of insectivores and primates. For each structure within each species an isoponderal index, i, was determined. The index takes into account the differential effects of body size (allometry) and is so weighted in each of the 63 species:

$$i = 100 \ k/ko \text{ and}$$
$$k = E/S^a$$

where k = coefficient of encephalization, ko the same for the structures' centroid, E = brain weight, S = body weight, and a = coefficient of allometry. The isoponderal indices were used to calculate interspecific distances and in turn the indices were used to generate species dendrograms. The dendrograms position *Tarsius* inside the prosimian group, while tree shrews are placed neither with primates nor insectivores but within a classification, Menotyphla (elephant shrews and tree shrews). Because of the effects of encephalization and specialized adaptations, dendrograms within insectivora, prosimians, and simians do not always agree with standard taxonomy, but the results should be taken into account by systematicians. In addition, the volumes of various brain structures were correlated with each other. The highest correlations were among structures known from experimental data to be interconnected. This knowledge help determine interconnections in taxa that are not available for expriments.

"Some Questions and Problems Related to Homology" (C.B.G. Campbell). Campbell's presentation was more theoretical than many of the other papers presented in the symposium. Campbell noted that establishing homologies is especially problematical in nervous tissue and he discussed criteria for determining homologies. Patterns of connectivity, topographic and topological positions of nuclear groups are important, and no one criterion will suffice. While the recognition of a homologous relationship depends on similarities, the phylogenetic definition does not. He made a convincing case for the *Aotus* MT and the rhesus monkey "motion area" being homologous as analyzers of visually perceived movement. They are both associated with the superior temporal sulcus and receive direct input, in part from the specialized solitary cells of Meynert. Their other connections also appear similar. The cells are sensitive to a particular direction of movement. The areas are obviously not identical, but the establishment of homology does not demand that they be so. Evolutionary concepts are not understood by most neurobiologigsts and in the future planned comparisons using identical methods must be done if we are to make strong phylogenetic inferences. Selection pressures may work through the brain's normal plasticity and thereby change connectivity patterns quantitatively. Because neurons are not likely to become unrecognizable, however, changes through this route are probably not profound.

The Paleoneurological Evidence for Human Brain Evolution" (D. Falk). Falk discussed the fossil record of natural australopithecine endocranial casts from South Africa. There are seven of these fossils which date from 2.5 to 3 million years B.P. One fossil (Sk 1585) is from a presumed robust australopithecine, the remaining six fossils represent gracile australopithecines. The most detailed endocranial cast is from the famous Taung "Baby" and Falk centered her discussion on this specimen. Sulcal patterns of all the specimens were compared to sulcal patterns of 17 human, 12 gorilla and 6 chimpanzee brains. Falk concluded that, contrary to the literature, South African australopithecines do not appear to be humanlike in the gross morphology of their cerebral cortices. Rather, australopithecine endocranial casts clearly resemble the cortical morphology found in African pongids, i.e., chimpanzees and gorillas. In particular, the position of the well-known lunate sulcus does not appear to distinguish australopothecine from ape brains as previously believed.

"Allometrical Determinants of Neocortical Fissurization" (H. Jerison). Jerison's paper was one of several that dealt with allometry of the brain. Jerison examined the relationships between brain size (volume), cortical surface area and amount of fissurization. Taking a sample of 48 species of mammals from the literature, Jerison discussed and quantified the well-known fact that cortex increases more rapidly with increasing brain volume than would be expected for a lissencephalic sphere, i.e., additional surface area is generated by fissurization in bigger brained mammals. Thus, the amount of fissurization in a brain of a given species is correlated with brain size, whereas the sulcal pattern is correlated with taxonomic factors. Jerison hypothesized that the positive allometric scaling of cortex surface area to brain weight may be to accomodate dendrites which he hypothesizes to have larger arborizations in larger brains.

"Encephalization and Obstetrics in Human Evolution" (W. Leutenegger). Leutenegger examined the relationship between obstetrics, encephalization, and morphology of the female pelvis during human evolution. These data, based on comparisons of modern humans and other living primates, showed that (1) human mothers bear relatively large-bodied neonates, (2) brain size in human neonates is only as large as would be expected for nonhuman primates, i.e., encephalization at birth is *not* higher in humans than in nonhuman primates, and (3) the large human neonatal cranium relative to the pelvic diameters is the result of the overall disproportionately large neonatal size. Leutenegger suggested that three-quarters of human brain growth is postponed until after birth because of obstetrical limitations. That is, in order to pass through the pelvic canal, human neonates must be born at relatively immature stages. There was selection against too high a degree of altricity and some pelvic remodeling.

"Allometric Approaches to the Evolution of the Primate Nervous System" (R.D. Martin). Martin discussed some of the statistical constraints that are particularly important when the allometry (nonlinear scaling) of brain size relative to body size is analyzed. Although standard least-squares regression analysis has been the favored technique used by paleontologists, Martin offered compelling statistical reasons for replacing this technique with a major-axis regression technique. This replacement would have the effect of increasing the slopes of baseline regressions. Thus, Martin suggests that the widely accepted slope of 0.667 for baseline groups should be replaced by a slope nearer to 0.75. The changed slopes have functional implications, perhaps related to metabolism. Martin's analyses for nonmammalian groups show that reptiles and birds scale differently and with lower slopes than do mammals. He also discussed the importance of carefully selecting taxa used to compile the baseline regression and of exercising caution when intervening variables, such as foramen magnum area or dental measurements, are used to predict body size in fossil species. According to Martin, use of a combination of different skull measurements, excluding foramen magnum area, provides a reliable basis for estimating body size in fossil primates.

"Quantitative Cytoarchitectonics of the Cerebral Cortices of Several Prosimian Species" (K. Zilles, G. Rehkamper, A. Schleicher). Cytoarchitectural analyses of brains provide important data for evolutionary concerns. Experimental procedures such as axonal transport studies cannot be used on rare and endangered species, so there is a need for a more powerful technique to analyze routinely stained brains. A computer-controlled scanning procedure and image analyzer were developed to quickly measure and plot different histological sections. Nissl-stained, serially sectioned cortices were studied with this new method. In this study the scanning procedure and measurements delimited regions within the isocortex (neocortex) of six prosimians *(G. demidovii, M. murinus, L. ruficaudatus, L. fulvus, T. syrichta,* and *D. madagascariensis).* Shifts in topological positions and morphological (predominantly laminar) structures were found to be useful in determining evolutionary changes. Regions observed to be similar in location and overall pattern included the primary auditory, visual, somatosensory, and motor cortices and the parietal-occipital-temporal cortex. Several interesting divergences were also noted. The striate (area 17) cortex appears more advanced in diurnal than nocturnal prosimians. The motor area of *Microcebus* and *Galago* are alike in their lack of Betz cells and these species are contrasted with *Lemur* whose motor area is divided into two parts of which the caudal division contains Betz cells. Computerized scanning is likely to be a powerful new tool for objective and quantitative cytoarchitectural studies.

Each paper dealt both with methods and concepts in primate brain evolution. Papers by Bauchot, Campbell, Martin, Zilles made particularly substantial contributions toward refining currently used methods. Bauchot elaborated a method of using neural volumes and the derived isoponderal indices to generate species dendrograms. Campbell discussed the crucial problem of how determine homologies in the nervous systems of different species. He suggested that a variety of data should be analyzed and that plasticity is not likely to interfere with drawing homologies. Martin's methodological contribution concerned proper statistical techniques for allometric studies and he made specific suggestions for refining these kinds of analyses. Finally, Zilles offered a new method for quantitatively studying cytoarchitecture (and by extension, other regions of the brain) based on computerized scanning procedures.

Papers that were especially conceptual dealt with internal brain allometries (Armstrong, Jerison) and evolutionary interpretations of primate brain evolution (Allman, Falk, Leutenegger). Armstrong's comparative work on allometry of thalamic nuclei suggests that the concept of mosaic evolution applies to changes within the brain (i.e., different nuclei evolved under different sets of selection pressures). Jerison's paper discussed brain cortex to brain volume allometry and its relationship to sulcal formations. Three participants emphasized the evolutionary implications of their work. Allman's extensive comparative work on primate visual cortices permitted him to describe common visual cortical areas in primates that were probably present in ancestral primates as well as derived structures. Falk's analysis on the natural australopithecine endocranial casts suggests that the widely held concept of brain evolution occurring early in human evolution needs reevaluation. Similarly, Leutenegger's work on encephalization and obstetrics in primates shows that selection operated during human evolution to produce relatively immature and not particularly encephalized neonates.

A second symposium on the same subject was held in Niagara Falls, U.S.A. The collection of manuscripts from both congresses will soon be published by Plenum Press in a volume entitled *Primate Brain Evolution: Methods and Concepts.*

Acknowledgments. We would like to thank A. Fasolo and B. Chiarelli for being so gracious and generous in arranging our accommodations. Dr. Fasolo was of the utmost help in providing space, in easing last minute perturbations and making the participants' stay scientifically exciting and socially enjoyable. We would also like to thank the Turin Academy of Sciences and the Institute of Comparative Anatomy for providing the rooms for our symposium.

The Effects of Drugs and Hormones on Social Behavior in Nonhuman Primates

A. KLING[1] and H.D. STEKLIS[2]

Introduction

This pre-Congress symposium at Pisa, Italy, in July 1980 brought together an international group of scientists actively working in the areas of pharmacology and endocrinology, with particular emphasis on social behavior.

Despite the vast amount of information describing the effects of psychotropic drugs on human behavior and psychopathology, there are few reports on how these agents manifest their influence within the context of the social group or larger social environment. Much of what we know about basic behavioral mechanisms has been obtained from rodent studies focused on motor, cognitive, or perceptual functions. In humans, the focus has been on the individual's taking a drug, particularly the effect on affect, perception, cognition, or the modification or induction of psychopathology.

Presentations in this meeting have shown the value of studying the influence of drugs within the broader social context, by focusing on the user or treated subject and other members of a social group, and evaluating the potential interactions between drug treatment and socio-environmental factors. Our present summary of the symposium papers is organized around four topics: (1) Influence of substance-abuse drugs on treated and drug-free members of a social group, (2) long-term effects of pharmacological agents on group stability and reproductive potential, (3) nonhuman primate models of human psychopathology, and (4) interactions between endocrine function and the social environment.

Influence on Members of a Group

Claus and Kling demonstrated that rhesus monkeys living in an established colony, when treated with Methaqualone (qualude), a hypnotic and commonly abused drug, may repeatedly show alterations in social rank and increased affiliative behavior (or aggression, depending on previous rank). Males regularly demonstrated autoerotic

---

1 Department of Psychiatry, Rutger's Med. School Univ. Heights, Piscataway, NJ 08854, USA
2 Department of Anthropology, Livingston College, Rutger's University, Piscataway, NJ 08854, USA

and copulatory activity, and even a sexually inexperienced male spontaneously developed copulatory behavior while under the influence of the drug. Thus, an aphrodisiac effect suggested by street users was indeed demonstrated. Untreated subjects temporarily rose or fell in rank depending on which subject in the group was treated. Radiotelemetered electrical activity from the brain showed spike activity in amygdala and hippocampus coincident with drug administration. The most intense activity occurred between 40 and 80 min post-administration and was correlated with the onset of sexual behavior.

Sassenrath reported on the short- and long-term effects of tetrahydrocannabinol (THC) administration to group-living rhesus monkeys. The short-term consequences of a 2.4 mg/kg dose (comparable to a human dose) were a decrease in arousal, exploration, and aggression, and an increase in sedation, stereotypical activity, and visual monitoring. Peak effects were obtained in 3 h. These acute effects were strikingly similar to those observed in man, including considerable variation in response by different subjects.

In both the Methaqualone and THC studies, dominant subjects tended to show diminished aggression, while lower-ranking members tended to display increased aggression. Thus, the behavioral outcomes of drug treatment may be dependent upon the subject's prior social history and status in a group.

Long-Term Effects

Another question raised in several papers was, how does the presence of drugged subjects and their prior social experience affect long-term stability and reproductive potential of the group. This is of central evolutionary importance and has implications for human social systems. For example, in the THC study by Sassenrath, the long-term effects of treatment were substantially different from the short-term ones. All treated subjects showed increases in aggression and eventually rose in rank. To quote Sassenrath:

> "The interaction of chronic drug exposure with psychosocial stress produces dramatic impairment of adaptive social behavioral responses of drugged individuals to nondrugged cagemates. This in turn, can result in disruption of group social structure and persistent elevated stress levels in nondrugged cagemates."

With regard to early experiential factors, infants of cage-reared mothers had higher cortisol levels than group-housed mothers, and when placed in a peer group, the former continued to have higher cortisol levels and lower rank. Similarly, social rank eventually achieved was related to the amount of previous social experience. Circulating levels of testosterone were higher in group-caged compared to individually-caged males.

In another study of long-term effects on social-sexual behavior, Steklis et al. studied semifree-ranging female stump-tailed macaques who were treated with a long-acting progesterone compound (Medroxy-progesterone acetate; Depo-provera). None of the treated females received copulation for a minimum of 60 days in the presence of untreated females. When all females in the group were treated, copulation diminished to less than 50% of pretreatment frequencies. This attenuation of female attractiveness suggests that human mating systems may be profoundly altered by the use of similar birth control agents.

Nonhuman Primate Models

A third issue addressed by these papers was the appropriateness of nonhuman primate social interactions as a mean of understanding the role of neurotransmitter systems in social behavior and its relevance to models of human psychopathology. Over the past decade experimental findings in this area have had a major impact on psychiatry and on our current insights, tenuous as they may be, on the role of dopamine, norepinephrine, and serotonin in the depressive and schizophrenic syndromes.

Munkvad reviewed his long-term studies on the effects of amphetamines in vervet monkeys, demonstrating that as little as 0.05 mg/kg eliminated most social behaviors including maternal-infant interactions. While neuroleptics were capable of modifying these effects, complete recovery was not obtained as neuroleptics independently tend to reduce social interactions. Munkvad remarked on the usefulness of amphetamine studies as a nonhuman primate model of schizophrenia, particularly with regard to the dopamine hypothesis.

Redmond reviewed the influence of substances which affect neurotransmitter systems emphasizing behavioral changes in untreated subjects. For example, treatment of *M. arctoides* with AMPT, which lowers brain cathecholamine, resulted in decreased social initiative, while such behaviors increased for untreated subjects. The same result was obtained with a more specific NE depleter, 6-hydroxy-dopamine.

Raleigh et al. summarized a series of studies which document the relative contributions of brain serotonergic systems to the mediation of vervet social behavior. Chronic drug treatment of adult animals with substances which enhanced central serotonergic activity (i.e., tryptophan, 5-HTP, and chlorgyline) produced specific increases in grooming, approaching, and resting. In addition, subjects were less solitary and less active. Opposite effects were obtained by treating subjects with substances which reduced central serotonergic activity (e.g., PCPA).

In addition, in drug-free animals, free and total tryptophan correlated positively with "approach", "groom", and "eating" and inversely with "avoid" and "be solitary". Whole blood serotonin correlated inversely with "avoid" and "be solitary" and, in addition, was significantly higher in dominant males than in others. Status also determined the behavioral response to tryptophan. The magnitude and rate of change were larger for dominant males.

Sigg discussed the influence of three classes of drugs on male-male dyads. He found that treatment with a beta blocker, neuroleptic, or antidepressant did not alter the development of positive social relationships between strange males. However, antidepressants tended to increase the amount of aggression, while beta blockers increased grooming between pairs.

McKinney described experimental mother-infant separation techniques and pharmacological manipulations as a nonhuman model of depression. Separation from the
mother results in two primary behavioral phases, i.e., "protest" and "despair". Despair
is characterized by decreased activity, social withdrawal, lower food and water intake,
EEG changes, and, if not treated, death. All of the above can be exacerbated by AMPT,
a catecholamine inhibitor. Opposite effects were obtained with fusaric acid and imi-
pramine. These experiments may provide a useful model for pharmacological studies
of depression.

In summary, those agents which lower central serotonin, dopamine, or NE reduce
or eliminate positive social interactions, such as grooming, approach, and proximity.
Raising these transmitter levels tends to enhance such behaviors; however, when dopamine is too high, as with amphetamine treatment, social behavior may be eliminated
and a syndrome resembling schizophrenia may occur. Similarly, behavioral changes
from social manipulation may be altered by drugs which affect these systems in the
"despair" reaction.

Interactions

The final group of papers addressed the interaction between endocrine regulation,
exogenous hormones, and the social environment. Slob et al. reported on the influence
of cyproterone acetate on the sexual behavior of male stump-tailed macaques. This
drug is used clinically to reduce libido in man. In contrast to the results in man, sexual
behavior was not significantly altered in laboratory male-female dyads treated for
6 weeks or more. Significant testicular atrophy was observed. Further experiments
with this drug are planned using a group setting rather than laboratory dyads.

The interactions between sexual and aggressive behaviors, social status, and hormone levels in talapoin monkeys were discussed by Keverne. In both dominant and
subordinate males, given access to estrogenized females in a mixed social group, testosterone levels increase. Previously dominant males, in addition, show decreases in stress-
related hormone levels of prolactin and cortisol. Subordinate males show increased
levels of prolactin and cortisol, perhaps reflecting their difficulty in coping with the
new social situation. Differences between high- and low-ranking females were also
found. Low-ranking females had higher levels of prolactin and cortisol and failed to
show an LH surge when challenged with estradiol. Social factors apparently inhibited
the LH surge, as indicated when removing a female from the group reinstated the LH
surge. This study clearly indicates some specific endocrine correlates of social rank.

The influence of social environmental factors on hormone-behavior interactions was
also stressed by Steklis et al. As previously mentioned, administration of Depo-provera
to female *M. arctoides* in a semifree-ranging setting abolishes copulation. However, the
same dose administered to one of a pair of laboratory-housed dyads, tested with one of
four males during 30 min sessions, failed to reduce copulation significantly. Copula-
tions with untreated females increased significantly, but no overall decline in copulation
rate was observed. The authors interpret this difference between the field and laboratory studies as a function of quality and quantity of male-female social-sexual contact.

It is clear from the results of this conference that the influence of a drug or hormone on behavior is not a unitary one, nor can its effects be characterized independently of the individual's social environment, previous social experience, or current social status. Because of these interactions physiological responses tend to differ between individuals in a social group, particularly in long-term studies where adaptation and coping ability may profoundly alter the entire group structure. Furthermore, because individual physiological states differ, the behavioral end points as a result of pharmacological intervention are likely to vary.

Finally, the least explored but perhaps most important issue was the effect of drug treatments on untreated group members. The preliminary findings presented would suggest the need for a long-term examination of this problem.

Report on Symposium Entitled: "Comparative Biology of Primate Semen"

A symposium entitled "Comparative Biology of Primate Semen" was convened as a satellite to the meeting of the International Primatological Society in Florence, 7–12 July 1980. The symposium was held in Siena, Italy, 25–28 June. There were 26 invited participants from nine countries. Of the participants, eight were from institutions which maintained primate colonies involved in the research reported at the symposium, while the remainder were scientists who used primate material supplied by others.

The use of nonhuman primate models for human reproduction is well established for the female, but less understood and utilized with regard to the male. This less-than-optimum use probably results from an imperfect understanding of male nonhuman primate physiology. This symposium was intended to evaluate our present knowledge and to direct attention to areas in which nonhuman primates could legitimately be utilized for future research. The presentations included a mixture of both review papers and reports of current work which provided a great deal of information in a short space of time. The overall standard of papers presented was high, and the volume of information presented made the two full days of scientific sessions very busy with more than 8 h of formal presentation and discussion on each day.

Following a plenary lecture by B. Afzelius on 25 June, entitled "Gustav Retzius and Italian Science", which pointed out the connections which have existed for many years between Italian and Northern European scientists with particular reference to the work of Gustav Retzius, the symposium began its scientific work with a review of sperm structure by Baccetti, and a presentation by Gould which directed the participants' attention to the responsibilities and reason involved in experimental use of the primates. He pointed out that it is important that great attention be paid to the validity of research to be conducted using nonhuman primates. It is evident, he said, that experimental models should provide statistical, physiological, economical, and ethical advantages over the use of human material. Statistical and economical advantages are readily obtained when using common laboratory animals, such as mice, rats, and rabbits, in which it is possible to involve large numbers of animals at a reasonable cost. Physiological advantages, representing either the existence of a mechanism demonstrated by the predominance of a specific structure or process in the experimental animal, can be sought and found in a variety of members within the animal kingdom.

---

1   Yerkes Regional Primate Research Center, Emory University, Atlanta, GA 30322, USA

When use of nonhuman primates is considered, however, statistical and economic advantages are harder to obtain, and this places increasing weight upon the demonstration of ethical and physiological advantages. It is therefore desirable that the papers presented in the scientific portion of this symposium be directed toward identification of features of the nonhuman primate species discussed which were uniquely suited for research investigation and which would provide legitimate models for future work. This concern for the ethics as well as economics of primate research was echoed by other speakers during the meeting.

The attention of the participants was focused on the testis by the review of M. Dym on testicular structure. He explained differences between primate species in the cell associations of the seminal epithelium and went on to delineate specific details of the amount and source of added androgen binding protein (ABP) in man. The information provided by Dym demonstrated clear differences in the process of spermatogenesis between the monkey (rhesus) and the human in all stages of spermatogenesis.

It has been shown, however, by studies conducted by A.K. Chowdhury, that in the chimpanzee spermatogenesis is irregular in a manner similar to that observed in man. An intermediate situation has been observed in the olive baboon *(Papio anubis)* where spermatogenesis is regularly irregular (more than one sequential stage in a cross section of the seminiferous tubule), and in that species the inappropriate presence of mature B-type spermatogonia in stage 7 cell association has been used to explain evolution of transitory or permanent multiple stages in localized areas of seminiferous tubules.

Dym continued to elaborate on the structural and functional significance of the Sertoli cells and provided eight functions for such cells including support and nutrition; release of spermatids; steroidogenesis (from C21 onward); phagocytosis; secretion of fluid, ABP, antimullerian hormone, and inhibin; metabolic coupling and cycle regulation; testosterone metabolism; and maintenance of the blood/testis barrier.

The presentation by Camatini directed attention to one component of the testis: the Leydig cell. In particular, she described new information on the development and maturation of the interstitial tissues and Leydig cells in the African green monkey *(Cercopithecus aethiops)* and demonstrated development of Leydig cells between 2 and 3 years of age in this species. At 2 years the interstitial tissue is principally composed of undifferentiated fibroblast-like cells with scattered differentiating Leydig cells present. By 3 years the Leydig cells are more numerous and developed, and all elements of steroid secreting cells are present. Lipid accumulation, however, is characteristic during this period, and has been associated in other species with lower than maximum testosterone secretion. Mature Leydig cells, while similar to those of other mammals can still be differentiated from those of the human. Camatini pointed out that the pattern of testosterone secretion and Leydig cell development in *Cercopithecus aethiops* does not parallel that of the human, and although the nonhuman primate can provide material for study of Leydig cell development, it cannot be used as a specific temporal model of the process as it occurs in man. Indeed it has been shown by other workers that the chimpanzee and gorilla exhibit the most similar pattern of hormone secretion during infancy and puberty to the human.

Stefanini continued our analysis of individual components of the testis and described the derivation and use of Sertoli cells and germ cells in mixed culture for analysis of hormone effects on cellular development and differentiation.

Graham presented a review paper on the endocrine control of the testis in primates. His paper focused on significant recent advances that had been made in understanding the endocrine control of spermatogenesis. He pointed out that relatively few primate studies have been done, and although we still lack a good deal of basic information, primates should be an important tool for such future studies. Graham drew on a very wide background of references, and provided us with an excellent background for this particular area of research. He reviewed the present status of the use of gonadotropins to control spermatogenesis, the interaction of LH and testosterone in the controlling process, and the role of androgen in supporting spermatogenesis. He emphasized the fact that there is a need for primate models of human endocrine infertility, in which the endocrine requirements for initiation and maintenance of spermatogenesis can be evaluated. Such models would be used for measurement of the value of treatment of infertility with gonadotropin releasing hormone (GnRH) and of treatment of infertility with gonadotropins and androgens.

He mentioned recent work involving immunization of marmosets and rhesus monkeys against GnRH. He further pointed out that although the antagonistic analogs of GnRH have been demonstrated to have antifertility action in male rodents, the activity of such analogs and of GnRH itself is not consistent among primate species. He made a case, in his presentation, for the use of the chimpanzee for certain portions of this work because of the demonstrated similarity in response to GnRH between the chimpanzee and the human. He concluded that although significant advances have been made in our understanding of the endocrine control of spermatogenesis, our knowledge in the primate is still based on relatively few species, and variability in endocrine control mechanisms will be a signficant factor in the selection of primate models for the study of the endocrinology of spermatogenesis.

Several papers in the symposium focused upon the epididymis from both structural and functional aspects. I. White provided us with a review paper concerning the epididymal compounds and their influence on metabolism and survival of spermatozoa. Epididymal fluid, which is in part derived from testicular fluid, contains several unusual compounds, and although little information is available on the composition of testicular fluid of the primates, the fluid of nonprimate species, such as the ram and boar, has been shown to contain high concentrations of inositol and certain amino acids. Analysis has, he said, been made of epididymal fluid collected from the cauda epididymis of the rhesus monkey, as well as from nonprimate species, but significant information on human material is lacking. The information available to date tends to imply that the composition of cauda epididymal fluid is similar between mammalian species, but differs considerably to that of blood, lymph and other extracellular fluids. This further implies that the environment of spermatozoa in the epididymis is highly specialized and contributes to the prolonged survival of the sperm in that organ, in part by providing substrates for spermatozoa metabolism. At least one antifertility compound (α-chlorohydrine) has been demonstrated to be concentrated in the epididymis subsequent to oral administration, and studies on laboratory animals, domestic species, and the human suggest that this compound inhibits enzymes of the glycolytic pathway. White concluded that detailed knowledge of the nature of testicular and epididymal fluid is very meager for primate species, and although such information as is available tends to suggest a similarity in composition between all mammalian species, further information should be obtained in the primate to identify the specific role of unusual compounds such as glycerylphosphorylcholine, carnitine, and inositol in epididymal fluid.

Analysis and investigation of epididymal structure was described in a paper by Moore and Pryor, in which the ultrastructure of the epididymis in several Old and New World monkeys was studied and comparison made to human epididymal tissue. They concluded that the general morphology of the epididymis was the same for all primates studied, and dissimilar to that of the common laboratory animals used presently. The ultrastructure of the epithelium, however, exhibited quantitative differences from that in man with the fine structure of the epididymal epithelium of the talapoin monkey and common marmoset being most similar to that of the human. They conclude that their evidence indicates that the nonhuman primate will be a more suitable model for epididymal tissue in man than rodents or other laboratory or domestic animals. However, they advocate caution when extrapolating findings from nunhuman primates to the human, since the idea that the human epididymis has a unique physiology is supported by studies on its microvascular structure and androgen-binding protein. The epididymis of the great apes has not yet been studied in this regard. They suggested that the marmoset in particular could prove to be a useful model for human reproductive physiology, and can be bred in captivity. Male marmosets do not show any seasonal variation in reproductive function, and so would be more acceptable as an animal model than the macaque which is presently most commonly used.

White's observation that limited data exists on the fluid composition of the epididymis in primates was addressed in the paper by Hinton and Setchell concerning analysis of fluid samples obtained by micropuncture techniques from the epididymis of the monkey and the baboon. The technique described by these authors permitted collection of 50–300 nl samples from several epididymal sites with subsequent analysis for inositol and for sodium, potassium, chlorine, osmolarity, carnitine, glycerylphosphocholine, and total phosphate. They showed that the microenvironment of the epididymis of the monkey, baboon, and human differs markedly from that of other species with regard to the concentration of certain compounds in the lumen fluid. In the human and monkey, the majority of the osmotically active compounds are inorganic ions, a finding in direct contrast to those reported by Hinton for the rat, hamster and boar, in which species organic compounds contribute more to the osmolarity of the fluid than do inorganic ions. Again, Hinton, in concert with other authors, concluded that further studies were legitimate and required in primate species in order to fully investigate the epididymal environment.

Several papers were presented at the symposium regarding structure and function of the male gamete product of the testis: the spermatozoa. Baccetti provided an overview account of those features of sperm morphology common throughout the animal kingdom, with special reference to the relationship of the structure of primate sperm to that of other species. Koehler elaborated on the specific structural changes involved in the capacitation of sperm and the development of the acrosome reaction around the time of fertilization. He reported on aspects of his own work which support the notion that there is an alteration in the surface components of spermatozoa at the time of capacitation, and this change is relatively similar between nonprimate and primate species. Continuing this discussion of surface changes of spermatozoa associated with fertilization. Young provided experimental data regarding the use of enzymatic iodination with lactoperoxidase and ^{125}I-sodium iodide to identify specific components on the human sperm cell surface and in seminal plasma.

The work of Young and Goodman demonstrated that the sperm surface and seminal plasma components are the same in molecular composition and that their work supported the hypothesis that some of the proteins on the human sperm cell surface are absorbed from seminal secretions. They pointed out that this approach was especially applicable to the studies of primate spermatozoa.

Talbot and Chacon described techniques which they have developed of specific application to human and nonhuman primate spermatozoa in which the acrosome has been difficult to identify. It is of interest to know the specific time at which the acrosomal membranes are lost from spermatozoa, and they reported on a triple staining technique which permits ready identification of the presence or absence of the acrosome together with the viability of the spermatozoon being observed. It was possible, using this technique, to distinguish the normal and degenerative acrosome reaction. Their observations showed that the human sperm undergoes the acrosome reaction in vitro in a time which depends upon the culture medium, perhaps as short as 1.5 h, and also that only a very small percentage of sperm underwent the acrosome reaction, a situation markedly different from that in most rodents frequently used in this type of study. Specific information on internal structure of spermatozoa was provided by papers by Pallini on the dyneins of the human and primate sperm and by Seuanez on the quantitation of DNA in individual spermatozoa. This latter technique provides a means for indentifying diploid sperm from the normal haploid population and has importance in the study of semen homogeneity between species. He points out that the greatest variability in ploidy is observed in the gorilla and in the human.

Linnett addressed the subject of sperm motility and by utilizing information on a chance observation related to the movement of agglutinated clumps of sperm under certain conditions of in vitro culture, he demonstrated that it was possible to derive information on the internal structure of the sperm midpiece and tail and to infer further information on the direction of the beat of the sperm tail, and suggested that his findings demònstrated a counterclockwise wave propagation toward the tip of the sperm tail. It also prompted the hypothesis that the tail curves in a given species of primates are concave toward the same fibers in the axoneme in every case.

Finally, there was a group of papers on functional aspects of male reproductive biology. Afzelius provided a comprehensive review of the various forms of abnormal human sperm, and provided some provocative hypotheses concerning the origin of these abnormalities. Phillips and Chalgee described their work on sperm/egg interaction utilizing interspecific (hamster ova and human sperm) fertilization to observe the patterns of association between sperm and ova. Their work demonstrated that in vitro morphological aspects of fertilization are different in ova in which the cumulus and zona are intact as compared to those of ova which have these investments removed. Their conclusion is that human spermatozoa associate with zona-free hamster ova tip first, but the observation that differences exist between in vivo and in vitro sperm/egg interaction suggests that caution must be taken in interpreting these results.

Much discussion was generated by the paper of Wickings, Zaidi, and Nieschlag concerning seasonality in rhesus monkey reproduction. Their laboratory has shown a marked seasonal change in reproductive parameters for the rhesus monkey *(M. mu-latta)* in Germany, with testicular volume between October and January being twice as great as in the same animals between March and June. They reported that electro-ejaculation was not possible in out of season animals — a finding in conflict with those of researchers working at lower latitudes. Data was presented on the annual changes in testicular and pituitary hormone function,, and their conclusion was that the rhesus monkey, under their conditions, could only be used for fertility control studies during the relatively short phase of testicular activity, but could also provide a model for study of the aspermatogenic state. They reported further on initial experiments on immunization on animals with gonadotropic hormones and on the use of LHRH super-agonists in attempts to induce functional sterility or infertility in primate species. Their conclusions showed that production of specific FSH antibodies only rarely resulted in suppression of spermatogenesis to the level of aspermia, but that the use of LHRH superagonists was not accompanied by change in testicular volume or in sperm count or motility. They suggested that nonhuman primates were essential for continuation of this work, and it is evident further work remains to be done in this area.

Information on the immunological consequences of vasectomy was provided in a review paper by Alexander in which she reported that up to 50% of animals retained circulating antisperm antibodies subsequent to vasectomy, although in man antibody development is less rapid than in the monkey. Histopathological studies revealed orchi-tis, aspermatogenesis, or both, in 92% of vasectomized monkeys compared to only 23% of control animals. She reported an apparent correlation of constant sperm anti-gen leakage subsequent to vasectomy with immune complex formation and subsequent development of arthritis and atherosclerosis. It is evident that this intriguing possibility must be pursued more fully.

A paper by Roberts reported on factors in seminal plasma which may have an immunosuppressive role effective in vivo during the fertilization and implantation phase of pregnancy. Roberts demonstrated quite clearly a need for continued use of non-human primates in this area.

A final paper by Tsong reported on the current status of the use of gossypol in inducing infertility in the human and in nonhuman species.

It is evident, then, that extensive reviews and important new information was pro-vided in the symposium on a variety of subjects appropriate to the study of reproduc-tive function in the male nonhuman primate. Discussion was very lively and prolonged after every presentation in the symposium and provided a groundwork for continued conversation throughout the entire meeting. After a brief round-table meeting, in which the present situation of funding of primate research and primate supply was presented, the symposium concluded with a dinner at the Collegio "M. Bracci" in Pontigano. Professor Baccetti is to be especially thanked for his conduct of the local arrangements. Almost all the visitors were housed in one hotel which aided in promo-tion of an atmosphere conducive to continuous work from 8 a.m. to almost 10 p.m. with a brief afternoon break to visit the city. The arrangements for meals, etc. also promoted continued discussion of the presentations. All participants agreed that Siena was a splendid site for such a meeting, and more than one suggestion has been made that a continuing series of such symposia be organized in the same setting.

Acknowledgments. We wish to acknowledge the sponsors of this symposium. They are the University of Siena, the Italian Research Council Project on Reproductive Biology, Menley and James Laboratories, the Yerkes Regional Primate Research Center of Emory University, the Division of Research Resources – NIH, Division of Research Services – NIH, and the Fogarty International Center – NIH.

Chromosome Banding and Primate Phylogeny. Inaugural Address

H.N. SEUÁNEZ [1]

The Early Days of Human and Primate Cytogenetics

Waldeyer was far from imagining in 1888 that the densely stained nuclear bodies he called "chromosomes" were to play a role in the transmission of heredity factors or to become a clue in the study of human phylogeny. Early in this century, Sutton and Boveri noticed, however, that the mechanical principles of Mendelian segregation and independent assortment applied precisely to the behavior of chromosomes in meiosis. Since then to now, cytogenetics has developed as a hybrid discipline, although its coming of age was far from being steady and easy. The diploid chromosome number of our species, *Homo sapiens,* was a subject of discussion until 1956, when Tjio and Levan on one side, and Ford and Hamerton on the other, independently reported that it was 46. Obviously, the dearth in the knowledge of human and primate cytogenetics was enormous at that time. However, as early as 1940, Yeager et al. had already observed the seminal epithelium of the chimpanzee *(sp. troglodytes),* and had correctly estimated the diploid chromosome number in this species based on the number of bivalents in the first meiotic division.

Significant advances in human and primate cytogenetics were achieved when simple techniques to obtain chromosome preparations from in vitro mitogenically stimulated lymphocytes became available. However, the study of human and primate chromosomes was initially limited to gross morphological and morphometric analysis, and to estimations of the diploid chromosome number of the species under study. In man, a few individual chromosomes were identified because they exhibited distinctive characteristics such as size, arm ratio, or the presence of a secondary constriction region, but the analysis of the whole chromosome complement was difficult. Some of these difficulties were partially overcome with autoradiographic techniques that allowed the recognition of homolog pairs within a group of chromosomes of similar morphology. A comparison between species, such as man and the great apes, for example, revealed obvious similarities at the chromosome level because many human chromosomes appeared to have similar counterparts in these species (Chiarelli 1962). One of these striking similarities was demonstrated by McClure et al. (1969) who found that a trisomy for a small acrocentric chromosome, similar to chromosome No. 21

1 Department of Genetics, Institute of Biology, Universidade Federal do Rio de Janeiro, Rio de Janeiro, Brasil

in man, was responsible for a pathological condition in the chimpanzee similar to mongolism in man. Thus, the production of a similar abnormality by an identical type of chromosome aneuploidy in two phylogenetically related species suggested that the small acrocentric chromosome carried the same genetic information in both man and the chimpanzee.

Chromosome Banding and the Recognition of Interspecific Homologies

Chromosome banding techniques were developed in the late 60s and early 70s, and they allow the unequivocal identification of chromosome pairs within the complement. This is because each chromosome pair is distinctively stained, showing a characteristic pattern of intensely stained regions alternating with pale regions across the arms which are called "chromosome bands". A large number of techniques have been reported, but the most commonly used are those called Q-, G-, R-, C-, and NOR-banding. Q- (quinacrine) banding was first observed when chromosomes were stained with the fluorochrome quinacrine mustard, a substance that is capable of alkylating pure DNA in solution. When applied to chromosomes, it was possible to distinguish a spectrum of fluorescent intensity across chromosome arms which ranged from regions of brillant, intense, medium, pale, and negative fluorescence. G- (Giemsa) bands were initially observed when chromosome preparations were incubated in saline solutions or after partial digestion of chromosomes with proteolytic enzymes followed by Giemsa staining. R- (reverse) bands were observed in chromosome preparations that had been subjected to heat denaturation and later stained with Giemsa or with the fluorochrome acrydine orange. C- (constitutive heterochromatin) bands were first observed in mouse chromosomes stained with Giemsa after alkali denaturation. NORs (nucleolar organizer regions) were demonstrated as areas of heavy grain density due to reduction and precipitation of silver in chromosome regions located in the vicinity of nucleoli.

These techniques have been extremely valuable in chromosome identification, and in showing a level of chromosome structure that was previously unrevealed. It must be pointed out, however, that there is no agreement on the nature of the substances that are specifically stained with these techniques or on the precise mechanisms by which chromosome bands are produced. This is because chromosomal DNA has different properties from pure DNA solution, a reason why it is stained in a different way by the same dyes. Chromosomal DNA is the main component of chromatin that contains histones and nonhistonic proteins as well. DNA and histones are organized in chromatin subunits, or nucleosomes, and these subunits are probably packed into larger subunits that form the chromatin fibers. Obviously, the spacial configuration of chromosomal DNA is different from pure DNA in solution, and this fact, together with the interaction of DNA molecules with other chromatin components, must account for the special properties of chromosomal DNA.

Another fact that makes chromosome banding difficult to understand is that the same kind of bands and the same pattern of chromosome banding can be obtained with totally different agents, some of which are completely unrelated to DNA. This is the case of G-bands that can be produced after incubation in saline solutions or after partial proteolytic digestion with trypsin or pronase. Incubation in saline solutions was thought to be operating on chromosomal DNA by allowing its reassociation after a previous treatment that was devised to denature it. Trypsin and pronase, on the other hand, can affect only chromosome proteins because they are well-known proteolytic enzymes that have no action on DNA. This finding suggested that chromosome banding might be dependent on the nature and the integrity of chromosomal proteins rather than on the nature or the composition of DNA. Sumner (1976) has demonstrated that most of the remaining proteins in fixed chromosomes are nonhistonic because fixation with methanol/acetic acid selectively extracts histones from chromatin. He has also demonstrated that there is a differential distribution of protein disulfides and free sulfydryl groups along chromosome arms, and that regions of positive G-banding are rich in disulfide bridges, as against regions of negative G-banding that are rich in free sulfydryls. Thus, chromosome banding would be a consequence of chromatin packaging, where positive G-band regions would correspond to areas of chromatin condensation, and negative G-band regions to regions of chromatin relaxation. It has also been shown by Sumner and Evans (1973) that the characteristic magenta color of chromosomes stained with Giemsa results from the formation of a magenta compound consisting of two molecules of thiazide and one of eosin. As the dye was found to be binding to DNA at two separate points, Sumner and Evans postulated that the differential decondensation of chromatin which is produced by saline incubation or by proteolytic digestion resulted in slight differences in DNA concentration. Such differences were in turn greatly exagerated by the binding properties of the Giemsa compound, as a result of which a sequence of intense and pale regions appeared across chromosome arms.

Although this is just one of the many hypotheses put forward to explain the mechanisms of chromosome banding, the fact that chromosome bands are not necessarily related to DNA is widely accepted. This is specially evident with Q-banding, where fluorescence intensity was initially thought to be related to a specific type of DNA composition. The initial hypothesis suggested that the intensity of fluorescence was positively correlated with the amount of AT-rich repetitive sequences. But in the mouse, for example, where AT-rich sequences are located in the near centromeric regions of all chromosomes except for the Y chromosome, there is dim fluorescence in these regions. The same regions, however, are more intensely stained in interphase nuclei, a fact that suggests that fluorescence intensity must be dependent on the stage of chromosome elongation or on the presence of different chromatin components in the cell cycle. Such variable components are likely to be proteins because there is no reason to believe that DNA composition is altered from interphase to metaphase.

A similar conclusion is valid for C-band regions when they are compared with the specific DNA composition of heterochromatin. C-band regions in different organisms may contain highly repetitive DNA, moderately repetitive DNA or nonrepetitive DNA. These bands, however, correspond to regions where DNA and nonhistonic proteins are tightly packed, and are therefore resistant to alkali denaturation. By measuring DNA content after alkali denaturation it has been shown that DNA and proteins are extracted from euchromatic regions while remaining almost intact in heterochromatic regions. Such resistance seems to be due to the particular packaging of DNA with proteins rather than to a specific type of DNA composition.

Nucleolar organizer regions (NORs) indeed contain 18S and 28S rDNA sequences, but the substance that precipitates silver is an associated protein that can be destroyed by proteolytic digestion, and that is present only when the 18S and 28S cistrons are transcriptionally active.

Although I am not trying to explain all the possible molecular mechanisms involved in chromosome banding, I think it must be stressed that the initial propositions suggesting that banding revealed specific kinds of DNA sequences are now untenable. For this reason, the obviously limitations of banding techniques must be taken into consideration in comparative cytogenetics and in chromosome phylogeny. This is because a comparison of banded karyotypes of phylogenetically related species of primates or between some primates and man has shown evident similarities at the chromosome level. Such similarities allow the recognition of presumed "interspecific homologs" because the morphological characteristics and the banding pattern of a chromosome pair in one species may be identical or similar to those of a chromosome pair in another species. Such similarities may sometimes not be so obvious, and a chromosome of one species may be "derived" from a chromosome of another by a presumed rearrangement that can transform one chromosome into the other. A derivation of this kind would implicitly suggest that a chromosome rearrangement has taken place during the phyletic divergence of the species under study, although it is impossible to know whether it occurred before or after speciation. But regardless of the time when rearrangements occurred, it is possible to envisage phylogenetic pathways for different chromosomes.

Most reports on primate chromosomes are confusing because each author uses his own criterium of chromosome nomenclature, and proposes different "interspecific homologies" between primate species or between nonhuman primates and man. This is because there are few standard criteria of nomenclature of primate chromosomes and for recognizing presumed interspecific homologies. The few species for which a standard criterium was first recommended are the great apes, viz. the chimpanzee, the pygmy chimpanzee, the gorilla, and the orangutan (Paris Conference 1971, supplemented 1975, Stockholm Conference 1977), while no similar criteria were then proposed for the rest of the nonhuman primates. This was due to the small amount of information on primate chromosomes available at that time. In most cases, reports on primate chromosomes are based on very few specimens because many of the extant primate species are under threat of extinction or are kept in small numbers in zoos and research centers. However, there is already enough information on the karyotype of the Rhesus monkey *(Macaca mulatta),* a species that has been widely used in laboratories. In view of the fact that *Macaca fascicularis* has an identical karyotype with *M. mulatta,* and the striking karyotypic identity between *M. mulatta* and *Papio papio* (Finaz et al. 1978), a standard criterium of chromosome nomenclature was needed for these species. Moreover, there was also valuable information on gene assignment in *M. mulatta* and *Papio papio,* a reason why a standard criterium of interspecific homologies was recommended in order to compare these species with man and the great apes [Human Gene Mapping 5 (1979)].

In the future, it will be necessary to undertake studies of primate groups or populations instead of restricting chromosome analysis to individual specimens. Population studies are most valuable when they are related to geographical distribution, behavior, and fertility, because the full implications of chromosome change within a species can be properly evaluated. This is the case of the owl monkey, *Aotus trivirgatus,* where seven different karyotypes, or "karyomorphs" have been described among animals of different geographical origin. The fact that specimens occupying different environments have fixed different chromosome rearrangements, such as pericentric inversions or translocations, suggests that we might be dealing with a case of rapid speciation. Ma et al. (1976), when analyzing some 300 specimens, have associated specific karyomorphs with specific phenotypes. Although phenotypic differences within *Aotus trivirgatus* are unlikely to have resulted from chromosome rearrangement, this observation is most valuable. The most important aspect is to clarify whether these populations are actually reproductively isolated from each other or whether crossbreeding might occur in natural conditions, in the event that they might be brought again into contact. A further step would be to look for differences at the biochemical level between *Aotus* karyomorphic groups, and correlate biochemical differences with chromosome change.

Another example of chromosome variation within a species is observed in the orangutan. This species is at present spread into two separate islands, Borneo and Sumatra, that have remained separate from each other for the last 8000 years. By comparing the karyotypes of Bornean and Sumatran orangutans it was possible to detect a small pericentric inversion that accounted for a slight difference in chromosome No. 2. As each population of orangutans carried a different type of chromosome No. 2 in the homozygous condition, the rearrangement has probably been completely fixed in the population where it originally occurred (Seuánez et al. 1979). In view of these findings, and taking into consideration that Bornean-Sumatran hybrid orangutans carry a heteromorphic pair No. 2, it is now possible to distinguish Bornean from Sumatran orangutans unequivocally whenever there is no precise data on their geographical origin. As the two orangutan subspecies are chromosomally distinct, it is probable that they might be more different from each other than we had previously thought. Bruce and Ayala (1979) have reported that the two subspecies show a greater genetic distance than that existing between two species of the same genus like *Pan troglodytes* and *Pan paniscus,* for example. This study, which was based on biochemical traits analyzed with electrophoretic migration, demonstrated that Bornean and Sumatran orangutans presented different electromorphs for the red cell enzyme adenosine deaminase, so that both subspecies could be distinguished from each other by a biochemical analysis. Although this biochemical difference is probably unrelated to the chromosomal difference between Bornean and Sumatran orangutans, both findings suggest that the taxonomic status of these subspecies might need some revision in the near future. Here again, the standard criterium of chromosome nomenclature of orangutan chromosomes must be revised because the original report of the Paris Conference (1971); supplement (1975) as well as that of the Stockholm Conference (1977) do not mention this chromosome difference, but propose identical karyotypes for the two subspecies of orangutan.

Banding Homologies vs Syntenic Homologies

It is evident that chromosome studies in the primates should not be based on banding patterns exclusively. As chromosomes are the vehicles of inheritance, they are actually linkage groups that transmit the genetic information from cell to cell, both during ontogeny and phylogeny. The main problem is whether the genetic information encoded in the DNA can be somehow revealed by chromosome banding if banding seems to be more dependent on protein denaturation than on DNA specificity. This question can be answered by comparing interspecific homologs, as recognized by banding, with data on gene assignment for the same chromosomes. Gene assignment can now be done by fusing cells of a primate line with cells of a rodent line. Fusion is usually followed by chromosome loss, and in most cases where human or primate cells are fused to rodent cells there is a preferential loss of human or primate chromosomes, and a preferential conservation of rodent chromosomes. By positively correlating the presence of a human or primate chromosome with the presence of human or primate structural loci in the cell hybrid, it is possible to assign a gene locus to a chromosome. It must be pointed out that chromosome banding is most valuable in this procedure because chromosomes must be unequivocally identified in the cell hybrid. This is possible because the banding pattern of a given chromosome is identical in all tissues, and each chromosome conserves its characteristic pattern even when it is present in a cell hybrid, in a different cellular background.

Experiments of this kind have clearly demonstrated that the principle of interspecific homology inferred by banding is generally confirmed by gene assignment. But we must clarify that this rule applies only to structural gene loci and not to repetitive DNA sequences like 18S and 28S rDNA or satellite DNA. And even in the case of structural loci the rule has exceptions. The best example of evolutionary conservation at the chromosome level is that of the X chromosome, not only among the primates but in mammals in general. The G-6-PD locus, for example, is X-linked in man, the chimpanzee, the gorilla, the orangutan, the Rhesus monkey, the African green monkey, the horse, the hare, the dog, and the kangaroo. And in many species such as man, the great apes, and the Rhesus monkey, the X chromosome is morphologically similar, with a constant banding pattern. This extraordinary conservation is what Ohno (1973) has called "an evolutionary frozen accident".

But the main question we must raise is how frozen other linkage groups have remained during primate phylogeny. Here again, the answer is not simple because there is good evidence that some primate chromosomes have been conserved as linkage groups, not only during human-ape radiation, but during Cercopithecoid radiation as well. This is the case of chromosome No. 1 in man whose origin may be traced as far as the time the Rhesus monkey, the baboon, and the African green monkey diverged from each other. But other human chromosomes like No. 9, for example, have no recognizable homolog in the gorilla, the orangutan and the Cercopithecoids. In this case, chromosome banding is useless to infer interspecific homologies, and such homologies can only be evident by gene assignment. In fact, data on gene assignment show that chromosome No. 9 in man contains the same loci as chromosome No. 13 in the gorilla, although this latter is completely different in morphology from chromosome No. 9 in man. On the other side, there are examples of chromosomes with similar morphology and banding patterns which do not contain the same loci. In the chimpanzee, for example, the *short* arm homolog of chromosome No. 2 in man contains the gene loci of the *long* arm of chromosome No. 2. Vice versa, the chimpanzee *long* arm homolog of chromosome No. 2 in man contains the same loci as the *short* arm of human chromosome No. 2. In this case, banding homologies are not coincident with syntenic homologies. A similar finding was reported in chromosome No. 20 in man and its interspecific homolog in the chimpanzee, the gorilla, and the orangutan. In man, chromosome No. 20 carries the locus of inosine triphosphatase (ITP), but in the great apes this locus is not contained in the corresponding homolog, in spite of the evident similarity between interspecific homologs. The locus for ITP has been actually found to segregate together with that of nucleoside phosphorilase (NP) which is in chromosome No. 15 in the chimpanzee. The NP locus was assigned to chromosome No. 14 in man, which is the homolog to chromosome No. 15 in the chimpanzee, and also to chromosome No. 18 in the gorilla, in spite of the fact that this chromosome is morphologically different from chromosome No. 14 in man or No. 15 in the chimpanzee (Winnipeg Conference 1977). The most possible explanation to the lack of coincidence between "banding homology" and "syntenic homology" is that chromosome banding is incapable of revealing a change in a single gene locus or the transposition of one locus from one chromosome to another.

The chromosome distribution of 18S and 28S rDNA sequences in the primates has also been found to be completely independent from the process of karyotypic evolution. This is why, in lower primates, the distribution of these sequences is multichromosomal as it is in some Platyrrhines, the great apes, and man. In other Platyrrhines and in Cercopithecoid monkeys, these sequences are contained in one chromosome pair (Henderson et al. 1977). The distribution of these repetitive sequences in the chromosome complement of the primates seems to be completely unrelated to the presumptive rearrangements that took place during primate phylogeny.

A comparable, though not identical, situation is found in the chromosome distribution of the four major human satellite DNA's in man and their homologous sequences in the great apes. The distribution of these repetitive sequences is unrelated to banding homologies, viz. interspecific homologs may contain different kinds of repetitive DNA or significantly different amounts of the same repetitive sequence. For this reason, chromosome banding is not informative on the chromosome distribution of satellite DNA in man and their homologous sequences in the great apes (Seuánez 1979). This is because the amplification of these repetitive sequences must have probably taken place independently *after* speciation, after the splitting of the Hominoidea into different species, and not in the common ancestor of man and the great apes.

In view of the obvious limitations of chromosome banding is it necessary to confirm presumptive interspecific homologies by gene assignment before further conclusions are extracted. Without data on gene assignment the postulation of presumed phylogenetic pathways and hypothetical chromosome rearrangements might be more fictional than real, because we would not know whether we are actually looking at the same linkage groups in the different species analyzed. The conservation of similar banding patterns during primate phylogeny might well depend on the conservation of few genes coding for the chromosome proteins that in turn are responsible for DNA packaging, and − indirectly − for chromosome banding. These loci could have been conserved in a similar way to structural loci, so that each chromosome would carry the genetic information for its own structural proteins. It is well known that isolated chromosomes maintain their original Q-, G-, R-, and C-banding patterns unchanged when they are put against a different cellular background. This is the case of a single human chromosome in a cell hybrid which contains a large number of mouse chromosomes. The remaining human chromosome has a G-banding pattern identical with that of the same chromosome in a normal human cell. A similar structural conservation could have occurred in primate phylogeny, and this could explain the common coincidence between banding homologies and syntenic homologies. In some other cases, however, the forces of disruption would overcome the forces of conservation. Point mutations could affect the loci of the chromosome proteins, and their interaction with DNA might have been affected. Moreover, chromosome rearrangements could have reshuffled the original karyotype of an ancestral species giving origin to many different karyotypes during phylogeny. We still do not know whether the maintainance of some linked loci in mammals has been adaptative or due to chance. For this reason it is difficult to understand what mechanisms have been operating in keeping gene loci unchanged or, alternatively, in producing karyotypic diversity. For all the above mentioned arguments, I think that chromosome banding should be carefully considered as one possible way of inferring presumed interspecific homologies, always to be confirmed by more precise procedures.

Chromosome Similarity and Phyletic Relationship

One of the main questions in comparative cytogenetics is whether similarities at the chromosome level as revealed by chromosome banding may be indicative of phylogenetic relationship. This is like asking to what extent the recognition of interspecific homologies in the primates is an indication of their common origin. Here the problem is actually twofold because on one side we have the origin of the primate species, and on the other, the origin of primate chromosomes. The origin of primate (and human) chromosomes has been tentatively explained by chromosome rearrangements that account for the karyotypic diversity of the extant primate species. The hypothetical pathways are based on the assumption that all primates originated from the same common ancestor whose original karyotype suffered substantial alterations.

I think, however, that the first question we must raise is up to what extent the genome, and not just the chromosome complement, is similar when species are compared. To answer this question there is no other way than analyzing homologies at the DNA level, and estimating the approximate amount of nucleotide substitutions that have occurred between the species that are compared. Present techniques for analyzing eukaryotic DNA allow the isolation of three different genome components based on their reassociation kinetics under standardized experimental conditions. This allows the isolation of one fraction with a fast rate of DNA reassociation which corresponds to highly repetitive DNA, a second with an intermediate rate of reassociation which corresponds to moderately repetitive DNA, and a third one with a slow rate of reassociation which corresponds to unique-copy DNA (Britten and Kohne 1968). Comparisons between species can be carried out with any of the three fractions, but the unique-copy DNA component is the one that will supply the most valuable information. This is because highly repetitive DNA has no known function, and most of the moderately repetitive DNA is not transcribed except for a few specific sequences like ribosomal DNA and histone DNA. As against these, the unique-copy DNA fraction contains the structural DNA sequences that are transcribed and translated, and which account for the functional genes of the genome. This is why it is important to look at this fraction and to estimate the amount of nucleotide replacements that have occurred in it as a consequence of the fixation of point mutations during phylogeny. This can be demonstrated by forming an interspecific hybrid DNA molecule, or heteroduplex, by incubating single-stranded DNA of one species with single-stranded DNA of another. Reassociation would occur if there is complementarity between the DNAs of the different organisms. The thermal stability of the hybrid molecule would be an indication of how well matched, viz. how similar, their DNAs are or how poorly matched, viz. how different, they are from each other. If the species thus compared have a common origin, the differences between their unique-copy DNA fractions can be extrapolated to the time when both species diverged from each other, so that the *rate* of nucleotide substitution can be approximately estimated.

An analysis of this kind has shown that man, the chimpanzee, and the gorilla are practically equidistant from each other up to the point that it is not possible to say which of these two species is man's closest living relative. On the other side, the orangutan stands out further away, in between man and the African apes on one side, and the gibbons and siamangs on the other.

This method is perhaps the best to infer phyletic relationships between species because it screens the whole unique-copy DNA fraction. Comparisons between species are carried out regardless of what mutations, viz. samesense, missense, nonsense, or frameshift, have taken place in the fractions that are tested. This approach is obviously more accurate than comparisons between species at the protein level. This is because structural loci account for only a fraction of the unique-copy DNA component, and also because a comparison at the protein level would not detect samesense mutations. Moreover, some of the methods used in protein comparisons, such as electrophoretic migration, would not detect amino acid substitutions unless they produced a change in the net charge of the molecule.

If man, the chimpanzee, and the gorilla are phylogenetically equidistant from each other, as is indicated by DNA reassociation studies, what does this mean in relation to chromosomes? A comparison of the banded karyotypes of man, the chimpanzee, and the gorilla shows that we can derive the human karyotype from that of the chimpanzee with fewer rearrangements than from man to the gorilla. Although this finding indicates that a greater similarity exists between man and the chimpanzee, at the chromosome level, than between man and the gorilla, this finding is irrelevant in terms of phyletic relationships between man, the chimpanzee, and the gorilla. This is because a comparison at the unique-copy DNA level had shown that these three species are phylogenetically equidistant from each other. The most logical explanation to this apparent contradiction is that chromosome change must have occurred at different rates during the phylogeny of man and the African apes. A similar situation has probably occurred in other groups such as the gibbons and the siamangs. These species are closer to man than man is to the Cercopithecoid monkeys, but few similarities are evident between man and the gibbons at the chromosome level. However, there are many similarities between human chromosomes and those of some Cercopithecoid monkeys such as the Rhesus, the baboon, and the African green monkey. Many of these similarities have been confirmed with gene assignment, so that many human chromosomes have been evolutionarily conserved for a period of at least 35 million years. This conservation at the chromosome level is, however, unrelated to the amount of nucleotide substitutions of the unique DNA component of man, the Cercopithecoids and the gibbons. The most logical explanation is that chromosome change must have been more rapid in one group, probably among the Hylobatidae, while it remained less drastic among the Cercopithecoids, the great apes, and man. For this reason, the simple assumption that two species may be phylogenetically closer to each other than to another one because of their greater similarity at the chromosome level is obviously untenable. If this assumption were valid, we would be postulating, perhaps unintentionally, that chromosome change has been constant during phylogeny, and as a consequence of being a simple function of time, it has been selectively neutral. If this had occurred, we would end up explaining chromosome change following a model of a "chromosome evolutionary clock", in which each "tick" of the clock would be marked by the fixation of a new chromosome rearrangement. This would imply that chromosome rearrangements were not adaptive, because only neutral mutations may be fixed at a constant rate regardless of different selective pressures. As most of the presumed chromosome rearrangements that occurred in primate radiation involved substantial amounts of euchromatic regions, it is unlikely that they could have been selectively neutral.

Although we are still far from understanding the actual significance of chromosome change in primate phylogeny, two alternatives are probable. One is that chromosome change, occurring *before* phyletic divergence, was the main factor in producing the splitting of primate populations into different subpopulations that became reproductively isolated from each other. The other is that chromosome change took place *after* speciation, and consequently, played no role in the emergence of new species.

Evolution seems to have been operating at different rates at the organic, chromosomal, and molecular levels. Although these three levels of complexity are indeed interrelated, we are still far from understanding the effects of molecular mutation and chromosome change on organic evolution. Within this context, the study of chromosomes is just one aspect of the whole problem.

References

Britten RJ, Kohne DE (1968) Repeated sequences in DNA. Science 161:529–540

Bruce EJ, Ayala FJ (1979) Phylogenetic relationships between man and the apes: Electrophoretic evidence. Evolution 33:1040–1056

Chiarelli B (1962) Comparative morphometric analysis of the primate chromosomes. I. The chromosome of the anthropoid apes and of man. Caryologia 15:99–121

Finaz C, Cochet C, Grouchy J de (1978) Identité des caryotypes de *Papio papio* et *Macaca mulatta* en bandes R, G, C, et Ag-NOR. Ann Genet 21:149–151

Ford CE, Hamerton JL (1956) The chromosomes of man. Nature (London) 178:1020–1023

Henderson AS, Warburton D, Megraw-Ripley S, Atwood KC (1977) The chromosomal location of rDNA in selected lower primates. Cytogenet Cell Genet:281–302

Human Gene Mapping 5 (1979): Fifth International Workshop on Human Gene Mapping. Cytogenet Cell Genet 25: No. 1–4

Ma NSF, Jones TC, Miller AC, Morgan LM, Adams EA (1976) Chromosome polymorphism and banding patterns in the owl monkey *(Aotus).* Lab Anim Sci 26:1022–1036

McClure HM, Belden KH, Pieper WA (1969) Autosomal trisomy in a chimpanzee. Science 165: 1010–1011

Ohno S (1973) Ancient linkage groups and frozen accidents. Nature (London) 244:259–262

Paris Conference (1971) supplement 1975. Standardization in human cytogenetics. Brith defects original article. Ser 11. National Foundation, New York, p 9

Seuánez HN (1979) The phylogeny of human chromosomes. Springer, Berlin Heidelberg New York

Seuánez HN, Evans HJ, Martin DE, Fletcher J (1979) An inversion of chromosome 2 that distinguishes between Bornean and Sumatran orangutans. Cytogenet Cell Genet 23:137–140

Stockholm Conference (1977) An international system for human cytogenetic nomenclature. Bergsma D, Lindstein JE, Klinger HP, Hamerton JL (eds) Cytogenet Cell Genet 21:313–409

Sumner AT (1976) The role of proteins and dyes in chromosome banding. In: Pearson PL, Lewis KR (eds) Chromosomes today, vol V. John Wiley & Sons, New York, pp 201–209

Sumner AT, Evans HJ (1973) Mechanisms involved in the banding of chromosomes with quinancrine and Giemsa. II. The interaction of the dyes with the chromosomal components. Exp Cell Res 81:223–236

Tjio H, Levan A (1956) The chromosome number of man. Hereditas 42:1–6

Winnipeg Conference (1977) 4th International Workshop on Human Gene Mapping. Cytogenet Cell Genet 22:1–714

Yeager CH, Painter TS, Yerkes RM (1940) The chromosomes of the chimpanzee. Science 91:74–75

Addendum: Symposium Results

The symposium was held from 4 to 5 July in Turin.

The contributions raised several new aspects of Primate Cytogenetics which are summarized as follows:

1. There is a trend to undertake chromosome studies in a larger number of animals of each species because reports on single specimens or on very few animals are less reliable, and could be misleading when used to make comparisons between species at the chromosome level.

2. It is important to analyze the intraspecific chromosome variability of primate populations. This approach is extremely valuable to understand chromosome change *within* species, which in turn could be illuminative in the understanding of chromosome evolution in primate speciation and radiation.

3. A higher number of primate karyotypes have been studied in greater detail than before with many of the presently available chromosome banding techniques. The DNA composition of some special chromosome regions (viz. NORs and heterochromatic regions) is a matter of special interest because it allows us to understand how repetitive DNA has evolved.

4. The study of nuclear DNA content in primates is no longer focused on merely estimating differences in DNA content between species, but in relating them to differences in specific DNA fractions of the primate genome.

It was generally agreed that primate cytogenetics should become a fully mature discipline that will provide useful information to conservationists and scientists who use primates in laboratory research.

The practical applications of primate cytogenetics in conservation are evident from the analysis of chromosome change within primate species. The existence of a discontinuous variation at the chromosome level between different populations within the same species is good evidence that each of them has been kept reproductively isolated from the other for considerable time. For this reason, separate breeding programs should be kept for each karyomorphic population because indiscriminate hybridisation might result in less fertile or even infertile offspring.

The recognition of different karyomorphic populations within the same species is also relevant to scientists currently using primates in biomedical research. As chromosome change might be just one of the very many differences existing between karyomorphic populations, one important conclusion is that not all specimens of the same species might be identical with each other at it is generally assumed. For this reason, it will be prudent to carry out a chromosome analysis of every specimen before data from different animals are compared.

The efficiency of chromosome research would be obviously enhanced if a standard criterium of chromosome nomenclature was adopted for a few species. Standard criteria should be recommended for all those species that have been extensively studied. Such criteria should not be anthropomorphic (viz. trying to use the human numbering system based on presumed similarities between human and primate chromosomes). On the contrary, individual criteria should be applied to each species regardless of how similar or how different their chromosomes might be from those of man.

The techniques and the materials used in the determination of DNA content in primate cells must also be standardized. Such standardization would allow us to carry out detailed comparison of data from different laboratories.

Finally, larger availability of material would be necessary to carry out much of this work, a reason why wider cooperation between primate cytogeneticists is recommended.

R. Stanyon, G. Ardito, L.E.M. de Boer, H. Seuánez
Institutes of Anthropology, Florence, Turin and Rotterdam and Department of Genetics, Federal University of Rio de Janeiro

The Participants at the Symposium Were: G. Ardito, Institute of Anthropology, University of Turin; G. Benedetti, Zoological Garden Turin; P. Bigatti, Institute of Anthropology, University of Turin; R. Corluy, Institute of Biomathematics, Free University of Brussel; R.M.R. Belterman, Cytogenetics Unit, Rotterdam Zoological Garden; L.E.M. De Boer, Cytogenetics unit, Rotterdam Zoological Garden; G.F. De Stefano, Institute of Anthropology, University of Rome; D. Formenti, Institute of Anthropology, University of Pavia; L. Lamberti, Institute of Anthropology, University of Turin; C.E.A. Pellicciari, Institute of Anthropology, University of Pavia; C.A. Redi, Institute of Anthropology, University of Pavia; H. Seuánez, Department of Genetics, Univ. Federal do Rio de Janeiro; R. Stanyon, Institute of Anthropology, University of Florence; C. Giani, Institute of Antropology, University of Florence; F. Piccotti, Ospedale Evangelico Valdese, Turin

Comparative Psychology Symposium: Introduction

P.A. BERTACCHINI [1]

I want to make the point that we are now living in a historical epoch that may again
see the study of man at many levels at the same time, including, but not restricted to,
those points of view characterized by socio-cultural, humanistic, and scientific contributions. In fact, I believe that there is no contradiction between science and humanism,
since man is a psychosomatic entity and not just a mixing of two heterogeneous substances in two different disjointed spheres. Man is an inseparable unity of psyche and
soma, as our sciences demonstrate, teaching us to find a new humanism that attempts
to understand the homologous needs of all organisms in an environment that reflects
a rational balance between biology and experience. To discover these laws of nature
one has to adopt a point of view that is illustrated by the comparative method. That is
to say, different organisms must be observed in the laboratory and in their natural
environment using a variety of testing procedures. This experimental method may
eventually lead to the uncovering of natural laws of behavior that are as accurate as
those that have been found by biologists and anatomists.

To be more precise, it can be said that Comparative Psychology can also study
biological phenomena, since behavior, both individual (intrapersonal) and social (interpersonal), is the effector of organic apparatuses, structurally constructed, functionally
supported, and synergically coordinated by macromolecules and specialized cells
integrated into a systemic unity known as the biological organism. Biologically speaking, it can be said that an organism undergoes a series of developmental changes that
can be described structurally and functionally. These developmental changes can
include physical growth, maturation, and the aging process, culminating in death.
This chain of events occurs casually, but may result from the influence of variables
that are perhaps causally and intrinsically interconnected to the autonomous organization of the living matter that is genetically referred to by Watson and Crick as the
action of DNA and RNA molecules. Therefore, it can be concluded that all known
living systems, ranging from single cells to man, manifest biological rhythms that can
be measured structurally and functionally.

The concepts and labels "phenotypic" and "genotypic" can be thought of and used
in a way different from how they are used by biologists, psychologists, and philosophi-
cal phenomenologists (see Lewin 1935 for a clear discussion of this argument). For
instance, if behavior is phenotypic in nature it must be considered in a different way

1 Università della Calabria, Dip. Scienze dell'Educazione, Via Busento, 87030 Roges di Rende,
Italy

from that behavior which is somaphenotypic in nature. On the one hand, this last point is the direct expression (i.e., first degree) of gene activity and is structured in a rigid manner in organs and tissues. On the other hand, behavior is the expression of the functionality in such organic structures and is manifested in much more plastic forms (or culturally learned patterns). Therefore, it can be concluded that behavior is of a very different phenotype and differs from somatic activity that was alluded to above, and must not be confused with it.

Such forms or patterns of behavior are often contrary to the specific genetic programs that are found in living systems: the single organisms and the single species that have been studied by biologists and psychologists thus far do not always succeed in accomplishing a behavioral adjustment that is truly adaptive to their environment, as it is the case with groups of somatic cells. It seems, however, that, statistically speaking, the ultimate minor causal project of the indefinite reproduction of DNA and RNA is not troubled by such small incidents, which are only general accidental fluctuations in the course of evolution. But some of these cultural incidents could be fatal to DNA, RNA, and to us, because it may also occur that our species will deny this general course of evolution. For instance, if a nuclear catastrophe were to occur, the destruction of life on our planet, notwithstanding the pool of altruistic genes with which our species is endowed, would result in a change in the course of evolution for many diverse species. With regard to the manifestation of behavioral phenotype, we find, for instance, that biorhythms, being extremely precocious organizers of the structures by which they are produced, are affected by the external environment. They are behavioral mechanisms which, when out of the genetic program, are realized in somatic structures and forms (morphological organismic phenotype), and develop and coordinate the somatic functions: those behavioral manifestations are not so rigid as the somatic structures.

In my opinion individual behaviors must be classified and considered as a second degree phenotype, because nongenetic factors act between genotype and the usual activity of the single individual by partly modifying the biological program, even when we exclude sociocultural interferences. Genes do not act in a vacuum because they are enclosed in a complex and varied environment of molecules and cells that are open to the influences of external factors. Genes and environment (internal and external to the organism) can be thought of as factors that influence the integrity of the continually changing (i.e., dynamic) field space, resulting in the expression of a behavioral act at the end of the final common (i.e., neuronal) pathway.

Genes never modify their messages, but the environment in which they are manifested can modify, distort, or inhibit the expression of these neuronally coded behavioral acts. The function of organic structures allows a given organism to vary its behavioral responses according to changing habitats. Therefore, the organisms belonging to diverse species may not react in a typical way even to the same environmental situation. This point is particularly important since it suggests that behavior is not the simple product of gene and environment interaction(s). Rather, it can be stated that while behavior is expressed through a neuronally determined final common pathway, it is also true that there are many forms of stimulus input that begin with the senses and are integrated in intermediate brain functions (i.e., limbic system, reticular activation system, sensory-motor areas, cerebral cortex, proprioceptive feedback loops, etc.).

Even for the individual organism it can be seen that behavior is the end product of many sources of influences (both biological and environmental) that at the human level reflect the emergent function known as intelligence or cognitive activity.

If we consider social behavior resulting from the interaction of two or more living systems or organisms instead of individual behavior, we may be able to see that at this level of analysis behavior is the direct expression of newer, more complex, vectors that mediate the influence of the genetic program (as modified or evolved in the context of the social field space) on behavior. It is assumed that these vectors operate to adapt themselves to an artificial ecosystem, not phylogenetically adapted. In fact, it may be that norms and cultural constraints, besides the limitations of physiological systems, may intervene in the cycle that is tied to the expression of a given behavioral act (i.e., perceptual definition by the organism of an externally defined stimulus, interpretation of this perceived stimulus by lower and higher cortical systems, the communication of this event in a form of a "meaningful" behavioral act in a socially defined act). Given the logical scheme that has been developed thus far, it can be seen that the mediation of habits, needs, individual predispositions within the limits set by the individual's physiology (i.e., Hullian theory), can be considered a phenotype of the third degree. The relationships emerging from gene, environment, and experience, for individual and dyads are shown in Table 1.

Table 1. The influence of biology and experience on behavior

Genes and molecules (DNA/RNA)	Degrees of influence				
	1st degree	Interferences	2nd degree	Interferences	3rd degree
Genetic blueprint or program (established at birth)	Effects on soma	Intra- and intercellular environment	Behavioral responses to physiological environment	Sociocultural environment	Interpersonal, sociocultural and behavioral responses

This table indicates that behavior can be conceptualized as the product of the biologically determined genotypic program on increasingly more complex organisms in diverse social systems. With this starting point in the study of behavior, the comparative method permits us to study several taxonomic categories; with each one of these groups there may be biorhythms that differ in some meaningful and systematic way. The term biorhythm is used here to refer to some orderly change that occurs in systems composed of one or more organisms. For example, this can include the rhythmic activity of the circulatory system for primates, or the occurrence of an exchange of behavioral acts (i.e., communication) between two living organisms — chimpanzees, gorillas, people, etc.

This table also indicates that the periodic and cyclic characteristics that describe the expression of organismic forms and functions may have the same general laws for all living systems. We can find this fact illustrated in biological clocks and alternating physiological states of the nervous system, as illustrated by REM-NREM phases in sleep and waking cycles. We know from research conducted on both men and other animals (Chance 1979, Buck and Buck 1976, Roberts and Church 1978, Binkley 1979) that there are some cells endowed with an oscillating activity, more or less integrated (according to the level in the taxonomic scale) into nervous circuits producing a spontaneous biorhythmic activity. For instance the cardiac rhythm of a crayfish is controlled by six cells linked to nervous excitatory or inhibitory fibers; the human cardiac rhythm is also governed by the autonomic nervous system (sympathetic and parasympathetic) which appears to have the same function.

There can be differences in the qualitative function of the diverse phenomena that are found among different species, and which are generally expressed at different taxonomic levels. These phenomena are perfectly homologous and the relative behaviors are comparable, since the differences are contingent on the three phenotypic phases discussed above. Since biorhythms have a primary function in carrying out social behaviors, it is of great importance that the comparative method should develop its own area of study, that will be defined within the context of a taxonomic classification and will study behavior by using the methodologies that have been developed by different sciences. Up to now it appears that the use of this "eclectic" approach to the study of behavior has generated some tentative conclusions regarding the physiological and psychological functioning of living systems. It seems that these phenomena are perfectly homologous and that the diverse behaviors are comparable, since the differences that are observed among behaviors may be contingent on the relationships described in Table 1.

Since biorhythms have a primary function in carrying out even complex social behaviors resulting from the interactions of many individual organisms, it is of great importance that the comparative method should develop its precise study area, defined with a "taxonomy" and a unitary transdisciplinary method, in order to give psychologists a possibility of classifying and studying different organisms in different environmental settings. In this way it will be possible to study the influence of the environment (i.e., experience) on the individual's behavior (i.e., mediated by anatomical and physiological pathways) in diverse social settings (i.e., one organism, two individuals, groups of three or more, etc.). For instance, phenomena such as ultradian rhythms and other microrhythms can be used to explain the distribution of some selected kinds of social behavior, including communication between primates, that occurs in social dyads. These meaningful behavioral units appear to be a function of the ultradian behavioral rhythms, when analyzed in terms of frequencies, durations, and refractory periods. It is of interest to note that Maxim et al. (1976) found a high correlation between the ultradian biorhythms and the daily distribution of social behaviors in groups of rhesus monkeys. These investigators concluded that both dominance and submissiveness, as well as the other behaviors of exploration, locomotion, self-grooming, and ingestion may be supported by a common arousal mechanism, that may coincide with one of the REM-NREM phases. It is of interest to note that the refractory phase for this biological rhythm lasts 40–45 min. This is of importance since social and individual behaviors showed higher frequencies of every 40–45 min, or every double phase (i.e., two consecutive time periods) of 80–90 min.

In a related study Maxim and Storrie (1979) found that rhesus monkeys that were given electrically induced brain stimulation in the pleasure center of the limbic system also showed the same general kind of periodicity in behavior, thereby suggesting a possible ethological basis to this behavioral phenomenon. Even in human behavior we can find ultradian cycles linked to the onset of REM phases. These rhythms were studied in adults almost exclusively during sleep, and yet we can say with sufficient certainty that waking behaviors are not independent from such factors, even if in civilized men a certain quantity of interferences seem to result from the influence of social reinforcements more than it happens in other primate groups. It has been observed that infants begin to smile with a certain rhythmic regularity. This ethologic-ally mediated behavior is believed to be initiated by the onset of cerebral activity, that occurs mostly during REM sleep, and is not a response that is generated by the sympathetic nervous system (Emde and Koenig 1969). Also when infants begin to react with smiling responses to external stimuli, they do it according to contingencies which may be related to some of the organismic variables that are mentioned above. However, it should be noted that the smiling response does not last a long time, even if the stimulation is persistent (Kagan 1971). In other words, it has been found that the smile responses show that after a few seconds an infant turns its eyes (and/or its attention) away before producing new responses, which when they occur, may be independent of external reinforcement. In older infants (3–5 years old) smile and laughter show noncasual oscillatory links, if the infants are observed preferring spon-taneous behaviors. Personal data from our laboratory, not yet published, that pertains to the spontaneous behavior of children's behavior, show that different distributions of the mean frequencies, durations, and refractory times of smile and laughter seem to have a predictable cycle. The behavioral patterns that were observed occurred in a testing situation that did not limit the behavioral spontaneity of the children. The temporal duration for laughter, smile, and play behaviors that were recorded for the children mentioned above are shown in Table 2.

Table 2. Frequency and duration of social interactions in children

Ethological categories observed	Behavioral evaluation of play			Sequential analysis of children's play		
	X frequencies	$\overline{X}$ duration (in s)	$\overline{X}$ interval length (in min)			
Laughter	1	3″	10.0′	** ** **		
Smile	3	6″	3.3′	** ** **	** ** **	** ** **
Play bouts [a]	Not constant in form			Not constant in form		

[a]   Ethogram of children's play bouts is shown below

In each case the behaviors were recorded for behavioral periods that lasted 10 min. In this table the behaviors of laughter, smile, and play are shown in the rows of the matrix, while frequency (F), duration in minutes (D), and cycle length are indicated in the three successive columns of the matrix. The ethogram on the right-hand side of this table clearly indicates that these three behaviors may be linked in a random manner, so that one behavior may function as a discriminative stimulus for the onset of the other. It may also be the case that the termination of one behavior may be a cue for the initiation and/or termination of another event. Differences in interval lengths may give us a clue as to the occurrence of diverse kinds of social interaction bouts that occur in groups of children, where play is not restricted. Thus, the behavioral regularity that is suggested by these data can be taken as evidence for the presence of a biological clock, that is probably found in the behavior of many diverse organisms, regardless of their ontogenetic or phylogenetic development. It is hypothesized that this kind of biological clock that can be inferred from behavioral patterns observed in children and adults may be indicative of a universally determined or biologically set mechanism that reflects the influence of different vectors — biology, environment, and experience. At the moment it is possible to study this complex matrix of biology and environment in organisms that are alone or in groups, and in systems that may be aligned along a preestablished phylogenetic scale. It is anticipated that this form of analysis will permit us to uncover the way in which sociological factors influence (directly or indirectly) physiological systems, thereby validating the use of the comparative method in the study of behavior and physiology.

The method used in analyzing phenomena cannot be the one adopted thus far to study incest taboo, since this procedure only relies on hearsay evidence and not systematic observation. As suggested in Parker's (1976) publication, cultural anthropologists may be that group of scientists that have not systematically used the comparative method well enough to satisfy psychologists. The most famous and authoritative example of this scientific pitfall is provided by Levi-Strauss (1967), who arbitrarily decided that incest prohibition (when considered as an exclusive norm of human behavior) could be used as a mean of determining the beginning of "culture" in some human societies. If we analyze the results found for many groups of nonhuman primates and other nonhuman vertebrates, we note that the young adults of all species show a fission behavior from the original group of family, thereby reflecting the presence of spacing mechanisms that may be related to the determination and defense of a new territory for a given organism. In this context, it should be recalled that under some conditions crowding can produce the presence of psychosomatic symptoms and pathological states of behavior in both human and nonhuman primates (Frank 1953, Hediger 1953, Calhoun 1958, Esser 1971, Edwards and Dean 1977, Heller et al. 1977, Kiritz and Moos 1974, Langer and Saegert 1977, Baum and Epstein 1978, Freedman 1979). Therefore, we can conclude that in the past selective pressure favored the fission of young adults from their original social groups.

Even if psychosomatic mechanisms producing pathologies in individual and social behaviors are not completely known (except for certainty with regard to the negative influence of the physiological pathology resulting from excessive secretion of the endocrine glands and their influence on behavior), it is a well-accepted fact that both men and other animals demonstrate a species-specific defense behavior pattern that may determine interindividual space in human and nonhuman primates. In this context, there are three levels of influence that can be described: (1) The personal buffer zone of every individual seems to determine the frequency and duration of interindividual contact. It is known that the buffer zone develops and changes ontogenetically from a value of zero to a few centimeters in the human infant, and at puberty, in a matter of weeks, there is consolidation and the attainment of behavioral baselines that reach stable levels. Both *Homo sapiens* and other animals distinguish themselves into "Ego et Alter". (2) There is also the concept of hierarchical space that determine for every person in the social group its status and role. This variable has also been found to influence spacing in nonhuman primates, as was mentioned above. This is so since this factor has been found to be related to the formation and maintenance of social groups. These variables also influence the maintenance of physical, social, and psychological security in individuals in the group, thereby permitting the gathering of food even at territorial boundaries. (3) Beside the species-specific biological mechanism of spacing, the neoadult organism will eventually attempt to increase interindividual spacing or contact and social exploration. Parker (1976) suggests that these behaviors may include fighting against unknown age peers and mating with partners that differ in terms of the familiarity variable.

At the present time it is obvious to us that at least three different classes of variables may determine the young adult's generalized fission behavior from their family groups: (1) The developmental need for a systematic social structure that eventually becomes more exigent during early maturity, both with regard to the periods of infancy, adulthood, and old age. Needless to say, as the organism develops cognitive mechanisms, they are later expressed behaviorally by that individual. The results of recently published studies indicate that there may be an imprinting-like phase which describes the development of new social structure motivation in animals, and in human primates as well. (2) The increase in cognitive capacities in human and nonhuman primates (and possibly in a few species of birds and a greater number of species of animals) can further facilitate the efficient formation and development of social groups. The benefits which can be obtained from intact physical environments and newly formed groups with "open" and flexible hierarchies are innumerable (for subjects endowed with developed cognitive capacities), like those found in human and nonhuman primates. (3) There is also the genetic factor that can influence both an organism's structure and function in a social group. This last argument has been developed in recent years by sociobiologists, thus it will not be discussed further.

Another important problem to be considered here in the area of social relationships is the one concerning the basic social unit which is structured within the general reproductive activity of an organism. In the past this particular social process has been oversimplified and labeled with the concept of "family", and compared to our monogamic nuclear family, i.e., without a relativity principle. Moreover, the monogamic family was considered to be a basic social unit, that may be monogamic in actuality, if one does not consider divorce, institutionalized adultery (i.e., prostitution), and noninstitutionalized adultery of swapping, which are statistically more frequent than monogamic mating, but may be limited to middle and high socioeconomic classes in western society. The study of social groups in both animals and humans is not without its problems. To put it simply, this is analogous to the traditional ethical concept synthesized by St. Augustine, who maintained that sewers were necessary to keep a city clean, as prostitution was necessary to keep girls chaste! However, all this does not agree with the influence of tools and technology on social behaviors. The monogamic couple is a historically recent acquisition for genus *Homo,* and certainly is not more ancient than the neolithic agricultural revolution. We must note that *Homo sapiens* adopted institutionalized individualism after the introduction of agriculture; this recent evolutionary trend in the history of man also evolved in monogamic institutionalized coupling. Yet, *Homo sapiens* could not give up actual polygamy. In some nonhuman primate societies this form of sexual behavior does play an important biological role in providing an adoptive social status, as well as in hunter-gathering human societies.

Another problem of social technology that can be studied using the comparative method concerns the general area of aggression, as it appears in human and nonhuman primates. While aggression may be controlled by ethological spacing mechanisms in apes and people, the development of guns and bombs as tools of aggression no longer permits its control in a natural manner. As a consequence, it is possible that man may destroy himself when the technology of war becomes slightly more sophisticated that it is today. To prevent this state of affairs, it is necessary that man's social structure should change to a point to permit the expression of ethological mechanisms that function to control aggression. The species-specific abilities in neutralizing aggression, which men seem to demonstrate, and contrary to what Lorenz (1963) has stated, seem to function to protect *Homo*'s genus during hundreds of millenia of wood, stone, and protometallic technologies, allowing only hand-to-hand fights. It may be possible to restore this check on aggression if man could be made to perceive this stimulus that derives from the other members of the social dyad or triad.

Sherif et al. (1961) demonstrated that the redirection of the aggressive tensions and competitive behaviors in two rival adolescent groups occurs if the members of the two groups pursue a common aim. In that case the intergroup conflicts cease and the members of the two groups cooperate in a friendly and congenial manner, thereby again establishing a social balance. Although this was a scientifically controlled study of group dynamics, we can see in these results the same social mechanisms that were found in an ancient human style of ruling. When inner tensions trouble a sovereign state, they are eliminated by declaring war on a foreign country or starting a new colonial enterprise. This is a kind of institutionally controlled fission of social, economic, and evil energies, which weakens the conditions for developing an inner conflict.

We can observe that cultural processes, though being biologically and physiologically based, were considered as discriminatory facts between men and other animals, which should have been considered only as instinctive mechanisms of machineries and not as organic systems. Therefore, thus far it seems to me that the living world has been dichotomized between biologism on inferior levels and sociologism on superior levels. Moreover the same concept of culture has become poor and ideologically deformed, since "culture" is a category that was used for men only, with a somewhat paranoic attitude. Even theologists and primatologists (Kawai 1965, Kawamura 1959) have coined the term "pre-culture" for nonhuman primates, as if, for instance, to mean a sociocultural acquisition of any learning of problem-solving techniques that can be passed on from one generation to another or among the members of a monkey group. With regard to the concept of culture it is clear that both nonhuman and human primates can transmit their own culture to other conspecifics. The socially learned behaviors include rules that govern role stratification, social interactions, and the use of tools. Studies that have started recently indicate that language-type skills can be learned and transmitted to other conspecifics by chimpanzees and gorilla.

From these data we can conclude that such cultural phenomena have phylogenetic roots, and are expressed in the behavior of human and nonhuman primates.

The research of historical origins brings us back to ancient times. In the recent classical prehistoric age we discover that as we observe man's development from *Homo sapiens* to Neanderthal, the existence of (lithic) stone cultures were characterized by the presence of very elaborate and functional techniques. We cannot exclude with any reasonable degree that such cultures included also a more or less elaborate language for verbal communication. Still going backward to our common ancestors of men and apes, we can find other cultures and social organizations that did not have any continuity. "Natura non facit saltus" or qualitative changes in evolution were documented from a study of Hominids and fossil sociocultures of some apes, and in the modern sociocultures of the living apes. Therefore, the phenomena of the acquisition of cultural behaviors seem not only to be a prerogative of our societies but also those of nonhuman primates as well.

Such behaviors were analytically observed, mostly in some colonies of free-living *Macaca,* or near some inhabited center, or in particular situations, such as ethological research exigencies that allowed for the direct interactions between the subjects and the observers (Asakura 1957). Daily food supply clearly influences the propagation of discoveries, defined by Japanese researchers as a "process of subcultural propagation", as discussed above. Field observations in Takasakiyama, Koshima, Minoo, Arhashiyama, Tsubaki and other places in Japan generally demonstrated that cultural propagation depends on the social rank and age of the subjects (Kawai 1965). It should be noted that females who had a higher rank in the social hierarchies played a more important role in the acquisition and social transmission of new habits and "customs". The newly acquired behavior patterns were first picked up by females, and males from their mothers. The females who lived at the extreme periphery of the group and the marginal group members were more influenced by their infants than by the inaccessible adult males. Moreover, the rate at which this kind of protoculture was propagated was higher in smaller than in larger monkey groups.

Before analyzing the propagation of culture in detail, we must also state that the hierarchical structure of a group influences the propagation of information and habits, at least in primate societies. When a community of primates acquires a new habit or makes a discovery, the propagation of this behavior pattern follows a course that is inevitably conditioned by the group social structure and the members' hierarchical position in the group.

The activity of potato washing is performed by Japanese monkeys in a similar way as it is performed by human individuals. The first monkey observed washing potatoes in September 1953 was a young female that was one and a half year old, named Imo. This behavior propagated slowly and gradually, since Imo was of a low hierarchical rank, so that in 1956 only 11 out of 50 monkeys had adopted the discovery inside the community.

The more we ascend the social hierarchical scale, the more frequently we find conservative neophobic resistances, in both nonhuman and human primates. In March 1958, 17 monkeys had adopted this new technique of potato washing, of which only 2 were adults and 15 under the age of 7 years old, with the following proportion of animals with regard to sex and age categories: out of 6 males and 5 females (11 adults on the whole), 15 young members, that is 78.9% of the young wash potatoes using the new technique. In December 1961, all the monkeys in the community had acquired the new behavior pattern, except the adults born before 1950. The old and the high-ranking members always remained inflexible to change. However, this finding should not be taken to mean that behavioral plasticity no longer occurs in adulthood. Rather, it may be that play must be the context in which change occurs. Thus, as the frequency of play decreases with age, one would expect that behavior patterns become established and are not as flexible as they are in young animals.

The factors that determine the adoption of the behavior of potato washing include (but are not necessarily restricted to) age, sex (or socio-sexual role), and the degree of kinship with the animal that has already learned the potato-washing behavior pattern. The acquisition and transmission of new behavior patterns occur and are transmitted as a function of the kind of social group that is present. Finally, I want to allude to the research programs that have examined the acquisition and transmission of language-type skills. They too seem to indicate that interindividual contact facilitates the learning of these new behaviors of communication.

We must now consider the taxonomic scale as a pyramid turned upside down. As the single cells integrate into groups (colonies of protozoa, for example), and then into more and more complex living systems and apparatuses, new elements (such as glandular structures, neural networks, etc.) add to the basic metabolic rhythms and regulator clocks, thereby incrementing this biological theme. The functional aims and goals remain, however, the same, regardless of the complexity of the organism being studied. In all cases the aims are to maintain and foster the energetic system, to develop the same system from a state of zero power to a state of reproductive power, and to transfer the reproductive potentiality into its effective use in actual situations. This process has to be accomplished through suitable alimentary and sexual behaviors; along the course of evolution the taxonomic forms, both somatic and psychic, and therefore sociocultural, multiply into structural and behavioral models differentially endowed with simple or complex apparatuses, the characteristics of which are biologically determined. It is sufficient to consider the psychosomatic aspect of ludic behavior in all superior animals. In any functional energetic system the relation between costs and benefits must result in favor of the latter. That is to say, benefits have to gain advantage over costs, at least with regard to the ultimate aim and goal of the future of the individual, group, and species (McFarland 1976).

Biopsychic systems seem to follow this general law at least since aerobic bacteria entered symbiotically into bigger cells, accomplishing a mitochondrial function which minimizes the costs of biological production of ATP. Such progress, on a merely energetic level, can result in the development of the sensory and motor organs, with their innumerable models that enrich this complex biological system.

On a more analytic psychological level it is found that efficiency is increased by multiplying and centralizing the methods and systems of intraorganismic and interorganismic communication, and of the communication between organisms and external environment. This process is based on an unconditioned reflex response to an internal or external stimulus, and then evolves into a symbolic response. At higher levels of symbolic communication is is found that attitudes may be phylogenetically and ontogenetically acquired by individuals, and that these experiences act by influencing the type and quantity of responding manifested by individual subjects, and less by the specific characteristics of the stimulus. The benefits of symbolic communication are found in the fact that it allows the individual to plan in advance the strategies to be used in the execution of a behavioral act, thereby eliminating the expansion of unlearned efforts and energies. For instance, trial-and-error learning results from immediate responses to simple stimuli, and also to more complex signals. Such "non-mediated" responses require a greater use of energy than that which characterizes relatively more complex problem-solving strategies of complex learning, and cognitive solutions to problems. Such an efficient increase of structures and functions occurred through the integration of processes and the bedding of cortical layers, where the higher levels are the result of the stratification of some elements in the immediately inferior layers (see the two criteria of Pettersson 1979 for details). So we can find qualitative differentiation depending on quantitative increments, and we can measure it using physical and mathematical methods, at least as long as the compound does not begin to transcend the specific properties of its components. In the case of comparative psychology we can find many of these qualitative changes, both in comparing human versus animal behaviors, animal versus animal behaviors, and human versus human behaviors, since the behavior patterns of the individuals within a group are always in a dynamic state (both in a primary and secondary group), and are also very different from individual behaviors.

Wilson's (1975) argument on this point does not take into account such phenomena, especially his discussion on the optimization of social structures (i.e., cost/benefits ratio). He says that social structures reside exclusively in those behaviors which increase the benefits for the group and not for the invidual within the community. Given this statement, Wilson (1975) maintains the following: (1) There exists a descending tendency of change from colonial invertebrates to vertebrates; (2) Social insect societies occupy an intermediate place that is superior to that position which is occupied by the other societies of vertebrates; (3) Man is the only vertebrate who, up to now, has inverted this hypothesized descending tendency. On this point I want to say that I do not agree with Wilson, because to consider a progressive tendency in the psychosomatic evolution of organisms one must also consider the benefits for single individuals, not only for aggregates, in calculating optimization. I do not think it is possible to consider societies and aggregates as homologous, since the latter forms are only subunits, when compared to the superior compound.

The vertical development complicates the whole system of a single organism, yet at the same time it still affords the advantage of rendering it more apt to perform plastic responses and to start more initiatives in its environment. In fact, rigid systems own a high probability of survival only in synchronous environments, whereas, in changing ecological situations, they are less favored than systems which own the characteristic of being plastic, and also do not lose a whole set of elementary basic functions. The wider systemic organization mentioned above serves phylogenetically advanced organisms by providing a basis for the expression of a biologically based mechanism for learning. We can say that culture, being a more or less coordinated whole of learning, is intrinsic to behavior, owning an adaptive value homologous to the one of the physiological process. Therefore, culture is, as defined, functional (or disfunctional) to the survival and reproduction of the single individual and of the whole species.

The history of human institutions seems to demonstrate to us, according to Lorenz (1975), an evolution that is at least analogous to the one of organic morphologic structures. Moreover, I want to stress that except for a part of the rising sociobiological theories, researchers have been always interested in the term "comparative psychology" because it includes the study of the phenomena of accumulating human behaviors and their comparison to those of other animals (Miller 1977). Such a theoretical position results from the analogy that is made to comparative anatomy, which confronts homologous structures in men and other animals. With regard to the study of behaviors, men, as we said earlier, have structures common both with animals and with other cultures, if we consider their interpersonal relations as well. Therefore, we can say on the one hand that it is useful to study a certain behavior, and to compare it with corresponding ones of other animal species; on the other hand, however, it would be useful to compare other behaviors to the ones of different sociocultural structures, using the results of anthropology, ethnology, and history (Bertacchini and Genta 1978). It would also imply the use of a diachronic and synchronic methodology, that may eliminate remaining doubts about comparing behaviors, as if they were homologous. Examples that illustrate this point include Goodall's data on the social behavior of free chimpanzees, Foot-White's data on the social behavior of the American Corner Boys, and the Levi-Strauss data on the social behavior of Brazilian Nambikwara. In other words, it is useful to make clear that the comparisons that are carried out with the comparative method, as used by scientists today, cannot be limited only to the sphere of biological evolution, even if it is of extreme importance for the progress of behavioral sciences. We cannot consider with a schizoid attitude social psychology as pertaining to another separate area, and we cannot ignore history as well. History, considered as evolution of human behaviors, is directly based on biological evolution, but it is still conditioned and influenced in an indirect way by many variables, and the underlying mechanisms may be autonomous from the points of view of the organismic system that regulates DNA and RNA synthesis.

The method that should be used to proceed should be analogous to anatomy that has given us satisfactory results, and as it is utilized by biologists and ethologists. I also want to propose here that we also realize that the fundamental contributions of the comparative method are that it tries to comprehend the sociocultural aspects of evolution and integration of the sociobiological and anthropological competence of man.

At this point we must also mention the problem of fixing the exigencies and limits of the methodology we are proposing. We can start by delimiting the phenomenologic area which is the object of study and of use of the comparative method. In particular, I believe that our phenomenology tries to first comprehend the heterotrophic relation that every organism establishes with its environment in which it lives, conditioning more or less directly its behaviors and even its somatic structures along the course of phylogenesis.

Our species, or better the genus *Homo,* offers a wide range of diversified examples; therefore, it is ideal as an initial starting point for us to select those varied behavior samples. In fact, in our study of *Homo* we could discover a series of subsistence strategies to reach a consistent basic survival level; some subsistence systems are "simple", or "complex", "advantageous" or "insufficient", some are harmoniously integrated into the surrounding environment and others, on the contrary, are more or less destructive of the surrounding resources and of conspecifics. In any case, together and within such subsistence systems, more systems can develop and become structures. The ones that pertain to the reproduction of the species (children), the establishment of intergroup relationships, and the ones concerning the world of fancy, ideology, religion, etc. can be studied to arrive at a description of these laws of biology and behavior. Leaving apart the attitudes of the various schools and disciplines that dictate the way behavior should be studied, it cannot be denied that such systems constitute the object of analysis of comparative psychology, which can then produce a body of knowledge to create social technologies.

Many crises and conflicts in the lives of modern men might be resolved if we opened up new horizons of knowledge and if we were to apply new social technologies to history and politics. Economic and political interests as well as moral values and prejudices influence our behavior, sometimes to a point that cannot be endured any longer. On the contrary, it should be possible, and not illusory and utopian, to afford a new perspective of resolution between public and collective interests on one side, and private needs on the other side, thereby forming the basis for a new social technology that is founded on a scientific background. Moreover, it should be desirable to oppose the growth of aggressiveness and aggression of ideological and moral taboos, and at the same time permit the dissemination of relativistic knowledge with wider horizons and a less absolutistic attitude than that which is found in today's society. In any case, ethology, comparative psychology, and sociobiology are using the comparative method with a variety of diverse and different groups of human and nonhuman primates. The use of a common methodology with diverse species of human and nonhuman primates will permit us to arrive at a precise description of individual and joint contributions of a diverse set of variables on the behavior of animals. It is hoped that the results of this symposium on comparative psychology will better define these issues for the scientists and scholars, as well as provide insights for the technologists of our society so that they may be able to improve the quality of man's life and possibly alleviate the inequities that are placed on us by biology and experience.

References

Asakura S (1957) A social relation between monkeys and man. Primates 1 (2):99−109

Baum A, Epstein YM (eds) (1978) Human response to crowding. LEA Publ, Hillsdale, New Jersey

Bertacchini PA, Genta ML (1978) Spazio psichico ed ecologia umana. Quad Dip Sci Educ 20:5−75

Binkley S (1979) Un enzima "marcatempo" nella ghiandola pineale. Scienze 130:24−29

Buck J, Buck E (1976) Lucciole che lampeggiano in sincronia. Scienze 97:84−93

Calhoun JB (1958) Space and the strategy of life. Reprinted in Esser AH (ed) Behavior and environment (1971). Plenum Press, New York

Chance MH (1979) Every 90 minutes, a brain storm. Psychol Today Nov: 172

Edwards EA, Dean LM (1977) Effects of crowding of mice on humoral antibody formation and protection to lethal antigenic challenge. Psychosom Med 39 (1):19−24

Emde RN, Koenig KL (1969) Neonatal smiling and rapid eye movement states. Am Acad Child Psychiatr 8:57−67

Esser AH (ed) (1971) Behavior and environment. Plenum Press, New York

Frank F (1953) Untersuchungen über den Zusammenbruch von Feldmausplagen *(Microtus arvalis)*. Zool Jahrb Syst 82

Freedman JL (1979) Reconciling apparent differences between the responses of humans and other animals to crowding. Psychol Rev 86 (1):80–85

Hediger H (1953) Les animaux sauvages en captivité. Payot, Paris

Heller JF, Groff BD, Solomon SH (1977) Toward an understanding of crowding: the role of physical interaction. J Pers Soc Psychol 35:183–190

Kagan J (1971) Change and continuity in infancy. John Wiley and Sons, New York

Kawai M (1965) New-acquired pre-cultural behavior of the natural troop of Japanese monkeys on Koshima Islet. Primates 6 (1):1–30

Kawamura S (1959) The process of sub-culture propagation among Japanese monkeys. Primates 2 (1):43–54

Kiritz S, Moos RH (1974) Physiological effects of social environments. Psychosom Med 36:96–114

Langer EJ, Saegert S (1977) Crowding and cognitive control. J Pers Soc Psychol 35 (3):175–182

Levi-Strauss C (1967) Les structures élémentaires de la parenté. Mouton, Paris

Lewin K (1935) A dynamic theory of personality. McGraw Hill, New York

Lorenz K (1963) On aggression. Bantam Book, Harcourt

Lorenz K (1975) K. Lorenz: the man and his ideas. Evans RI (ed). Harcourt, New York

Maxim PE, Storrie M (1979) Ultradian barpressing for rewarding brain stimulation in rhesus monkeys. Physiol Behav 22:683–687

Maxim PE, Bowden DM, Sackett GP (1976) Ultradian rhythms of solitary and social behavior in rhesus monkeys. Physiol Behav 17:337–344

McFarland DJ (1976) Form and function in the temporal organization of behaviour. In: Bateson PPG, Hinde RA (eds) Growing points in ethology. Univ Press, Cambridge

Miller DB (1977) Roles of naturalistic observation in comparative psychology. Am Psychol 32 (3): 211–219

Parker S (1976) The precultural basis of the incest taboo: Toward a biosocial theory. Am Anthropol 78 (2):285–305

Pettersson M (1979) Vertical taxonomy: for certain social, biological and physical structures. J Soc Biol Struct 2 (4):255–267

Roberts S, Church RM (1978) Control of an internal clock. J Exp Psychol Anim Behav Proc 4 (4): 318–337

Sherif M, White BJ, Hood WR, Sherif CW (1961) Intergroup conflict and cooperation: the robber cave experiment. Norman, Univ Oklahoma Book Exchange

Wilson E (1975) Sociobiology: the new synthesis. Harvard Univ Press, Cambridge

The Present and Future Status of Comparative Psychology: Proceedings of the Corigliano Calabro Symposium

J.T. BRAGGIO [1]

Introduction

This chapter will summarize the papers on the program of the pre-Congress symposium on Comparative Psychology that was held at Corigliano Calabro (Cosenza, Italy) on 3–5 July 1980. The Comparative Psychology symposium convened prior to the start of the VIIIth Congress of the International Primatological Society in Florence, Italy, on 7–12 July 1980. As the organizer of the VIIIth Congress of the International Primatological Society, Prof. B. Chiarelli of Florence decided to have many small study groups of about 10–20 scientists meet somehwere in Italy to discuss specialized areas related to the fields of anthropology and primatology. One of the specialized groups to be formed by Chiarelli was this one, on Comparative Psychology.

Some of the papers submitted to the program committee of the VIIIth Congress of the International Primatological Society that were concerned with some aspect of comparative psychology were grouped together. Then Chiarelli asked J.T. Braggio (Asheville, North Carolina, U.S.A.), P.A. Bertacchini, and A. Tartabini (both from Cosenza, Italy) to organize the pre-Congress symposium on Comparative Psychology. During the winter and spring of 1980 the three of us sent out letters of invitation, selected a meeting site, and printed the final program.

The 3-day meeting was finally held in an ancient and historically rich castle located in a lovely seaside resort town in Southern Italy. During the 3-day pre-Congress symposium all of the participants had an opportunity to combine the best of both worlds – the excitement of participating in a scientific symposium and of traveling to another country. In fact, all of us met in a region of Italy that was known for its traditional values of scholarship, and as a place that provided the world with a model for a refined and joyful life. On behalf of the participants and my two Italian colleagues, I thank Professor Chiarelli for his help in making the initial plans for this pre-Congress symposium on Comparative Psychology to be held in the picturesque town of Corigliano Calabro.

1 Department of Psychology, The University of North Carolina at Asheville, P.O. Box 8467, Asheville, NC 28814, USA

Introduction to the Comparative Psychology Symposium

The theme of this pre-Congress symposium was succinctly stated by Prof. P.A. Bertacchini, our gracious host from Calabria University. His theoretical paper was entitled "Introduction to the Comparative Psychology Symposium". In this paper (preceeding the present report) Bertacchini explains the utility of the comparative method. He proposes the use of a field theory or gestalt approach to the study of nonhuman and human primates. Field studies provide valid measures of species-specific response tendencies such as tool use, nest building, mother-infant interaction, etc. Laboratory studies, on the other hand, provide reliable measurements of preselected behavioral categories. By combining the results of field and laboratory studies it may be possible to provide a synthesis of valid and reliable measurements of the behavior of nonhuman and human primates.

In Bertacchini's theoretical paper on the comparative method, it was concluded that there are some phenomena that should be examined in detail. The papers that were on the program on the pre-Congress symposium on Comparative Psychology provided experimental support to the arguments in his theoretical paper. To summarize the laboratory and field studies that were on the program at Corigliano Calabro, they will be grouped into three categories: (1) Assessment of Perceptual, Intellectual, and Cognitive Capacities, (2) Social Interactions in Nonhuman and Human Primates, and (3) Physiological and Incentive Factors. Once each of the papers in these three sections have been summarized, the implications of these studies for the present and future of Comparative Psychology will be outlined in the last section of this chapter.

Assessment of Perceptual, Intellectual, and Cognitive Capacities

Dr. K. Swartz (West Lafayette, Indiana, U.S.A) presented a paper entitled "A Comparative Perspective on Cognitive and Perceptual Development: What the Human Infant Can Tell us About the Nonhuman Infant and Vice Versa". The general aim of this theoretical paper was to emphasize the importance of understanding perceptual mechanisms of nonhuman primates. It is only after the functional stimulus is defined with reference to the perceptual mechanisms of nonhuman primates that it will be possible to evaluate the intellectual and cognitive capacities of the various animal species in the order Primates. Swartz mentioned that the experimental paradigm to be used here should be analogous to the one employed by Franz and other child psychologists who have already studied the perceptual mechanisms of the human infant. Perceptual mechanisms of nonhuman primates should be examined with reference to an information-processing model. That is to say, processing functions such as intelligence and cognition may be a direct result of the information that is perceived by the nonhuman primate. Even though similar stimuli are presented to a variety of nonhuman primate species, what is in fact perceived, will depend on the following factors: First, visual scanning strategies may enable a nonhuman primate to fixate on some stimulus feature such as color, form, etc. Also, visual scanning strategies change developmentally, and may differ as a function of the species variable. Second, the actual stimulus pattern which is detected by the eye or another receptor is changed to a biochemical-electrical impulse as it is processed by the nervous system, and then to the brain. The transmission of this neuronal signal becomes altered as meaning is assigned to it by perceptual mechanisms. Swartz suggests that the proposed research program in this area should begin with an emphasis on the study of the habituation phenomenon. It is possible that the nonhuman primate, like the human infant, may demonstrate that the habituation phenomenon forms the basis for the appearance of relatively more advanced psychological functions of intelligence and cognition. That is to say, advanced psychological functions may only appear once the perceptual mechanism has fully developed. Swartz proposed that the habituation process and form constancy may precede the occurrence of more advanced behavioral phenomena of categorization, seriation, conservation, or other cognitive capacities that have been observed in the human infant, and described by Piaget. In conclusion, the importance of this work is that the results of this kind of research program may form the bulwark for the understanding of the intellectual and cognitive capacities that have been found to occur in nonhuman and human primates.

Dr. R.K. Thomas (Athens, Georgia, U.S.A.) presented a review paper that was entitled "The Assessment of Primate Intelligence". The purpose of this paper is to attempt to construct an ordinal scale of primate intelligence. In his presentation Thomas summarized prior attempts at the evaluation of the intelligence of nonhuman primates. In his review he summarized the work of Harlow, Razran, and others who have attempted to use Piagetian theory and methodology to evaluate the cognitive capacities of nonhuman and human primates. According to Thomas, prior attempts at the evaluation of nonhuman primate intelligence may have failed because of inadequate controls for species differences that are due to sensory, motor, or motivational differences. In addition, he mentions that most of the studies which have used Piagetian type tasks to evaluate the cognitive capacities of nonhuman primates have been carried out without adequate experimental controls. As a result, it is impossible to determine if the obtained species differences are in fact indicative of variations in intellectual functioning, or simply due to the influence of uncontrolled performance factors.

Having reviewed several attempts at evaluating the intellectual capacities of nonhuman primates, Thomas outlines his rationale for constructing an ordinal scale for evaluating the intellectual capacities of nonhuman primates. His proposed system contains eight levels that are arranged hierarchically. The various levels include the following: (1) Habituation — a learned decrement in responding that results from the repeated presentation of a nonreinforced stimulus, (2) signal learning or Pavlovian conditioning, (3) stimulus-response learning, (4) chaining — a chain of two or more stimulus-response connections, (5) concurrent discrimination learning — two or more stimulus-response connections that are learned independently and concurrently, (6) affirmative concepts — involve the logical operation of affirmation and its complement, negation, (7) conjunctive, disjunctive, and conditional concepts — these determine relationships among elements, at least one of which must be an affirmative concept, and (8) biconditional concepts. The final aim of this research program is to assign a number that reflects the intellectual attainment of a given species. The advantage of the proposed ordinal scale of intelligence is that each one of the eight levels can be subdivided into even smaller levels. In addition, intellectural functioning at the higher levels is reflected in the complexity of the conceptual relationships which can be used by a nonhuman primate to solve a given problem or respond correctly on a task. In his theoretical paper Thomas reviews some of his published studies that seem to provide empirical support for his proposed ordinal scale of intelligence. Based on the derived intellectual hierarchy, Thomas reviews the work of some of the comparative psychologists that have attempted to teach great apes to use language-type skills.

In opposition to the view expressed by Thomas and his students, we have taken the position that it is possible to use Piagetian theory and methodology to evaluate the cognitive capacities of nonhuman and human primates. For the last several years my colleagues and I have attempted to administer a 2 by 2 Piagetian-type multiple classification task to juvenile chimpanzees, retarded children, and children of average intelligence. This Piagetian-type task purports to evaluate the simultaneous use of two stimulus dimensions of color and form in order to solve a series of unique multiple classification problems. Each 2 by 2 stimulus matrix is formed by combining two different colors with two different forms in order to generate a unique matrix. On each problem a nonhuman primate or a human child is required to select one of the stimulus items from the response tray and place it inside the incomplete stimulus matrix which always lacks one stimulus object. Analysis of our data indicated that both the juvenile chimpanzees and the human children (retarded and normal groups) were able to solve the series of unique multiple classification problems presented by selecting stimulus objects which were correct on both color and form cues. These results suggest that juvenile chimpanzees, like retarded children and children of normal intelligence, are able to use two stimulus dimensions simultaneously to solve a series of unique multiple classification problems.

In a recently completed study (see below for details) we found that the three groups of nonhuman and human primates still differed from each other in terms of their multiple classification performance. The children of normal intelligence attained near perfect performance on the multiple classification task in about 26 problems, while the retarded children required almost twice as many problems. For both groups of human children the stimulus objects that were correct on both color and form were selected nearly 100% of the time by the end of testing. However, the juvenile chimpanzee group was able to select the stimulus objects which were correct on both color and form cues at values that were above chance, but still below the performance shown by the two groups of human children. Four of the juvenile chimpanzees were able to select the stimulus objects which were correct on both color and form cues so that their strings of correct responses in a block of problems was significant at the 0.05 level. A fifth juvenile chimpanzee was able to select the stimulus object that was correct on both stimulus dimensions in a long enough run of correct responses in a block of problems so that the attained level of performance was significant at the 0.10 value. Also, the best juvenile chimpanzee was able to select the stimulus objects that were correct on both color and form cues nearly 70% of the time by the end of testing. In our work we wanted to determine if it was possible for nonhuman and human primates to demonstrate similar cognitive capacities on a Piagetian-type task, but still differ in their expressed level of performance. In other words, the two studies to be described below attempted to determine if there are performance factors that can attenuate the performance of juvenile chimpanzees (or other groups) on a multiple classification task.

Our first study was co-authored by J.T. Braggio, A.D. Hall, J.P. Buchanan, and R.D. Nadler (Asheville, North Carolina, U.S.A.) and entitled "Logical and Illogical Errors Made by Apes and Children on a Cognitive Task". In this report we examined the problem-solving strategies used by nonhuman and human primates on a 2 by 2 Piagetian-type multiple classification task. We reported that there was an inverse relationship between absolute error rate and attained levels of performance on the multiple classification task. Absolute error rate was highest for the juvenile chimpanzees, lowest for the children of normal intelligence, and intermediate for the mentally retarded children. It should be recalled that the multiple classification performance was best for the normal children, worst for the juvenile chimpanzees, and intermediate for the retarded children. Analysis of the features of the task indicated that the juvenile chimpanzees had a strong preference for color cues, while both groups of human children had a strong preference for form cues. We hypothesized that the presence of language in the human children, and not in the juvenile chimpanzees, may have been the factor which permitted the former group to suppress these preestablished response tendencies to a greater degree than the latter group.

In this study logical and illogical errors made on the first and second trials of each multiple classification problem were also analyzed. Operationally defined, logical errors involved the selection of incorrect stimulus items which led to more information about the task on the second trial than what was available on the first trial. On the other hand, illogical errors involved the selection of incorrect stimulus items which provided less information on the second trial than what was available on the first trial. It was found that for all three groups of nonhuman and human primates tested, the proportion of logical errors remained constant at about 80% throughout the experiment, save for a value of zero for the normal children for the third Part of Testing. This outcome was due to the fact that the human children did not make any errors, thus generating a value for the logical errors that was also zero. The data of this study suggest that nonhuman and human primates may demonstrate similar cognitive capacities on a Piagetian-type task, even though there may be differences in performance. Specifically, it appears that nonhuman and human primates may manifest an unlearned capacity to use two stimulus dimensions simultaneously. It is assumed by Piaget that it is this kind of cognitive capacity which forms the intellectual bulwark that permits human children to acquire language. It is possible that it may be the presence of this cognitive capacity in nonhuman primates that permits chimpanzees to acquire and use language-type skills, as well.

In our next study by A.D. Hall, J.T. Braggio, J.P. Buchanan, and R.D. Nadler (Raleigh, North Carolina, U.S.A.) and entitled "Partitioning the Influence of Level and Rate Factors on the Performance of Children and Apes on a Cognitive Task", we wanted to derive mathematical formulas that permitted us to evaluate the degree to which performance factors may function to limit the performance of nonhuman primates on a cognitive task, and not bias the performance of human children tested on the same task. That is to say, we wanted to examine the assumption that under some circumstances phylogenetic standing may not be directly related to performance, since the rate factor may limit the expression of the level's variable. This question was asked by analyzing changes in selected parameters in the 2 by 2 multiple classification performance of juvenile chimpanzees, retarded children, and children of normal intelligence. In doing this analysis it was possible to derive a series of mathematical equations that permitted us to assess the differential influence of both level and rate factors on performance. A computational formula was derived to denote the relative compression of mathematically defined planar surfaces that reflect the influence of rate and level parameters. The input of the formula, i, is a performance index that varies from a minimum value of 0.00 to a maximum value of 1.00. Values close to unity of 1.00 reflect the expression of the levels variable which occurs at an accelerated rate, while values that are less than unity may be indicative of an underestimation of the levels variable. The degree to which the computed performance index underestimates the levels variable can be evaluated by solving the formula for different relative and absolute constants. In this paper the conditions illustrating the differential influence of rate and levels factors on the multiple classification performance of nonhuman and human primates are illustrated mathematically and graphically. Two objectives can be attained by analyzing the multiple classification performance of nonhuman and human primates in this manner: Firstly, it is possible to quantify those performance factors that can operate to suppress the accuracy of nonhuman primates tested on a cognitive task. Secondly, it is possible to generate a specific numerical value for each group tested. In this paper we showed that the computed indices for the three groups were 1.507×10^{-2} for the juvenile chimpanzees, 1.135×10^{-1} for the retarded children, and 2.501×10^{-1} for the normal children. Therefore, these numerical values may be indicative of the facilitative effects of human language on the performance of human children of normal intelligence tested on a Piagetian-type multiple classification task.

The last paper in this section of the Comparative Psychology symposium was co-authored by J.T. Braggio, S.M. Braggio, L. Friedenberg, and W. Bruce (Asheville, North Carolina, U.S.A.) and entitled "Piaget vs Vygotsky: Clarification of the Language-Cognition Controversy in Nonhuman and Human Primates". In this theoretical paper we tried to show that the results of the language training programs with primates deal with the classic controversy of language and cognition as outlined by Piaget and Vygotsky. Piaget has argued that cognitive structures must precede the emergence of language development. In opposition, Vygotsky believed that experience gained through language forms the bulwark for symbolic thought. Up to now this controversy has resisted resolution, even though the issue is fundamental for an interpretation of the cognitive functioning of normal children, exceptional children, and the language-training programs with nonhuman primates.

This theoretical paper examined the controversy within a comparative-developmental framework. By using this point of view it was shown that verbal, visual, and/or tactile "labels" can be learned through association. However, the internalization of these language labels as "words" only occurs if the subject possesses biologically inherited perceptual and/or cognitive mechanisms, as is the case for species in the order Primates. Based on a review of the literature on the acquisition and use of language-type skills in nonhuman primates (e.g., Gardner and Gardner, Rumbaugh and Savage-Rumbaugh, Premack, etc.), and the psycholinguistic literature of normal and exceptional children, it can be concluded that experience forms the basis for the attainment of optimal levels of linguistic or cognitive competence only when there is parallel development between biological and psychological structures. This review suggests that under conditions of normal development there is a correlation between linguistic competence and cognitive development. However, in some groups of exceptional children it is found that cognitive development follows linguistic competence. It is concluded that Piaget's interpretation of the language-cognition issue may only apply to normal children, while Vygotsky's view may be applicable to both normal and exceptional children. It was further suggested that the mathematical equations developed by Hall and associates (Hall, Braggio, Buchanan, and Nadler, and mentioned above) may be useful in describing (and possibly predicting) the interactions of biological and experiential factors, and determining their influence on the cognitive development of nonhuman and human primates. The possible use of Piaget's or Vygotsky's theory in teaching language to exceptional children is also discussed.

Social Interactions in Nonhuman and Human Primates

Dr. Swartz's (West Lafayette, Indiana, U.S.A.) second paper was entitled "Issues in the Measurement of Attachment in Nonhuman Primates". This theoretical paper reviewed some of the issues that pertain to the measurement of attachment in nonhuman primates. This paper is an important contribution to this symposium since it served as an introduction to the other laboratory and field studies that examined social interactions in nonhuman and human primates. Swartz begins her discussion on attachment by considering it as an organizational construct, as a trait, and as an intervening variable. Each of these definitions provided a basis for the definition of attachment. She then concludes that an operational definition of attachment can be derived from theoretical definitions of this construct. Operational definitions of attachment should use multiple measures combined to describe behavioral patterns. Once attachment behavior is defined in this manner, it is then possible to examine the quantity and quality of attachment behaviors as they change developmentally for animals of different species, and for animals in different social groups. Dr. Swartz concludes by saying that it is the result of studying the mechanisms of attachment that we can begin to understand the basis for social organization in nonhuman and human primates.

Dr. A. Tartabini, Professor P.A. Bertacchini, and Dr. M.L. Genta (Cosenza, Italy) presented a paper entitled "Mother-Infant Interaction, Social Grooming, and Agonistic Behavior in Rhesus Monkeys *(Macaca mulatta)*". The purpose of this paper was to clarify the basis of mother-infant interactions in rhesus monkeys. The authors mentioned that up to now it is known that the mother's hierarchical status does influence (indirectly) the infant's behavior in the social group. However, there are other factors that influence the behavior of the infant in the social group, these include the age of the infant, the mother's dominance, and the mother's biological state of being pregnant or not pregnant. When these factors were considered in this study, it was found that older infants tended to have more body contact with their mothers than the younger infants. Also, the frequency of body contact between the mother and the infant was higher when the mother had a higher social rank in the group, and lower when the mother had a lower social status in the group. The status of the infant in the group was determined during the first few months of life, and appeared to have been a function of the dominance of the mother in the social group. Further, mothers who were more socially dominant groomed their infants more than those mothers who were less socially dominant. These investigators also found that aggressive attacks were directed toward the infants only after they were 4–5-months old. The biological state of the mother was important in determining the frequency of mother-infant interactions. Pregnant mothers kept their infants away from them more than did nonpregnant mothers. Tartabini and his associates concluded that the status of the infant rhesus monkeys in the social group was determined early in the infant's life, and was a function of the dominance of the mother. This study underscores the importance of mother-infant contact since it later permits the infant to establish itself in the social group.

Dr. G. DeJonge and H. Dienske (Rijswijk, The Netherlands) presented a paper entitled "The Importance of Parent-Infant Interactions in Infant Development in Rhesus Monkeys and Man". The purpose of this study was to determine the effects of environmental deprivation on the quality and quantity of mother-infant interactions which occurred in rhesus monkeys and human children. For the nonhuman primates tested it was found that environmental deprivation resulting from partial isolation and less mother-infant contact had profound effects on the adaptive behaviors of the nonhuman infants. In the deprived group of infants the pathological behaviors observed by DeJonge and Dienske ranged from overprotection to little or no body contact with the mother or other conspecifics. Also, the absence of contact with the mother or a conspecific was expressed as a delay in the appearance of play, exploration, imitation, and cognitive development. DeJonge and Dienske noted that cognitive development in the rhesus monkeys seemed to be a function of exploration and imitation of the mother.

These primatologists also examined the records of human psychiatric patients to determine if there were analogous effects of deprivation on human children. They found that parental neglect was later manifested as less body contact, less educational guidance, and poor school performance. The infants of these psychiatric patients also appeared to be passive in social situations, and more aggressive than other human children. These investigators suggested that by comparing the effects of environmental deprivation which occurs in nonhuman primates with those effects that are found in human children it may be possible to eventually understand how early social experiences with the mother contribute to adult development. From these data it is clear that nonhuman and human primates require specific sets of high-quality social experiences with the mother in order to assure that cognitive and social skills develop. This study also illustrates the value of using nonhuman primates as animal models to study the influence of experience in psychological and behavioral development.

Dr. P.R. Ojha (Naguar, Rajasthan, India) used a combination of the ethological method and a field theory approach to examine the reactions of nonhuman primates to the infant-leaving behavior of rhesus monkeys in an experiment entitled "Infant-Leaving Behavior in Rhesus Macaque — *Macaca mulatta*". This observational study was conducted in Mahoth Village (District of Nagaur, Rajasthan, India) during the months of May and June 1972, and June—July 1976. When the study was carried out the estimated ages of the infants ranged from 3 weeks to about 3 months. The data consisted of 38 cases of infant-leaving behavior during which time the mother remained away from the infant for a few minutes to over 1/2 h. The results indicated that in nine cases when the infant was first separated from the mother it cooed continually. Even when the infant ran toward the mother, the mother ignored the infant. In 23 cases the infant continued to play with other conspecifics or engaged in exploratory behavior when it was separated from the mother. In six cases the lone infant first started to play, then began to emit cooing sounds, and on two occasions the subadult animal went to other members of the social group for support.

Ojha also described instances where group members tried to protect the infant when it was separated from the mother, and it was cooing. In this instance group members moved toward the infant, aunts (i.e., females other than the mother) sat near it, groomed the infant, and allowed it to play with other infants. Juveniles also sat near the cooing infant when it was separated from its mother. Some group members protected the cooing infant by threatening the observer when he approached the isolated infant. In 20 cases the mother retreated from the infant. In eight cases the mother only retreated from the infant after high-pitched vocalizations or other sounds were heard, or when danger calls were emitted by the alpha male. Ojha also noted that the distance which the mother allowed the infant to travel away from her increased as the infant became older. Therefore, this field study clearly documents the reaction of the mother, other adults, and juvenile animals to the occurrence of infant-leaving behavior, thereby illustrating the kind of experiences that form the basis for later socialization of the developing infant in the social group.

In a study carried out at the Yerkes Regional Primate Research Center, Atlanta, Georgia, we examined in detail those variables that may determine the formation and termination of dyadic social bonds in captive-reared juvenile chimpanzees and orangutans. This study was co-authored by J.T. Braggio, S.M. Braggio, A. Weber, and R.D. Nadler (Asheville, North Carolina, U.S.A.) and entitled "Degrees of Familiarity Differentially Influence the Quality and Quantity of Social Interactions in two Species of Great Apes". We previously established that ethological mechanisms such as the presence of the play-face and low-pitched vocalizations set the occasion for the occurrence of nonaggressive social interactions between two conspecifics. In this study we wanted to determine how familiarity influenced the duration of nonaggressive social interactions in juvenile chimpanzees, and orangutans. The familiarity variable was operationally defined as either keeping two animals caged together (familiar pair) or not caging two animals together (nonfamiliar pair). It was found that the frequency of approaches, rough-and-tumble play, and retreats were higher among nonfamiliar pairs of animals than between familiar pairs of animals. This laboratory study provides evidence that may explain the formation and termination of dyadic social bonds formed by captive-reared juvenile chimpanzees and orangutans. These data suggest that it is the presence of some ethological cues such as the play-face and low-pitched vocalizations that account for approach behaviors, and the initial formation of a dyadic social bond. After a time the ethological cues habituate (i.e., when they are not followed by reinforcement such as reciprocal grooming) and are not able to maintain two nonhuman primates together. At that point the social bond is weakened, and the two animals retreat from each other. This cycle of the initial formation and eventual termination of a dyadic social bond is repeated enough times so that all of the animals in a group have interacted with every other animal in the social group. This proposed mechanism of behavior may be useful in accounting for the transmission of newly acquired behaviors such as nest building, tool use, use of ladders in captivity, etc. from one animal to another, and the establishement of a linear social hierarchy. It may be that the proposed mechanism may also account for the establishment of a protoculture, even in nonhuman primates that have had limited interactions with the mother or other adult members of the social group.

Dr. P.R. Ojha's (Naguar, Rajasthan, India) second paper on the program of the Comparative Psychology symposium is entitled "The Rhesus Macaque: Food Feeding in the Indian Desert". In this study Ojha attempted to demonstrate the adaptability and survival skills of rhesus monkeys in extreme environmental conditions where there is little or no food available. In his introduction Ojha states that few studies (if any) have examined the behavior of nonhuman primates in a desert habitat. This field study was carried out at Maroth Village, in the district of Naguar (Rajasthan, India). A total of 93 nonhuman primates were observed in a climate that was arid, dry, and had a low annual rainfall. Under these conditions Ohja found that the nonhuman primates he observed showed unusual eating habits. Specifically, he noticed animals eating deciduous thorny trees, and grasses during the rainy season. Rhesus monkeys were also observed taking food from the houses of the local inhabitants; the foods that were taken included corn, fruits, vegetables, etc. The nonhuman primates also consumed "numar", a substance used as stone building plaster. The monkeys were also observed eating bird eggs, on rare occasions. Foraging usually lasted about 5 h, but less time was spent looking for food during the winter months. Therefore, Ojha's observations suggest that the specific eating behaviors of rhesus monkeys are not species-specific, but may depend on the ecological and climatic conditions that determine the kind of vegetation that can serve as a food staple.

Dr. S.C. Makwana's (Jodhpur, India) paper entitled "Field Ecology and Behavior of the Rhesus Macaque *(Macaca mulatta)*. V. Foraging and Ranging", attempted to investigate the variables that influence the adaptive significance of the home range in foraging and traveling groups of rhesus monkeys. Makwana noted that all of the nonhuman primates observed restricted their foraging and ranging activities to a home range that was operationally defined as a measurable circumscribed geographical area. All of the nonhuman primates described in this study were observed in the Dehra Dun forests, located in Northern India. The results of this study indicated that the majority of the rhesus monkeys in the various groups traveled between 900 and 2100 m per day. Analysis of the order of progression indicated that 75% of the time females initiated and led the group, while the alpha male (8.3%) and other males (16.6%) led the group occasionally. Also, the alpha male tended to be located in the middle of the group 70.8% of the time, and at the end of the group 20.8% of the time. However, the rate of progression was not correlated to group size, or to the length of the home range. There were more anticlockwise turns than clockwise turns during foraging. In his discussion of these field observations, Makwana suggested that the obtained results on home range reflect group cohesion or a general "social tonus" factor. If groups of rhesus monkeys are observed over a long period of time, it may be possible that this "social tonus" construct may correlate to ranging patterns or other measures of the home range.

Physiological and Incentive Factors

Up to now there have been few studies that have examined systematically the effects
of physiological and incentive factors on the behaviors of nonhuman primates. In one
sense Ojha's study on the feeding behavior of rhesus monkeys in the Indian desert is
an example of the examination of an incentive variable on the behavior of nonhuman
primates in their natural habitat. In this section I will describe other studies that were
on the program of the Comparative Psychology symposium at Corigliano Calabro.
All of the studies in this section either altered the physiology of the subject (i.e., lesion),
or varied incentive factors. The effects of these independent variables were evaluated
by looking at behavioral baselines or measures of complex learning (i.e., Harlow's
Wisconsin General Test Apparatus or Rumbaugh's modified Discrimination-Reversal
paradigm). While these studies are only descriptive and not explanatory, they still
provide experimental evidence that experience can only influence behavior or performance within the lower and upper limits established by biology.

The first paper to be presented in this section of the pre-Congress symposium on
Comparative Psychology was co-authored by a distinguished primatologist and his
research associate, Prof. A.J. Riopelle and Dr. D.C. Hubbard (Baton Rouge, Louisiana,
U.S.A.) and entitled "Vestibular Function and Development of Swimming Behavior in
Rhesus Monkeys". In presenting this paper, Riopelle mentioned that this study is one
that has evolved from an ongoing research program that is aimed at studying the development of rhesus monkeys following pre- and postnatal protein and mineral deprivation and antibiotic insult to physiological systems. Specifically, there were three main
purposes to the present study: First, Riopelle and Hubbard wanted to determine how
early the rhesus monkey was capable of maintaining itself above water, and of propel-
ling itself to a specific (if not distant) point. Second, these primatologists tried to
determine the speed and direction with which proficiency in swimming behavior developed. Third, this study tried to determine if the vestibular system was involved in the
mediation and development of swimming behavior in infant rhesus monkeys.

In this study a total of 14 infant rhesus monkeys were used as subjects. All of the
nonhuman primates tested were born in the laboratory, and removed from their biological mothers at the day of birth. Throughout this study the nonhuman primates
were housed in individual cages having a terry-cloth floor and a transparent partition
which enabled the animals to view a nonhuman primate neighbor. There were eight
nonhuman primates that served as control animals (in the sense that their vestibular
system was intact). In the three remaining groups of two animals each the vestibular
sense organ was altered by using one of three procedures. In the first two groups
(for a total of four animals) the vestibular sense organ was damaged by the introduction of prenatal and early postnatal manganese deprivation or a combination of manganese and zinc deprivation. The last group (of two animals) was given nearly daily
doses of streptomycin, beginning on day 6 (for the first animal) and on day 24 (for the
second animal), and continued throughout the duration of the study.

The infant rhesus monkeys were tested for swimming behavior in a water tank that
was 330 cm long, 100 cm wide, and 50 cm deep. On all test trials water depth was
always greater than 35 cm. Each test trial lasted 10 to 30 s, and each animal was given
up to three test trials per day.

The results of this study indicated that normal infant rhesus monkeys at birth or within a few days after birth are capable of treading water or swimming for a distance of 5–6 m and as long as 30 s, the longest length of time allowed on each trial. Specifically, the first finding was that all of the control and experimental animals tested during the first few days of life could at least tread water. The second finding was that four of the no-manganese nonhuman primates and both of the streptomycin subjects failed to swim between the 23rd and the 60th day of age. Failure to swim occurred on at least six trials for these animals.

Riopelle and Hubbard also described the ontogeny of swimming behavior in both groups of animals tested. These primatologists noted that self-clutching was a stereotyped response that was particularly incompatible with swimming. In the discussion section of the paper the ontogeny of swimming behavior that occurs for the human infants was compared to the swimming behavior that was observed in the infant rhesus monkeys. Based on these human and nonhuman primate comparisons, the tentative explanations for the failure of the occurrence of swimming behavior in nonhuman primates may be the result of the following: (1) the persistence of conflicting responses such as self-clutching, turning aimlessly, flaccid body position, etc., (2) normal reduction in reliance on swimming reflexes, and (3) degeneration of the sensorineural pathways of the vestibular system or the otoconial structure. Which one of these three proposed hypotheses (or possibly others) accounts best for the failure of the onset of swimming behavior in nonhuman primates that sustained damage to the vestibular system has to be determined experimentally by future research on this important topic.

Dr. S.D. Hill (New Orleans, Louisiana, U.S.A.) presented a paper entitled "The Development of Cortisol Rhythmicity and Social Responsiveness in Protein-Deprived and Nondeprived Rhesus Infants". The overall aim of her research program is to use the nonhuman primate as an animal model to investigate abnormal physiological functioning in human children. It is already known that in human children malnutrition produces severe impairments in attentional, arousal, and motivational systems, rather than in disrupting basic learning abilities per se. In the human literature this cluster of symptoms has been attributed to a category known as hyperactivitiy or hyperkinesis. As malnutrition continues, it has been hypothesized that there are severe disruptions in the normal rhythms of the hypothalamic-hypophyseal-adrenocortical system(s). Because of ethical and legal considerations, it is neither desirable nor possible to carry out in depth studies involving long-term protein deprivation in human children. An alternative strategy that is available to the comparative psychologist is to use nonhuman primates such as rhesus monkeys as animal models to study these phenomena. The primary aim of the present experiment was to examine the development of diurnal cortisol rhythms in protein-deprived and nondeprived infant rhesus monkeys. A secondary objective of this study was to compare protein-deprived and nondeprived infants on the development of early social responses, such as lip smack, grimace, fear responses, etc.

A total of six infant rhesus monkeys were used as subjects in this study. One infant was placed on a 13.4% protein diet, two infants were on a 6.7% protein diet, and the remaining three infants were on a 3.35% protein diet. The infants lived apart from their biological mothers in cages that permitted visual access to a neighboring infant living in an adjoining cage. Cortisol levels were taken from blood samples beginning in the first week of life and continuing through day 240 of the experiment. Behavioral measures were made of the infants in their home cages, as well as in specified test areas.

The results of this study indicated that there was an inverse relationship between degree of protein deprivation and amount of weight gained. That is, the animals on the most severe protein restricted diet (i.e., 3.5%) gained very little weight over the 4-montth period of diet restriction. Analysis of the cortisol levels indicated that the expected diurnal variation (i.e., rise in the early morning, followed by a decrease in the early afternoon, and a low at 3 a.m.) was disrupted after the 7th week for the rhesus monkeys that were on the severely restricted protein diets. However, cyclical variations in cortisol levels continued for the entire 8 months of the study for the infant rhesus monkeys that were not protein deprived. Hill also noticed that there were delays of about 5–7 days in the appearance of some of the species-specific behaviors, such as facial displays, lip smack, and grimace for the protein-deprived animals as compared to the non-protein-deprived animals. These results were interpreted to mean that the adreno-cortical system may be fully developed at birth, even though the characteristic diurnal variation in cortisol levels can be altered by diets low in protein content. Therefore, these data suggest that there is a clear relationship between protein deficiency, body weight, species-specific behavior, and cortisol levels.

In another study co-authored by S.M. Braggio, J.T. Braggio, T.C. Cochran, and P. Ellen (Asheville, North Carolina, U.S.A.) and entitled "Discrimination-Reversal Performance of Normal and Septal Rats", we attempted to evaluate rodents on Rumbaugh's Transfer Index (TI) task, and determine the contribution of the limbic system (i.e., septum) to the expression of an operationally defined measure for complex learning. The reader should note that Rumbaugh's discrimination-reversal paradigm has been used to test the complex learning skills of great apes, Old and New World monkeys, but not rodents. A total of eight Long Evans hooded rats were tested in a modified operant chamber that had two levers, one below each one of the two projectors used to present the two dimensional patterned stimuli that differed from each other in hues and/or brightness values. The stimulus item pairs and the automated testing apparatus used in this study were identical to those already utilized by Rumbaugh and his associates to test many species of nonhuman primates.

Each one of the rodents was given a sufficient number of trials to attain the 67% and/or the 84% Acquisition criterion. Once the preestablished Acquisition criterion had been reached by the subject, ten reversal trials were administered. The eight subjects received between 30 and 60 problems each. It was found that for the 67% Acquisition criterion the TI values for the entire group was 0.79 (range was 0.67–0.92). No significant effects were found between the normal rodents and the rats that sustained lesions in the septum, in terms of the number of trials needed to attain the 67% Acquisition criterion (Normal = 22.9; Septal = 21.9), and TI values (Normal = 0.82; Septal = 0.80). These results suggest that the rodents tested seemed to attain TI values that were similar to those of the squirrel monkeys, save for the fact that only the nonhuman primates were able to attain the 84% Acquisition criterion.

We also examined the TI performance of the entire group of rodents to determine how it is possible for animal subjects to attain optimal performance on Rumbaugh's modification of the discrimination-reversal task. To do this, we examined the performance of the four subjects with the highest percentage correct on the second reversal trial (High Group or HG) and the four subjects with the lowest percentage correct on the second reversal trial (Low Group or LG). As expected, we found that the HG had a higher percentage correct on all reversal trials than the LG. The HG, and not the LG, demonstrated a win-say and loose-shift strategy on the first and second reversal trials. Further analysis of the data indicated that single errors (i.e., incorrect responses that were not reinforced) served as the cue(s) that set the occasion for the occurrence of the loose-shift response strategy.

Several implications can be reached based on these results. Our findings indicate that the septum is not involved in the mediation of those complex learning skills that are tapped by Rumbaugh's modified discrimination task. For the rodents tested, the task may have been reduced to a black-white discrimination where stimulus intensity was the salient feature. This was probably due to the fact that rodents do not have color vision like nonhuman primates. Under these circumstances rodents were also able to demonstrate relatively complex problem-solving strategies (i.e., win-stay and loose shift) that resembled those that have been found for nonhuman primates (especially rhesus monkeys).

The last paper in this section was co-authored by Prof. C.W. Hill and A.J. Riopelle (Baton Rouge, Louisiana, U.S.A.) and entitled "Reward Frequency Versus Amount During Discrimination Learning and Preference Testing with Rhesus Monkeys". In the introduction to their paper these authors state that when two objects in a discrimination problem are rewarded with a differential frequency, some species respond by selecting the object that is rewarded most frequently. However, under these cirumstances frequency of reward may be confounded with amount of reward. The purpose of this study was to use a factorial design to evaluate the differential effects of reinforcement on discrimination learning. In experiment 1 the rhesus monkeys first learned to respond to relevant frequencies (i.e., 100%, 50%, 33%, and 25%) and amounts (i.e., 1, 2, 3, and 4 units) that were associated with four of the stimulus objects presented two at the time. In experiment 2 the specific combinations of frequency and amount were counterbalanced by using a pretraining procedure that consisted in presenting each stimulus object by itself, with its associated frequency and/or amount for reward, for 12 consecutive trials.

In both experiments a total of eight rhesus monkeys *(Macaca mulatta)* were used as subjects. There were an equal number of male and female subjects in the two groups. The ages of the animals in both experiments ranged from 4 to 6 years. All behavioral testing was carried out using a standard Wisconsin General Test Apparatus. Gellerman series were used to randomize the location of the differentially rewarded stimulus objects in the response tray of the WGTA. Also, the various rewarded frequencies were programmed by using a table of random numbers.

The results of the first experiment indicated that rhesus monkeys were more likely to respond on the basis of greater frequencies of reward, especially when the greater frequencies used were close to 100%; in addition, the rhesus monkeys were unlikely to prefer the alternative of greater amounts of reward per trial, despite the fact that the same total amount of reward could have been obtained by utilizing either one of the two consistent response preferences mentioned above. When the frequency ratios were below 2:1, the rhesus monkeys were more likely to divide their selections equally between the two options presented. Under these circumstances, it was noted that the rhesus monkeys demonstrated the use of a position habit.

In general, it can be concluded that the results of the second experiment indicated that there was an interaction between kind of group (i.e., experienced vs naive) by Problem Block (i.e., first or second). The experienced group consisted of the four rhesus monkeys tested in the first experiment, and four additional rhesus monkeys assigned to the naive group. The animals in the naive group were not included in the first study, but also had extensive testing backgrounds with the WGTA. In the second experiment the pretraining factor consisted in presenting each stimulus object by itself, with its associated frequency and/or amount of reward, for 12 consecutive trials. Two consecutive test sessions of 48 trials each were administered to both the experienced and naive groups of rhesus monkeys. It was found that all eight subjects tested (i.e., both experienced and naive groups) appeared to be influenced by the preliminary training, so that the animals were more likely to select to a greater extent the object associated with the larger amount of reward. However, this preliminary training effect seemed to influence the rhesus monkeys in the naive group more than those in the experienced group. In the discussion section of this study Hill and Riopelle alluded to the potential significance of these results by suggesting that this pattern of performance may be indicative of some form of generalized rule learning. Whether this is indicative of the learning of a specific or a generalized rule cannot be answered by this study. This issue (regarding the kind of rules used by monkeys in solving discrimination problems) has to be addressed by future research on this topic.

Round-Table Discussion

The moderator of the Round-Table Discussion was Prof. C.W. Hill (Baton Rouge, Louisiana, U.S.A.). Several themes emerged from the discussion which included the participants of the Comparative Psychology symposium, and the visitors from Calabria University and the town of Corigliano Calabro. As a retrospective analysis of these discussions, it was my impression that all of the points that were made at this meeting can be summarized around the theme of the "Present and Future of Comparative Psychology in the 1980s". I will try to expand on these comments for each one of the three areas mentioned in this paper, as well as allude to the future direction(s) of comparative psychology.

The papers that were grouped together under the general heading of the "Assessment of Perceptual, Intellectual, and Cognitive Factors" were seen as making an important and timely contribution to comparative psychology. All of the participants agreed that it is important to continue to analyze the intellectual and cognitive capacities of nonhuman and human primates. On this point, Thomas raised an important issue regarding the need to use rigorous experimental designs in order to evaluate the behavioral basis of these psychological capacities in nonhuman and human primates. Other primatologists pointed out that comparative psychologists should not underestimate the contribution of other disciplines to this area of study. That is, under some circumstances scientific advancement occurs when a novel idea is discovered or a different methodology is used to study nonhuman primates. In this regard, it is possible that such conceptual or methodological advancements may help us to better describe and explain the intellectual and cognitive capacities of animals in the order Primates.

In opposition to the methodological arguments proposed by Thomas, our work in this area has attempted to determine the usefulness of the contribution of Piagetian theory and methodology to the study of cognition. Our initial results suggest that the cognitive capacities of nonhuman primates (especially the great apes) are similar to (but not identical with) the cognitive capacities of normal and mentally retarded children. This conclusion is based on the laboratory results that the great apes can use two stimulus dimensions simultaneously, and can also solve Piagetian-type conservation tasks. It is further recognized that even though there may be similarities between nonhuman primates and human children, they are expressed more efficiently by human children (both normal and retarded) than by nonhuman primates. This difference may be due to the presence of human language. It is further suggested that the presence of human language in human children, and language-type skills in nonhuman primates, can both function to minimize the negative effects resulting from the appearance of previously unlearned response tendencies that can operate to impair the performance of nonhuman primates on cognitive tasks.

Even though there were some differences of opinion regarding the approach that should be taken to study the intellectual and cognitive capacities of nonhuman and human primates, all participants agreed that there is need for a systematic study of the perceptual mechanisms of the nonhuman primate infant. Swartz's work on this topic suggests that before intellectual and cognitive capacities can be fully understood, it is necessary to be able to define the physical and functional stimuli perceived by the nonhuman primate infant. This area of research should involve the use of the methodologies which have been employed by child psychologists to investigate the development of perceptual mechanisms in human children. In addition, an investigation of perceptual mechanisms in nonhuman primates should also include a psychophysiological and neurophysiological study of how experience and brain factors interact to determine the process by which a compound stimulus matrix (and its sensory analog on the retina) is perceived by the nonhuman primate infant. It is likely that perceptual mechanisms may change as a function of development (i.e., biological and experiential).

The second group of papers were discussed under the heading of "Social Interactions in Nonhuman and Human Primates". These papers seemed to clarify some of the behavioral mechanisms that may form the basis for the appearance of socially organized groups in human societies. Some of the participants underscored the importance of investigating mother-infant interactions to determine how subadult animals eventually become socialized in a group. The study by our Italian colleagues Tartabini, Bertacchini, and Genta, indicated that the nonhuman primate mother does play an important role in determining the rank of its infant in the social group. Also, these primatologists stated that it is a result of the mother's influence on the infant that eventually determines the quality and quantity of social interactions that involve the infant with conspecifics in the group.

The papers contributed by the two scientists from India provide a valuable source of information regarding the behavior of nonhuman primates living in their natural habitat. These three papers suggest that nonhuman primates live in social groups that reflect the reciprocal interaction of animals that differ in age, gender, experience, social roles etc. It was of interest to notice from the results of Ojha's study that it may be the reaction of the animals in the social group to infant-leaving behavior that may form the basis for the eventual socialization of subadult animals into the group. Further, Makwana's field study along with Ojha's report on the eating behaviors of rhesus monkeys in the Indian desert seem to suggest that even unlearned behavioral repertoires can be influenced by ecological and climatic conditions.

The last group of papers on the program at Corigliano Calabro concerned the general area of the influence of physiological and incentive factors on behavioral and performance measures of nonhuman primates and rodents. Thus far not enough attention has been directed at a systematic investigation of physiological and incentive factors. It may be that the only way that it will be possible to examine the interactive effects of biological and experiential factors on the behavior of nonhuman primates is to examine the influence of different brain states (i.e., lesions) and physiological states (i.e., hormones, manipulation of drive) on behavior. It was the feeling of the participants of the Comparative Psychology symposium that this issue should be examined further in the future.

Aside from these general points that are directed to the three areas that were included in the pre-Congress symposium, there were three additional trends that were raised during the Round-Table Discussion. First, it was stressed that there should be a greater emphasis on the use of the ethological method, along with the comparative method, in order to study the behavior of primates, both in the laboratory and in the field. The advantage of this approach is that it is possible to examine the degree to which there are species-specific response tendencies that may form the basis for the appearance of more complex intellectual, cognitive, and social behaviors in nonhuman primates. Second, many of the papers presented at this symposium underscored the importance of using nonhuman primates as animal models for the study of phenomena that are of primary importance to mankind. It will be recalled that the paper by DeJonge and Dienske from the Netherlands examined the effects of environmental deprivation on the psychological and social development of nonhuman and human primates. Our own research in this area has emphasized the use of the results of Piagetian studies on the acquisition of language-type skills by nonhuman primates as a basis to better understand and remediate the educational handicaps of exceptional children. S.D. Hill and Riopelle of the United States have tried to use nonhuman primates as animal models to study the effects of protein deprivation on diurnal cortical levels and the development of swimming behavior in nonhuman primate infants that sustained damage to the vestibular system. A third and final issue concerned the general agreement that if comparative psychology is to continue in its study of nonhuman and human primates, there must be a far greater awareness on the part of federal funding agencies and private foundations regarding the importance of these laboratory and field studies in answering both theoretical and applied questions that pertain to primatology. The participants were in general agreement that research support is needed by all scientists, from all over the world, who are studying nonhuman and human primates in the laboratory and their natural habitat.

Behavioral Ecology and Sociobiology

ISSN 0340-5443   Title No. 265

Managing Editor: H. Markl, Konstanz

Editors: B. Hölldobler, Cambridge, MA; H. Kummer, Zürich; J. Maynard Smith, Brighton; E. O. Wilson, Cambridge, MA

Advisory Editors: G. W. Barlow, Berkeley, CA; J. Brown, Albany, NY; E. L. Charnov, Salt Lake City, UT; J. H. Crook, Bristol; J. F. Eisenberg, Washington, DC; T. Eisner, Ithaca, NY; S. T. Emlen, Ithaca, NY; V. Geist, Calgary, Alberta; D. R. Griffin, New York, NY; W. D. Hamilton, Ann Arbor, MI; D. von Holst, Bayreuth; K. Immelmann, Bielefeld; W. E. Kerr, Ribeirão Preto, SP; J. R. Krebs, Oxford; M. Lindauer, Würzburg; P. Marler, New York, NY; G. H. Orians, Seattle, WA; Y. Sugiyama, Inuyama City, Aichi; R. L. Trivers, Santa Cruz, CA; C. Vogel, Göttingen; C. Walcott, Ithaca, NY

The electric eel *(Electrophorus electricus)*, often reaching a size of 8 feet and a weight of 200 pounds, electrocutes its prey with a burst of high-voltage electricity, about 500 Hz. In contrast, the eel locates prey using an electric organ emitting low-voltage impulses. An important question confronting behavioralists and physiologists alike is whether these two electrical systems are interrelated in some fundamental way. An answer will not only reveal important information on the physiology of *E. electricus,* but will also further our understanding of the basic relationship that exists between the environment and neuron function.

In response to this and other ecological and sociobiological questions Springer-Verlag initiated the publication of **Behavioral Ecology and Sociobiology** in 1976. Since then the journal has become a major forum, publishing original research on the functions, mechanics, and evolution of ecological adaptations and emphasizing social behavior. An international board of editors and advisors, aided by numerous reviewers, guarantees the very highest standards.

Contributions are welcomed from scientists the world over; publication is almost exclusively in English. Topics treated cover a broad range, including orientation in space and time, communication and all other forms of social behavior, behavioral mechanisms of competition and resource partitioning, predatory and antipredatory behavior, and theoretical analyses of behavioral evolution. Quantitative studies carried out on representatives of nearly all major groups of the animal kingdom are published, from spiders through fish, birds, primates, and other mammals. Empirical studies on the biological basis of human behavioral adaptations are also welcomed.

Behavioral Ecology and Sociobiology has become an important source of information on the progress in animal behavior research. Scientists, researchers, and graduate students will regularly find papers of significance in their particular fields of interest.

For subscription information or sample copy write to:
Springer-Verlag, Journal Promotion Department,
P. O. Box 105 280, D-6900 Heidelberg, FRG

Springer-Verlag
Berlin
Heidelberg
New York

Primate Evolutionary Biology

Selected Papers (Part A) of the VIIIth Congress of the International
Primatological Society, Florence, 7–12 July, 1980
Editors: A. B. Chiarelli, R. S. Corruccini

1981. 73 figures. IX, 119 pages
(Proceedings in Life Sciences)
ISBN 3-540-11023-2

The contributions selected for inclusion in this volume provide a
compact, yet comprehensive and up-to-date review of current
investigations into primate evolutionary biology. They cover in particular functional morphology, evolution and paleontology, with the
papers on evolutionary morphology arranged in logical succession
from lower primates to apes and man. Although the contributors
represent many different disciplines, their diverse points of view
and emphases lend an extraordinary measure of depth to the studies
presented here.

Primate Behavior and Sociobiology

Selected Papers (Part B) of the VIIIth Congress of the International
Primatological Society, Florence, 7–12 July, 1980
Editors: A. B. Chiarelli, R. S. Corruccini

1981. 65 figures. IX, 182 pages
(Proceedings in Life Sciences)
ISBN 3-540-11024-0

The papers of this volume cover in particular social aspects which
reflect or affect primate biology, one of the more important sub-
disciplines in primatology today. Although the contributions were
drawn from both field and laboratory studies, special emphasis was
laid on work done in the field as a more reliable reflection of social
groups of monkeys and apes.

Springer-Verlag
Berlin
Heidelberg
New York

H. N. Seuánez

The Phylogeny of Human Chromosomes

1979. 49 figures, 10 tables. X, 189 pages
ISBN 3-540-09303-6

"**Interesting reading for researchers** involved in genetics, cytogenetics, evolution, primatology, and anthropology ... Dr. Seuánez
is at his best when describing his own studies with the orangutan...
In this volume he has included a wealth of information about a
broad range of techniques. Students and those interested in primate
cytogenetics will find the volume useful because it **includes a large
number of references...**"

Nature